THE GOLD HUNTER'S FIELD BOOK

Also by Jay Ellis Ransom

Gems and Minerals of America

Fossils in America

A Range Guide to Mines and Minerals

The Rock-Hunter's Range Guide

Petrified Forest Trails

Arizona Gem Trails and the Colorado Desert of California

THE GOLD HUNTER'S FIELD BOOK

How and Where to Prospect for Colors,
Nuggets, and Mineable Ores of GOLD
by Amateur and Serious Followers
of Jason and the Golden Fleece

JAY ELLIS RANSOM

1817

HARPER & ROW, PUBLISHERS

NEW YORK, EVANSTON, SAN FRANCISCO, LONDON

FIRST EDITION

Designed by Sidney Feinberg

Library of Congress Cataloging in Publication Data

Ramsom, Jay Ellis, date
    The gold hunter's field book.
    Includes index.
    1. Prospecting—Handbooks, manuals, etc. 2. Gold
ores—Handbooks, manuals, etc. 3. Gold mines and
mining—North America—Handbooks, manuals, etc.
I. Title.
TN271.G6R36 1975   622'.18'410973   74-20409
ISBN 0-06-013511-5

75 76 77 78 79 10 9 8 7 6 5 4 3 2 1

For Today's Gold Hunters

Contents

Acknowledgments

In compiling this range guide to prospecting for gold, special thanks must go to the directors and geologists associated with the various state water resource agencies and state departments of mines and minerals. Their titles and addresses are appended to the introductory remarks for the states and provinces covered in Parts II and III. Many of these interested officials and agencies have contributed information freely as an assist to gold hunting ventures within the areas of their authority.

America needs new sources of gold, and the United States Geological Survey (USGS) is actively engaged in encouraging amateurs and professionals alike in finding it. Similarly, Canadian mineral and resource agencies are interested in helping to develop a new era of gold prospecting and mining across their immense country. Thus, in general, wherever a gold hunter tries his hand at prospecting, he will find cordiality and assistance on every hand. In many states and provinces, the old-time practice of "grub-staking" a knowledgeable prospector is being reinstituted, reminiscent of the gold rush years of the nineteenth century.

The USGS provides detailed coverage of every significant mining area. Among its publications are map indexes; geologic maps; and mineral resource and topographic maps on which every gold mine is pinpointed by section, township, range, and degree and minute of latitude and longitude. Descriptions of these publications, most of which are available also in major public and university libraries, and information concerning their purchase may be found in the catalog *Publications of the Geological Survey* and its supplements. The catalog may be obtained without cost by writing to the Branch of Distribution, United States Geological Survey, Washington, D.C. 20242.

Although all those who have helped to provide the information from which this book is compiled are individually too numerous to credit by name, I am sincerely grateful to each and every one.

JAY ELLIS RANSOM

The Dalles, Oregon

Introduction

Seventy years ago C. W. Purington wrote a brilliant monograph* on placer mining in Alaska, in which he observed: "The winning of gold from alluvial material is a business difficult both to learn and to conduct successfully." The same observation goes doubly for extracting gold from deep within the earth, where hard-rock veins require machines and technologies far more expensive than those used in placer mining.

In pioneer days gold gleamed in stream gravels in full view of the passerby, or it appeared at the grass roots in the form of solid nuggets. It was placer gold prospecting that opened the frontier West and the Far North to settlement. Not really until the more easily obtained alluvial gold had been pretty well cleaned out did miners begin searching for the original sources of the gold, buried nearby within the earth, usually in quartz veins. Thus, following the era of placer mining, came the era of lode-gold mining.

These two means of winning gold from the earth complement each other. Showings of placer gold in the form of colors, flakes, grains, or nuggets mean that somewhere nearby—"upcountry," as the Westerners phrase it—there must be a vein or outcrop from which the alluvial gold was washed down over eons by freshets and snow melt runoff. Going in the opposite direction, the modern gold hunter can expect that wherever

---

*Methods and Costs of Gravel and Placer Mining in Alaska (Washington, D.C.: U.S. Geological Survey, 1905), Bulletin 263.

a lode-gold mine has been developed, there is a likelihood that somewhere "downcountry" from it he may find some remaining evidences of placer gold to color the bottom of his gold pan. Perhaps in the general vicinity he may discover an as yet unknown lode vein or quartz stringer that might encourage him to try his hand at developing his own small gold mine.

Placer gold is still very much around. In Sierra County, California, an amateur gold hunter found a 28-ounce solid gold "grass roots" nugget in March 1973, a good 125 years after the great California gold rush. An even larger nugget, weighing 25 pounds troy and possibly the largest nugget ever found in California, was picked up by another amateur in October of that same year. This gold-quartz nugget was gleaned from the Yuba River near Downieville by a skin diver. He estimated the worth of its gold content as $22,500 at the then official price of $42 a troy ounce, but the nugget was clearly worth even more as a museum specimen. In almost every gold district in America, as well as in Canada, gold nuggets are still being found, literally at the grass roots, either overlooked by previous generations of prospectors or newly weathered out of the surrounding auriferous formations.

Lode-gold mining suffered a serious setback in 1942, when the War Production Board issued its famous Order L-208, forcing the closure of all gold mines in the United States for the duration of World War II. Almost none of the mines was reopened after the war ended. The miners were gone, their mining skills diverted to industry, where wage levels had risen to levels above those gold mining could afford.

At the same time, many old mining districts were paved over and converted to housing subdivisions, especially in California; the mine properties themselves became worth much less as producers of gold than as real estate. These factors, on top of the effects of an order issued in 1934 by President Franklin D. Roosevelt severely limiting the amount of gold (200 ounces) that any American could possess, contributed to the almost total demise of the gold-mining industry.

At long last, on March 17, 1968, the United States Treasury discontinued buying and selling domestically mined gold and ceased to control its price. Since that date, the gold prospector and miner has been allowed to sell his product to any willing purchaser at prices determined by supply and demand. Thus ended a long period of government control over the price, sale, and purchase of gold. The last restriction on the 200-ounce limit of possession ended on December 31, 1974.

When Treasury buying ended, the official fixed price of gold was $35 a troy ounce 1,000 fine—the price set in 1934. With government controls lifted, the price began rising, slowly at first as the United States began devaluing her currency, and then dramatically as the great London gold market—the world's largest—began trading gold bars and coins with other world centers, such as Zurich, Paris, Frankfurt, Milan, Hong Kong, Bombay, and Beirut. By 1975 the world price per troy ounce of pure gold (not including collectors' valuations for individual gold coins, which are much higher than for gold bullion) had risen above the $175 an ounce mark, with wide and rapid fluctuations listed daily in the world's press.

This sharp price rise has had a considerable effect on gold prospecting and mining throughout the world. Even though many pre-World War II mines will never be reactivated, the increased price of gold now permits many other old and often marginal mines to be reopened, cleaned out, and further developed at a profit. Lower-grade deposits may now be explored and mined. Already we see a new surge in gold hunting, especially by amateurs—weekenders enjoying themselves along the nation's streams looking for placer gold. Others are vacationing in long-abandoned mining districts, picking over old mine dumps for specimen ore containing raw gold. The more professional prospectors have taken to leasing particularly promising old mine dumps, where the low-grade detritus thrown away in past generations may now be worth reworking.

New mining companies are being formed by both Americans and Canadians. Leases or options have been taken out on many old properties, and active exploration and development programs are under way. In a number of once-great mines, where the gold veins were exhausted within their original profit requirements, systematic examinations, including diamond drilling, are being conducted to determine the feasibility of mining them by lower-cost methods. Such deposits include extensive zones of alteration known to contain disseminated gold, large bodies of auriferous schist and massive greenstone, and districts containing extensive vein systems or zones too low in gold content to have interested the original prospectors.

The marked increase in the activities of weekend and vacationing amateurs from all parts of the country—scuba divers, "snipers," and "pocket hunters" from every walk of life—has, indeed, led to new discoveries of gold. Understandably, the widespread newspaper and television publicity following such new finds has served to stimulate further prospecting, especially around century-old gold-mining ghost towns.

It is the purpose of this range guide to assist the man or woman without technical mining education to prospect intelligently, and to suggest the means for equipping himself or herself with the tools and elementary techniques for developing an auriferous deposit at the least cost in terms of labor and out-of-pocket finances.

As every experienced gold prospector and miner knows, and every state geologist and mining consultant insists, the places to begin looking for new sources of gold are the old places where the metal was first discovered and mined. To assist the new generation of gold hunters and to remind the older generation still hunting for gold, all the major and most of the minor auriferous districts, mines, and potentially productive areas recorded in the gold-mining literature are listed or described in Parts II and III of this volume. From the known to the unknown is an adage of science that applies likewise to the search for gold.

Note: In the interest of conciseness, a number of non-standard abbreviations have been used for place names in Parts II and III.

PART I

Introduction to Gold Hunting

1

About Gold

We are reasonably certain from excavations of New Stone Age burials in Mesopotamia and the Near East that gold was the earliest metal gathered by human beings. Although the gravel bars of the streams descending southward from the Caucasus Mountains of present-day Soviet Georgia were rich in gleaming nuggets of placer gold, earlier men had found no use for it; gold is much too soft for making functional arrow or spear tips. By Neolithic times, however, its nonutilitarian value became appreciated, and people began hammering it into ornaments. Some of these have been dated to fifteen thousand to twenty thousand years ago.

The Physical Nature of Gold

Gold is a comparatively rare native metallic element, ranking fifty-eighth in abundance among the 92 natural elements in the earth's crust. Chemists have given gold the chemical symbol Au, from the Latin *aurum*, "gold." In its massive or nugget state, gold is sun-yellow and shining with metallic luster. Gold has a specific gravity of 19.3 when pure, a density exceeded only by that of platinum. Despite its softness of 2.5 to 3 on the Mohs scale of hardness, gold has a relatively high melting point of 1,063°C (1,945°F) and a boiling point of 2,600°C (4,712°F), both critical points being somewhat below those of iron.

Although gold is almost chemically inert, it does possess valences of

3

1 and 3, so that it occurs very occasionally in chemical combination with tellurium in the "telluride" minerals calaverite, $AuTe_2$, and sylvanite, $(Au, Ag)Te_2$; and also with tellurium, lead, antimony, and sulfur in the mineral nagyagite, Pb_5Au $(Te, Sb)_4S_{5-8}$. Gold dissolves in mercury to form an amalgam (an alloy used in filling teeth).

Pure gold is the most malleable and ductile of all metals; it can be beaten or hammered to a thickness of one three hundred thousandth inch without disintegrating. A single ounce of gold can be drawn into a hairlike wire sixty miles long. It is the beaten gold leaf that has modern decorative value in certain types of gilding and in the bookbinder's art. The fine wire has minor applications in space-age electronics.

In the field, gold is chiefly recognized by its "heft"—that is, its great weight relative to other minerals—and by its malleability and color. Gold is normally found in varying shades of yellow, but it may be silver-white to orange-red, depending on the associated content of silver and, frequently, copper and iron. Finely divided gold is black, like most other metallic powders. Colloidally suspended gold varies in color from deep ruby-red to purple—hence its use in the manufacture of the finest ruby glass. Variations in the colloidal color derive from differences in the size of the suspended microscopic particles.

So chemically inert is gold that its dust and nuggets and vein occurrences have survived climatic and geologic changes throughout the earth's 4.5 billion years of upheavals, erosion, and deposition. Gold is not affected by air, heat, moisture, or by any ordinary solvent except mercury. A gold ring, therefore, should never be polished with "quicksilver" to make it shinier; mercury is rubbed off with its dissolved, or amalgamated, gold, and thereby pits and mars the ring surface.

Gold will dissolve chemically in only one acid, aqua regia, a mixture of three parts concentrated hydrochloric acid and one part concentrated nitric acid, to form chlorauric acid, from which a chemically important compound, gold chloride, may be obtained. The only other commercial solvent that will dissolve gold is a dilute solution of sodium or calcium cyanide. The application of this solvent to an extractive process was discovered in Australia in 1897, where it was used to remove finely disseminated gold from pulverized rock. The "cyanide process" is the only known method of profitably treating massive low-grade gold ores. Utilizing the cyanide process, auriferous rocks carrying as little as one part gold to 300,000 parts of worthless material can be treated successfully.

Measuring Gold

Gold, because of its durability, radiance, and common occurrence in its pure state, has always been valued by man. The legendary standard of beauty and worth, gold is also a highly practical commodity—it has, throughout history, been accepted in return for services and goods—and the one that serves today as a key element of the international monetary system. In earliest times gold was used for jewelry, but this use was, in turn, the basis of a value greater than the metal's ornamental qualities. The Bible reports that Joseph put "bundles of money" in his brothers' sacks, meaning weighed rings of gold and other metals and bags of gold dust that had a recognized value.

To carry on their active trade with neighboring states, the ancient Babylonians introduced a standard value for a given weight of gold. Actually there were two standards—the heavy and the light. A heavy gold shekel weighed 252 ⅔ grains, a light one, 126½ grains. The heavy gold mina was 12,630 grains, and the light was half of that. The largest unit, the talent, had a heavy weight of 758,000 grains, with a light talent of half that weight. (A grain is a very ancient measure, equalling the weight of one kernel of wheat—or barley or corn or whatever was grown locally—taken from the middle of the ear; officially today it is .0648 grams.)

The word "to weigh"—*stater*—gave its name to the daric or stater, a Persian gold coin that circulated throughout the Middle East from 500 to 300 B.C. The etymology of the word "pecuniary" reflects another of gold's ancient measures. *Pecu*—"cow" in Latin—was the accepted value of one gold standard.

Coins, of course, were not the only way to measure gold. As early as 1100 B.C. gold circulated in China in small cubes that people understood, by their weight and size, to be worth a certain amount. In classical

Mexico and in the civilizations of West Africa gold, presumably in dust form, was stored in transparent quills. The most flexible standard of value, however, has surely been that of many Indian maharajas who, upon ascending their thrones, received from their subjects their own weight in gold.

The value of gold, no matter the form, appears to be intrinsic. When Greek tyrants needed coins, they ordered their subjects to melt down statues and ornaments, but there are also many instances, for example in medieval Europe, of the coins of the pious being donated and melted down to make ornaments and statues. Gold itself is the value, and even today jewelers in many parts of the world determine the cost of their products in large part by placing them on a scale.

Throughout the Middle Ages various measures for the valuation of gold were used. From Florence came the florin, the fine gold coin of Italy, weighing 48 grains. The Tower pound—5,400 grains—was the standard of England until it was replaced in 1527 by the troy system, believed to have originated at the great international trade fairs held in Troyes, France. Today troy weight is the system used by jewelers and, incidentally, apothecaries, as well.

Troy Weight	Avoirdupois Equivalent	Metric Equivalent
1 pound = 12 ounces	0.823 lb ·	0.373 kilograms
1 ounce = 20 pennyweights	1.097 oz.	31.103 grams
1 pennyweight = 24 grains	0.0548 oz.	1.555 grams
1 grain = $1/5760$ lb	1 grain	0.0648 grams

Fineness and the Price of Gold

Gold never occurs 100 per cent pure in nature, but is nearly always contaminated slightly with other metals. The most common contaminant is silver, with which gold forms a complete solid-solution—that is, gold ions can substitute for silver in any proportion in a natural alloy. When natural gold contains more than 20 per cent silver, the alloy is called electrum and has a pale-yellow color. Pure gold, which for the sake of providing finer distinctions is termed 1,000 fine, is a product solely of the gold refiner's laboratory.

The term "fine gold," meaning pure gold, is very important to the gold trade. It means that no matter what the going price on the national or world market may be, whatever gold you bring in for sale will not bring you that price but only a percentage of it. Let us suppose that you bring in a cigar box full of mixed gold finds that you have panned out of a particularly productive creek bed, or gleaned from an abandoned gold-mine dump. There are likely to be bits of quartz attached to some of the grains and nuggets, or your grains of gold may be embedded in bits of quartz. Besides these obvious contaminants, the gold itself is bound to contain some small percentage of silver, perhaps tellurium, or lead, antimony, and sulfur. Therefore, a chemical assay must be made, and the actual content of 1,000 fine or pure gold present must be determined.

In the very richest deposits, the raw gold may assay up to 950 fine—that is, it would be 95 per cent pure gold. In another deposit the gold may assay only 600 fine. In every gold-producing district there will be a high and a low degree of fineness, which can be averaged to a "midrange," so that purchasers can go by a general rule of thumb and pay a gold prospector at the midrange percentage rate. Variations within any particular district will depend upon the creek or the source of the gold in each paystreak, but generally the gold in any one deposit will show a consistent fineness, as well as a characteristic grain size and shape of particles.

To find the market value of gold which, for example, assays out to 846 fine, simply multiply the fineness by the current world price quoted daily in the financial section of most newspapers. This will give you the dollar value per troy ounce. Thus, if the going price is pegged at $190 an ounce, 846-fine gold will be worth $160.74. Of course, practically all gold found today by amateur and semiprofessional prospectors will bring a price considerably higher than the quoted world prices, when gold is sold as specimen or jewelry gold.

Specimen gold is any bit or piece of gold-bearing rock or ore which prominently reveals raw native gold, such as colors, grains, flakes, single or tangled wires, leaves, octahedral crystals, or nuggets still encased in the matrix. Also considered specimen gold are individual nuggets and small concentrations of colors or grains, which must usually be kept in a vial. Jewelry gold is raw native gold of any type sold by prospectors directly to a jeweler or lapidary for manufacture into jewelry or decorative objects. It is used in its original form and not melted down. The actual fineness of specimen or jewelry gold is secondary to its aesthetic value for display purposes.

Jewelry manufacturers are by far the largest single industrial users of gold, accounting for an estimated sixty to seventy per cent of the world's total industrial usage. However, because of the softness of gold, the industry rarely uses it in its pure form, that is, as "jewelry gold." Instead, gold is alloyed with other metals to increase its wearability. The percentage of gold in the alloy is expressed in carats. The term "carat" refers to a 24th part, used to denote the percentage by weight of gold relative to the alloying metal. Pure gold would be 24 carats, while a 12-carat piece of jewelry would contain fifty per cent gold and fifty per cent other materials.

While the world price of gold represents the value only of gold bullion refined into bars and held by banks or national repositories such as Fort Knox, specimen gold, even when contaminated with quartz or other material, carries the price tag of "whatever the buyer is willing to pay." The price will always be higher than the value of the raw gold content itself, within the saturation point of potential sales. Museums and private collectors constitute the major buyers of good specimen gold and the more spectacular the specimen, the higher the prices paid.

2

Placer Sources of Gold

Prospecting for gold is something that probably everyone dreams of trying at least once. If you are concerned with gold hunting primarily as a vacation diversion, prospecting offers a special excitement. There is a constant hope that the next pan of sediment will contain pay dirt. No other thrill can compare with that experienced when you see even a few tiny flakes ("colors") of gold glittering in the black sand at the bottom of your pan. Very often, the search itself is its own reward. However, the would-be full-time prospector hoping for substantial financial gain should carefully consider all the facts of hunting gold before deciding to set off on a prospecting expedition.

The Natural Distribution of Gold

Even though gold is a rare natural element, it is widely distributed in small amounts in all igneous rocks and in varying degrees in ocean water, streams, groundwater, and even in the tissues of plants and animals, according to a report of the U.S. Geological Survey. The amounts in water, still not financially practicable to remove, average one part gold to 500 billions parts water, with considerable variation. For instance, the waters of the North Atlantic Ocean contain ten times the amount of gold found in the South Atlantic, and the total amount of gold in seawater worldwide is estimated to be around 27.5 million tons.

From the time of Thomas Jefferson's first mention in 1782 of gold

9

finds in Virginia, the total gold production of the United States (including Alaska) has been a seemingly whopping estimated 307,182,000 troy ounces. Converted to avoirdupois, however, this turns out to be only 11,666 tons, or 968.6 cubic feet—barely enough to fill a one-car garage alongside a tract home. This is not very much when compared to the production of silver, lead, copper, iron, or, for that matter, almost any other common metal except platinum.

It is a conservative estimate that fewer than one in every one thousand prospectors who opened the West to settlement during the nineteenth century ever made a strike. Most of the gold-mining districts were located by the pioneers, many of whom (in America, at least) were experienced gold miners from Alabama and Georgia. But even in pioneering times, when gold glittered brightly in very many streambeds or could be picked by hand out of the grass roots of hillsides, only a small proportion of the tens of thousands of avid gold seekers actually discovered valuable deposits.

The old-time prospectors were very thorough. Placer gold was their primary objective, and they were unlikely to have passed over many near-surface bonanza deposits. The rewards of most of today's gold hunters will probably be modest: colors or a few nuggets of small size from bench and stream gravels that were too lean to attract the original hand-miners, or small concentrations of placer gold in underwater crevices that can be probed only by a skin diver. While new discoveries are still entirely possible, a great deal more effort will have to be expended to obtain gold today than was the case in the boom periods of the past.

Even in definitely auriferous districts, gold-bearing placer gravels will vary in gold content somewhere between 25 and 75 cents per cubic yard. A dollar-a-yard placer deposit is considered rich. By selective panning, you might be able to find gravel worth several times as much, not counting a few rare nuggets that periodically add to the overall average worth. Consider, however, that one cubic yard of sand and gravel is 180 large panfuls and that it takes at least five minutes of careful panning to work down the concentrate in a pan without losing it.

As an experiment, try panning with sand and fine gravel in a tub of water or in the nearest stream. An ordinary round-point shovel well filled is a panful. Mix iron filings or buckshot with the sand to simulate gold, be careful to save it all as if it were really gold, and separate it from the dirt. Time yourself for half a dozen pans and calculate how long it would take you to pan out a cubic yard. Other than enjoying such a vacation activity, would you be happy with, say 75 cents' worth of raw gold, which might be your total day's reward for all this effort? During the Great Depression, thousands of unemployed men streamed into long-abandoned gold fields and tried making a "day's wage" of $1.00 to $5.00 per ten-hour stint with a gold pan. Few were ever able to support a family panning for gold.

The great lode mines also operate on ores of extremely low gold content. The largest and richest gold mine in the United States is the Homestake Mining Company operation at Lead, South Dakota. In 1969 the mine produced 593,000 ounces of gold from ore that had a mill-head grade of 0.324 ounce per ton. The second largest gold-producer in America is the mammoth Kennecott Copper Corporation's open-pit copper mine at Bingham, Utah, from which gold is produced solely as a by-product of copper refining. The Carlin Mine of the Newmont Mining Corporation at Carlin, Nevada, discovered in 1965, is America's third-largest gold mine and the largest found in the last century. In 1969 the mine produced 212,000 ounces of lode gold from ore that ran only 0.289 ounce to the ton of rock.

Gold Sources

There are three principal gold sources: gold placers—that is, sites where metalic gold washed out of its ores appears mixed in with alluvial soils; gold ore, which occurs in hardrock deposits known as lodes; and base-metal ores, such as those of copper, lead, silver, and zinc, which yield gold as a by-product of refining. By and large, the casual gold hunter will only be interested in placer deposits, or in the waste gangue and tailing piles of a long-abandoned mine in which the gold lost by the original owners or that which was too low in grade to interest them may be considered as a form of artificial placer deposit.

From time to time, however, a lucky modern gold hunter prospecting upcountry will discover a new ledge or outcrop in which lode gold occurs. In such a case, he may find it worth his time to stake a claim and begin developing his own small lode mine, at least to the point where he can attract a well-financed buyer or a mining company to purchase the property.

GOLD PLACERS

Most gold placers were formed by the erosion of a surface with outcrops of gold ore (lode gold). Weathering decomposes all rock formations that project above the general level of the land, and runoff waters carry the disintegrated rock down the ravines and gullies to the creeks and rivers and oceans, where boulders, rocks, gravel, sand, and clay are finally deposited in descending order of size. Thus the farther from the original outcrop a sand or gravel deposit lies, the finer will be any gold contained in it. The origin of many placers is traceable directly to auriferous veins and stringers, lodes, or replacement deposits which, in most instances, were never of very high gold content.

Because gold's specific gravity is so high, when it is freed by weathering from its rocky host, it sinks in running water and lags behind, while the greater part of the sand and gravel is swept on. Hence, gold is concentrated by natural mechanical means, and a placer deposit is formed.

Placer deposits are most likely to occur where the carrying power of a stream is reduced, as in sand and gravel bars or in low points or pockets and crevices in the streambed. Irregular surfaces of bedrock beneath the stream flow aid in trapping gold particles and serve like the riffles in a sluice box or long tom often used in the hand recovery of gold (see Chapter 5). Fine particles are carried out to ocean beaches at the mouths of streams, where the combination of stream washing and wave action settle out the tiny gold fragments, usually in an abundance of black sand that also carries a much greater quantity of ilmenite, magnetite, hematite, chromite, and smaller quantities of platinum.

Prospecting for placer gold is not expensive, and a deposit may be found and worked with very little capital other than that normally expended on a weekend or vacation trip. Almost the only tools needed are a shovel, pick, mattock, gold pan or batea, and plenty of elbow grease. The presence of water is desirable but not entirely necessary, for in waterless regions many placer deposits can be successfully worked on a small scale by winnowing in the wind or building a dry washer consisting of screens, catchment plates with riffles, and a bellows blower.

Dredging operations of a large-scale deposit are expensive and better left to a major mining company. But if there is an ample supply of water, hydraulicking does not ordinarily require the expenditure of any considerable sum of money for equipment, unless a ditch or pipeline to bring water in has to be laid for any considerable distance (a ditch in Siskiyou County, California, brought water to the Yreka placers along a 100-mile circuitous route).

The best conditions for concentration of gold into placers are those where deep decay of the rocks has been followed by a slight regional uplift. The best placer concentrations probably occur in rivers and streams of moderate gradient (about 30 feet per mile), under well-balanced conditions of erosion and deposition. Except where gravel bars may form in the slower reaches, particularly within the arcs of curves, very little gold concentration will take place in gorges or steep-sided canyons. Ancient gravel bars, through further deepening of the channel, may be left as elevated benches or terraces, and in auriferous areas these geologically older sands and gravels may be rich in gold content.

Most of the gold in a placer generally rests on or near bedrock. In some instances, the coarser gold may be scattered through the lower 4 to 20 feet, or the gravel may be richest a few feet above bedrock, but never is the richness equally distributed vertically.

Among the best types of bedrock are compact clays, somewhat clayey decomposed rock, and slates or schists whose partings form natural riffles. Smooth, hard material does not catch or retain alluvial gold effectively. Gold also works down for some distance into minute crevices in hard rock, for one to 5 feet into pores of soft rock, and for many feet along solution cavities in limestones. The high insolubility of gold in most surface waters means that flake or flour gold may be carried by rivers of moderate gradient for hundreds of miles. The Colorado River is one example, where showings of gold all the way to the head of the

Gulf of California represent gold derived from Rocky Mountain sources in Wyoming and Colorado.

The fineness of placer gold is generally higher than that of vein gold from the same district. The increase in purity, which is proportional to the distance that the placer material has been transported and to the decreasing size of the particles, has been shown to be due to solution and abstraction of silver by the transporting waters.

Types of Gold Placers

There are eight easily recognizable categories of placers. In any one state or region, only certain categories will be found, as can be inferred from the following descriptions:

Creek placers are gravel deposits in the beds and intermediate flood plains of small streams.

Bench placers are ancient stream gravels standing from 25 to several hundred feet above the present streams. They were formed by a general uplift of the terrain and a sharper down-cutting of the channel by runoff waters.

Hillside placers are gravel deposits lying between creek and bench placers, where the bedrock slopes. This category may include normal mantle or alluvial-type placers weathered out of veins located upslope from the deposits.

River-bar placers occur as gravel flats in or adjacent to the beds of large streams or rivers.

Gravel-plain placers occur primarily in coastal or other lowland plains. In the western and southwestern desert states, gravel-plain placers may cover hundreds of square miles usually notable for a complete absence of water. Many well-known gravel-plain deposits listed in Part II have not been worked because of a lack of water, or at best only here and there with dry-washing equipment.

Sea-beach placers are reconcentrations of coastal-plain sands by waves along the ocean shore. Many of the black-sand beaches along the Oregon and Washington shoreline have produced gold and platinum in commercial quantities. Such placers are renewed annually by the continuous deposition of new silts and sands and by winter storms at sea.

Ancient-beach placers are deposits found on the coastal plain along a line of elevated benches and terraces, created by the uplift of the continental land mass within recent geologic times.

Lake-bed placers are gravel accumulations in the beds of present or ancient lakes that may have been dammed by landslides or glacial morains.

Gold Particle Sizes

Placer gold is found in a variety of forms and sizes, most of it small to microscopic. At the top of the list and most eagerly sought by all gold hunters is the *nugget*, a term that prospectors reserve for any bit of free, native gold larger than one grain in weight.

Gold *grains* are smaller than a grain-sized nugget. In a placer deposit grains of gold are often much smaller than the grains of associated minerals.

The term *coarse gold* refers to gold that will not pass through a 10-mesh screen. *Medium gold* will pass through a 10-mesh screen but not a 20-mesh screen; it averages about 2,000 particles to a troy ounce. *Fine gold* will pass through a 20-mesh screen but be captured by a 40-mesh screen; it averages twelve thousand colors to the troy ounce.

Very fine gold, averaging around forty thousand colors to the troy ounce, can be screened through a 40-mesh screen. Finer yet is *flour gold*, which is not well defined but is obviously in still finer particles than very fine gold. Flour gold tends to float right out of a gold pan and over the riffles in any type of mechanical or sluicing operation. Such extremely fine gold has a very large surface area in proportion to its weight and so cannot break the surface tension of water.

Pan Sampling a Gold Placer

The ability to estimate accurately the amount of gold in a cubic yard of gravel from the colors saved in a gold pan is one of the most useful and practical accomplishments that the gold hunter can learn. An old method of calculating the worth of a placer deposit is actually based on gold panning and counting or estimating the sizes of colors, a technique that improves with experience. By taking the gold fineness at a general average of 850, and a gravel swell of 25 per cent, you can utilize a standard gold pan by including all rock or making an allowance for whatever portion of a boulder that is properly a part of the sample.

You count 181 "struck" pans as equivalent to one cubic yard (one ton) of material in place. A struck pan is one filled level with the edges, not heaped up, and this number of pans allows for gravel swell (a rock is compact and smaller in cubic measure than an equivalent weight of sand). The value per yard of dirt is calculated from the average weights of the five sizes of gold colors ordinarily found in a pan but does not include flour gold.

Very fine gold	0.05 milligram
Fine gold	0.20 milligram
Small particles	1.00 milligram
Medium particles	2.00 milligrams
Coarse gold	4.00 milligrams

Nuggets larger than coarse gold are weighed or estimated separately and added to the total counted above.

For rapid computation, count or appraise the number of each category of color saved in your pan, reduce them to milligrams, and add the total weight of all five categories. If you multiply the total gold weight in milligrams that would be in 181 pans (actually panned out or calculated by averaging several pans) by 85 per cent of the current price per troy ounce of gold, you will obtain an excellent approximation of the dollar value of the gold per ton in the placer deposit. The total amount of rock and gravel in the deposit itself can be calculated from standard tables in a mining engineer's handbook or estimated from a rough measure of the area of the deposit multiplied by its average depth, thereby providing a sound estimate of whether it would be worth your time to stake a claim and develop a small placer mine.

Panning for placer gold is by far the easiest procedure for garnering a few colors, grains, or an occasional nugget of gold. Many gold hunters simply prefer to pan for gold as a part-time hobby. Nevertheless, a deposit that seems to afford a certain amount of wealth is not one that the normal prospector wants to let go to somebody else. If you do discover a profitable placer, your next step is to stake a claim and begin the annual assessment work required by mining law to give you full authority to reap from it as much gold as your ingenuity and mining capabilities allow. Now comes the hard work.

3

Staking a Claim

Mining law is a complex subject, and although based on federal regulations, it also includes special laws that separate states have imposed and that differ from one state to another. A gold hunter who suspects that he has found a potentially productive placer or lode gold deposit must obtain a copy of all laws affecting that location and abide by their differing provisions. In general, both federal and state regulations are on file with the larger libraries in each state.

Locating a Placer Claim

Many minerals are open to "location"—that is, a mineral deposit (in this case, gold) may be appropriated by filing a mining claim. In general, all unreserved and vacant public lands belonging to the United States and to many separate states are open to mineral location. Reserved lands include national parks and monuments (except in Alaska, where prospecting and mining are still permitted), Indian reservations, and military reservations. To locate a valid placer claim in most states, you must:

1. Make a valid discovery of gold or other mineral.
2. Erect a discovery monument and post thereon a notice of location (each state provides standard forms).
3. Mark the boundaries, unless the claim is taken by legal subdivision.
4. Record in the county recorder's office a true copy of the location

18

notice together with a statement of the marking of the boundaries (if done) and including section, township, range, and the base meridian within which all or any part of the claim is located.

(In the past, very many mining claims have been located in particularly attractive areas, most being on U.S. Forest Service lands, but never used for mining. Use of the provisions of mining laws to appropriate public land for recreational or residential purposes is perpetrating a fraud and is likely to lead to legal trouble.)

A Valid Mineral Discovery

You are not permitted to guess that a valuable mineral, like gold, is present. A valid mineral discovery means that you know and can prove the presence of gold (or any other mineable mineral) within the boundaries of your claim.

Discovery Monument and Location Notice

A conspicuous and substantial monument must be erected at the point of discovery, and the location notice must be posted in or on the monument. As legally defined, the monument may be a wooden post or stone structure at least 3½ inches in diameter or a metal post at least 2 inches in diameter, projecting at least 3 feet above the ground. Most locators simply pile up a 3-foot cairn of rocks, inside of which they bury the claim notice, often in a tobacco can.

The location notice to be posted must contain: (1) a name for the claim; (2) the name and address of the locator or locators; (3) the date of location—that is, the date of posting; (4) the number of feet or acreage claimed; and (5) a written description of the claim referred to some natural object or topographic feature, or area section marker if one can be found, so that the claim may be clearly identified on a quadrangle map.

Boundaries

The boundaries of a placer claim not located by legal subdivision must be marked so that they can be readily traced, and a conspicuous monument must be erected at each corner. Each such corner monument shall be labeled so as to designate the corner—for example, northwest corner, southeast corner, etc.—to which it applies, together with the name of the claim.

The maximum size of a placer claim is 20 acres for an individual and 160 acres (quarter section) for an association of not less than eight persons or, if the association comprises fewer than eight, 20 acres per member. There is no limit to the number of claims you may hold, so long as you can prove the existence of a valuable mineral on each one and meet the other requirements.

A placer claim must be located upon the ground in such shape (usually 1,500 feet long by 300 feet wide) and position that it will conform as nearly as possible to the lines of a public survey—that is, it should have north–south and east–west parallel boundaries. In a legal subdivision a limit of 10 acres is placed on any one claim, and the plot must be square.

Recording the Claim

Within ninety days after posting a placer claim notice of location, you must record its true copy in the office of the county recorder. However, only thirty days are allowed in which to record a claim in any National Forest Wilderness area, and that with the district forest supervisor or ranger having jurisdiction over the land.

You should bear in mind that state statutes supplement the United States mining laws. Therefore, before formalizing your claim, and after checking the status of the land to be claimed as to whether it really is open to mineral entry, you must obtain the supplemental information and directives from state officials and abide further by the specific regulations in these documents.

Retaining Rights to the Claim

At least $100 worth of labor or improvements must be performed each year on or for the benefit of each claim you declare. This is called the "annual assessment work." This work must be completed on or before noon every September 1 in order to prove your active interest in the claim. This must be done in order to maintain your right to the claim against having it "jumped"—that is, taken over by somebody else who has that right the moment your assessment work lapses. You should also record in the recorder's office each year a statement that you have performed such work.

You may not build a house or cabin or other improvements such as tool sheds or ore storage bins, etc., on any mining claim, unless such structures are reasonably necessary for your use in connection with actual mining operations. Remember, an unpatented mining claim is to be used for mining purposes only.

PATENTING YOUR CLAIM

A patented mining claim refers to a piece of ground for which the federal government has given a deed or has transferred its title to an individual. An unpatented claim is one under which an individual, by the act of valid location under the mining laws, has obtained only the right to remove and extract minerals from the land, but where full title has not been acquired. You may pay no county or state taxes on an unpatented claim; you do on a patented piece of land. Your rights under each type of claim differ somewhat, and it is not necessary to have a patent in order to mine and remove minerals from a valid claim.

If you establish a valid claim, perform and record the annual assessment work, and meet all other requirements of federal or state mining laws and regulations, you establish a possessory right to the area covered by your claim. You may sell, inherit, or be taxed if your claim is on state or county land, but no other person can mine the minerals you have claimed without your consent.

But until you obtain a patent to the claim, you do not hold full title to the land. At any time, the government may challenge your claim on one of several grounds: that it lacks a bona fide mineral discovery; that the minerals have been mined out; or that the claim fails to meet other provisions of the law. If the government's contentions prove out, the claim may be canceled and you have no further rights to the land; you can even be charged with trespass.

Once you have obtained a patent for the claim, however, you become the legal owner of the land in accordance with the regulations in Title

43, Code of Federal Regulations, Part 3400, a copy of which you can obtain from any office of the Bureau of Land Management. Each application for patent must be accompanied by a $25 filing fee. The application must show your right of possession to the claim and must state briefly the facts constituting your right to patent. This means that you have to show discovery of a valuable mineral deposit within the limits of the claim and that you have performed not less than $500 of assessment work, an amount that can be expended all at once or at the rate of $100 per year for five years. If a final certificate for the claim is granted, no further annual assessment work is required, but you do become immediately subject to county and state property tax assessments.

Finding a Lode-Gold Deposit

Lode gold originates at great depths in the earth's crust. It is carried, along with other minerals, by rising, hot liquids emanating from a source such as a cooling body of magma. Under considerable pressure, the hot mineral solutions follow natural cracks or crevices in the crustal rocks and travel upward until they encounter cooler rocks. Here, where temperatures and pressures are less, the minerals crystallize out of solution and accumulate along the fracture surfaces and in voids along the fractures to form veins.

Commonly, the principal constituents of a fracture filling or vein are quartz and calcite, but the vein material generally also contains some copper, lead, iron, silver, and other metal-bearing compounds. In many veins, brasslike iron sulfide or pyrite is abundant. Pyrite and iron-stained mica are often mistaken for gold by the mineralogically uninitiated, especially when found as small particles in a gold pan—hence the name "fool's gold."

The deposition of gold is rarely uniform, and it may occur in bands, in a series of lenses separated or connected, or in irregular shapes like splashes of ink on blotting paper. Visible gold is not common in an average ore, and it is necessary to sample and assay frequently and to plot the gold assays on an accurate mine map to guide the mining of the deposit.

Lode-gold deposits are difficult to discover, and the history of gold mining is full of tales of fortuitous circumstances that have led to famous discoveries. Lode deposits do not always crop out, and when they do only a small area may be exposed, and that area may be hard to distinguish from the surrounding rock. The presence of placer gold in a stream bed usually indicates a source in a vein or lode deposit somewhere in the vicinity, always uphill from the placer showings

PROSPECTING FOR A QUARTZ VEIN

The first requisite in all prospecting is to learn to use a miner's gold pan efficiently, or in the absence of one, a small grease-free frying pan with sloping sides. The gold pan is simply a sampling tool that can lead you from the first show of colors to the source lode that they came from.

A gold-bearing vein may or may not be visible on the surface of the ground, for it may be buried beneath rubble and vegetation. During the slow breakdown of a mother lode through weathering, gold is freed from its matrix rock and becomes scattered in the soil, usually close to or on the bedrock below the vein. The movement of gold eroded from a vein is like a slow downward seepage of water.

For example, visualize a small vein of gold ore occurring well up on a hillside and running in a direction nearly parallel to the base of the hill. If, at a point about 20 feet below and on a line parallel to the vein, a number of samples 5 to 10 feet apart are gathered by digging down to or nearly to bedrock, they are likely to yield gold colors in panning. On a line 40 feet below the vein and still parallel to it, samples taken in the same manner may also yield gold colors, but they will probably be fewer in number in the pan. At 60 feet or perhaps as far as 200 feet below the vein, colors might still be obtained, but sparser yet in number per pan. In searching for a lode-gold deposit, conditions are simply reversed. The source is unknown, but the finding of colors is an indication that a gold-bearing vein exists at some higher point.

In prospecting a hill, holes are dug along its base at intervals of 50 feet or more, and the alluvium down near the bedrock is panned very carefully. When colors are found, the prospector ascends the hill perhaps 20 feet, where he digs a series of holes, similarly spaced, along another line parallel to the first. He pans samples along this line and then climbs another 20 feet higher and repeats the process, beginning his first hole above that section of the lower lines where he obtained his best color showings. What he is actually attempting to do is to follow the gold flow across the face of a narrowing triangle to its source at or near the apex by picking up and noting the number of little yellow flecks of gold concentrated in the bottom of his pans. As the number of colors increases in his pan, he shortens the distance between each hole. As he nears the apex of the triangle, the number of colors he finds near bedrock will be much greater than at the base of the triangle along the bottom of the hill.

If the prospector's topmost row of holes shows no colors, then he can be certain that he has crossed the vein into barren ground. Once he has more or less outlined the lode vein, he can stake a claim along 1,500 feet of its length and extending above and below its center for 300 feet. He can stake as many claims along the length of the vein as may be necessary to enclose it completely, but very few gold-bearing veins extend to lengths of more than a quarter mile.

This method of prospecting is called "post-holing." If the prospector wishes to be thoroughly methodical about it, he can map his progress in a notebook, labeling the separate rows with Roman numerals and assigning Arabic numerals to the holes within each row.

From Creekbed to Lode Vein

In auriferous districts gold can usually be found on the bedrock of creeks or gullies. To search for gold in a dry creek, find a place in the watercourse where the bedrock is exposed or nearly so. Gold lodges under large rocks and in cracks in the solid foundation. Find a fracture in the bedrock. Pry it open with a pick or bar. Your pan filled with water should be handy, even though in the aridity of the creek or gulch you'll probably have to haul your water with you.

Lift out the rocks as they are broken and wash them in the pan, scraping off any adhering clay or sand. Scrape up all the sand from the crevice and place it in the pan. A small paintbrush, a spoon, and an old table knife or putty knife are useful in scraping up all the fine sand that might be lodged in a crevice or under a boulder. Scrape the bedrock vigorously and brush up the sand and dust carefully, for gold particles may have sunk deeply.

Sometimes three or more scrapings from different parts of the watercourse may be obtained for a single panning test. Pan very carefully. If gold is found, ascend the watercourse and continue to pan the bedrock materials at spacings of 50 feet or more. When a point is reached where panning no longer yields any colors, or the number of colors greatly diminishes, go back to where the last gold was obtained and "post-hole" a triangle across the face of the hillsides, first on one side of the watercourse and then on the other.

With paper and pencil you can outline the probable course of quartz or float from a vein and plan the finding of the deposit by tracing the cast-off flakes of gold or pieces of quartz.

Disintegrated quartz that has separated from its vein and become scattered follows a downhill course from its point of origin (the apex of the triangle), spreading out laterally just as does gold, only the quartz is likely to travel and spread much farther. Quartz is often found right on the surface of the hillside, and you can trace the float backward to its source by careful observation without the necessity of digging postholes to bedrock. Of course, many quartz veins do not carry any gold at all, and there is no use tracing back quartz that has not panned out traces of gold after being crushed in a mortar.

In testing quartz for gold, pulverize a small amount in a mortar by pounding and grinding it to fine sand with the pestle. Or use a 1½-inch length of 2-inch iron pipe upended on a flat rock; drop in a few pieces of quartz, crush them with one end of your prospecting pick, and scrape every particle into your pan. You may have to use a large pan filled with these fines, and your panning must be very carefully done.

Quartz, feldspar, and calcite all look very much alike to the uninitiated. Before attempting to locate a quartz vein (feldspar and calcite rarely contain gold), learn to recognize it from any handbook or gemstone hunter's field guide.*

When you find quartz that you suspect may contain a trace of gold

---

*E.g., J. E. Ransom, *Gems and Minerals of America* (New York: Harper & Row, 1974).

and have pulverized it in your mortar, use a magnet or magnetized knife blade to stir around in the powder to remove possible fragments of iron. Have an expert identify for you any unknown heavy mineral you find in your panning concentrate. It might be platinum, which is even more valuable than gold. Learn also to identify the black grains of iron oxides, ilmenite, chromite, or other heavy mineral constituent of black-sand deposits. Use a pocket magnifying glass to study your concentrate, almost grain by grain. Have any quartz that contains fine-grained pyrite or lead and silver minerals (for example, the cubic gray crystals of galena) assayed for gold.

Claiming a Lode Discovery

Before making any substantial investment of time or money in a mining claim, it is good business for a prospector to check out the status of the land on which he hopes to record a claim. The land status records located in a state's land office or in the office of a county recorder are one of the prospector's most important tools, surely equal to his pick and shovel.

Even though the lands may lie within the public domain, they may have been withdrawn from mineral location. A withdrawal may be made to reallocate the land for a wide variety of uses, such as for a Defense Department training area, for national parks and monuments, for power sites, for Indian reservations, or for reclamation purposes. A withdrawal usually closes the land to entry under the public-land laws. The important thing is to remember that not all public lands are truly public.

You do not need a detailed check on every piece of land you may look at. Failure to check on the general area surrounding your projected claim can, however, result in the loss of a substantial investment in the time and effort to perform your original prospecting.

Once you have determined that a lode deposit exists and wish to establish a valid claim, you must follow the four basic procedures described for locating a placer claim (see page 18). There are some differences, however. As with placer sites, within sixty days of the date of location, you must distinctly mark out the entire lode claim, but you must place additional substantial monuments at the center of each end line.

The maximum size of a lode claim is the same 20 acres, but the side boundaries must not lie more than 300 feet on either side of the ore vein, which may actually twist and curve as it is traced out on the ground. Only the end lines need be parallel, to provide extralateral rights. Moreover, a lode claim may be smaller than the maximum size allowed. The sides do not have to be parallel, but may follow along any irregularities in the vein itself. The discovery point on which the location monument and claim notice are posted may be anywhere along the vein within the claim boundaries. Procedures for recording the claim are the same as for placer sites, and the assessments are the same.

Abandoned Mines

In practically every mining district in the nation you will find apparently abandoned mines in every stage of decay. Many old mines will show every evidence of disuse dating back fifty or a hundred years or more. Nevertheless, since old-mine dumps and subsurface gold deposits may be valuable today for minerals other than gold, or because modern techniques and power sources can make mining low-grade left-over ores profitable, such abandoned mines may be worth reclaiming and reopening.

But remember, an inactive mining claim has not necessarily been abandoned. Even though a claim may look inactive, the relatively small amount of work valued today at $100 for assessment—which may be going on underground or on an adjacent claim—will satisfy the annual legal requirement to maintain possessory rights in a claim. Furthermore, if a mine—especially a developed, obviously once productive mine—is inactive, there is probably a good reason. Perhaps the ore is mined out, or costs of production have jumped far above the profit margin; perhaps the current price of that mineral commodity (other than gold) has dropped too low for a prudent man to make a reasonable living; or perhaps expensive machinery is required for further development work but hasn't arrived. Despite these obstacles, the original claimant may have retained his possessory interest.

In mining law, abandonment of an unpatented claim most commonly requires that there be some deliberate, positive action to demonstrate intention to relinquish all possessory rights. Few mine claimants are aware of this regulation, or else they simply ignore it and move on elsewhere, once they have gleaned as much gold as they can from a claim. In prospecting through many mining districts, you are likely to find posted claims that show no evidence of any annual assessment work performed within the year prior to your rediscovery of the claim. In such circumstances, if you wish to "jump" the claim and take it over in your own name, the burden of proof that the annual assessment work has not been done by the previous claimant is upon you.

In some states, however, failure to file an annual statement of work performed is taken as prima-facie (that is, self-evident) presumption of the act and intention of the owner to abandon such claim. The burden of proof that the required work has been done then lies on the original claimant.

The history of practically every mining district anywhere is a record of legal disputes (some tying up mining for years on end), brawls over claim jumping, and court battles of epic proportions. In all too many cases, the attorneys have ended up with all the gold.

ANNUAL ASSESSMENT WORK

Annual assessment work, also called "annual expenditure," is yearly labor required by federal law to hold possession of an unpatented mining claim. The law is not generally specific as to the exact nature of acceptable work, but it implies that the work be directed toward production of an economically valuable mineral or toward exploration or enhancement of the economic mineral value of the claim. Only work or labor can be applied; the cost of equipment does not count. It is left to the courts to decide the applicability of any work that is challenged.

Assessment work may be done either underground, as in a lode mine; or in a drift tunnel in a buried-channel placer claim; and even off the claim entirely (for example, in building a road for access), if the work being done is for the benefit of mineral production in time. In determining the amount of work done on a claim for purposes of representation, the legal test is as to the "reasonable value" of such work, not what you may have paid somebody else to perform it.

Geological, geophysical, or geochemical surveys by qualified experts may be considered "labor" for no more than two consecutive assessment years per claim. In order to count at all, such technical work must be verified by a detailed report filed in the county recorder's office specifying, among other things, the name, address, and professional background of the person or persons performing the survey.

SEARCHING THE MINING RECORDS

On the face of it, locating the history and records of ownership of an old mine or claim, especially if the latter was never patented, is a nearly impossible task. Most mining districts are 50 to 150 years old. Tens, perhaps hundreds, of thousands of prospectors have filed literally millions of claims since the first gold discoveries were made, and many of these claims may overlap one another in a literal tangle of potential legal confusion.

Also, in the early days, counties were often very large. As population increased, these big counties were subdivided. For example, trying to locate the records of a claim that your grandfather may have filed in 1850 "somewhere in Mariposa County, California," might require a detailed search of the county recorder's books in all of the present-day counties of Mono, Merced, Madera, Fresno, Inyo, Tulare, Kings, Kern, and San Bernardino.

The search of a county recorder's records can be time-consuming indeed, because of the vast number of documents on file. Very few recorders maintain cross-references and indexes to records of mining claims. Many recorders, or their clerks, simply recorded all mining claims chronologically, just as they came in year after year, perhaps for a century or more.

There are generally no maps available from official sources showing the locations of unpatented mining claims; the boundaries of such claims were rarely surveyed—as is necessary to obtain a patent—and they cannot be plotted accurately. Nevertheless, the recorder's office in any county is the best source of names and addresses of current or recent owners of unpatented claims.

The county assessor can provide maps, called plats, showing the location of all patented claims, names and addresses of those to whom the property is assessed for taxes, and tax records filed with his office. This readily available record is your best lead to taking out a lease on an old mine or its dump, if you believe that it would be to your profit to work it anew.

New Forest Service Regulations

Late in 1974 the U.S. Forest Service proposed certain new regulations affecting prospecting and mining on any of the 140 million acres of public land it supervises, giving some leeway to casual prospectors and gem and mineral hunters. The revision, certain to become law, was aimed at clarifying an original proposal made by the Department of Agriculture in accordance with the demands of ecologists. Basically, the new regulation requires that all those who prospect for or mine any mineral must be held responsible for leaving the area just as they found it, or as near as possible to its natural state.

The regulation will not affect the weekend or vacationing gold hunter armed only with hand pick, gold pan, and pickup truck, although scuba divers will be held to general regulations prevailing in some areas relating to stream pollution. However, if you decide to move large equipment over public land in order to work a particularly profitable claim, you will have to file a notice of intent with the Forest Service.

If the Forest Service determines that your operation might cause significant disturbance to the environment, you will be required further to submit a plan showing how you expect to conduct the prospecting or mining, and this plan will require approval by the Forest Service before you begin operations. Furthermore, you will be held responsible for restoring the mined area or its access routes and will have to certify to that responsibility by posting a bond sufficient to cover the costs of such reclamation.

4

Elementary Placer-Mining Methods

Although buried-channel placers are found in many auriferous regions, the amateur prospector will be interested primarily in surface placers. These are of two types: residual placers and transported placers. These are of two types: residual placers and transported placers.

When quartz or other gold-bearing rock becomes disintegrated by weathering, some of the lightweight rock fragments may have been removed by erosion, leaving the heavier gold concentrated as a residual placer. Transported placers are formed when both the gold and the rock fragments are moved from their source, as was described in Chapter 2. Of the various categories of gold placers, stream deposits are the easiest to work and are usually the most productive, where the greatest concentrations of gold are found close to or in crevices in the bedrock. It is advisable, therefore, to work the bedrock to a shallow depth in order to obtain complete gold recovery.

Placer-Mining Methods*

The basic tools for the placer miner are few and inexpensive: a long-handled, round-pointed shovel for digging; a 16- or 18-inch gold pan or batea; a hand magnifying glass for inspecting tiny gold particles;

*Abstracted from *Basic Placer Mining* (Washington, D.C.: U.S. Geological Survey, 1974), Special Publication 41; revised from *Mineral Information Service*, Vol. 10, No. 6 (Sacramento, Calif.: California Division of Mines and Geology, August 1957). An expanded and classic detailed procedure for placer mining is Wm. F. Boericke, *Prospecting and Operating Small Gold Placers*, 2d ed. (New York: John Wiley & Sons, Inc., 1936).

a prospecting pick; and a full-sized pick and a mattock. In addition, a compass, notebook and pencils, maps, and sample sacks come in handy. The complete prospector is likely also to equip himself with measuring tape, dip needle (used in magnetic prospecting), geochemical kit, carbide lamp, and camping-out supplies and equipment. A sampling moil and hammer, mortar and pestle, and small shovel are other useful tools. Much equipment for prospecting in one part of the country may be useless in another section. It is often useful to the beginner to find out from old-timers or resident prospectors what types of equipment they may recommend for exploration in their areas.

Wet Placer Methods

Gold Pan and Batea

Of all the prospector's tools, the simplest and most useful are the gold pan and the batea, both types of pans serving to clean gold-bearing concentrates and to hand-work rich deposits. Gold pans can be purchased in almost any hardware store in a mining region; bateas, used mainly in the tropics and in the Philippines, where they are considered superior to all other pans, may have to be put on order. However, a large wooden salad bowl makes an excellent substitute.

The gold pan is a shallow pressed-steel pan 15 to 18 inches in diameter at the top and 2 to 2½ inches deep, with sides sloping at about 30 degrees. It weighs 1½ to 2 pounds and has a rim turned back over heavy wire for stiffening. A skilled operator can pan out a cubic yard of gravel (about 3,400 pounds) in ten hours. In practice, a good miner can handle about six pans an hour of mixed sand and gravel.

The object of panning is to concentrate heavier materials by washing away the lighter debris. The pan is filled about three quarters full of gravel to be washed, then submerged in water. The large gravel is first picked out by hand, and any lumps of clay are broken up. Then the operator raises the pan to the surface of the water, inclining the pan slightly away from him, and moves it in a circular motion combined with a slight jerk, thus stirring up the muck and light sand and allowing it to wash off the lower edge.

This is continued until only the heavier materials remain, such as gold, black sand (usually in considerable abundance), and other substances having a high specific gravity (for example, platinum and stream tin). These heavier fine-grained materials, called concentrates, are saved until a large quantity accumulates in a pile or in a container. The larger particles of gold may be extracted with a pair of fine-tipped tweezers and placed in a medicine vial or bottle partly filled with water for show. Smaller gold particles may be amalgamated with a small amount of mercury, preferably in a copper-bottomed pan.

If the separation is difficult and the grade and quantity justify, the concentrates may be shipped to a refiner or smelter. Processing charges for this service are outlined in Chapter 7.

Panning may best be learned by watching an experienced operator at work, learning various tricks from him. Incidentally, a clean 6- or 8-inch frying pan makes an effective prospecting or clean-up pan, but it must be absolutely free of grease. It is well to burn out an iron pan after having used mercury in it, and then scour it thoroughly with a soft rock or piece of brick; otherwise, it may be impossible to see small colors or flakes of gold.

The batea is a shallow, cone-shaped wooden pan that performs essentially the same function as a gold pan, although techniques of using it differ somewhat from the conventional gold pan. Most bateas are from 15 to 24 inches in diameter, with an angle of 150 to 155 degrees at the apex. A 22-inch batea holds about one third of a cubic foot of gravel, and a skillful operator can handle from 80 to 120 cubic feet per day. Many experienced gold hunters claim that wood will hold fine gold much more effectively than will metal.

The Long Tom

A long tom is an inclined trough used to concentrate gold from auriferous earth. Used singly or in a series of several troughs feeding successively one into the other, a long tom has a greater capacity than a rocker (see page 34), but it also uses more water in operation, because flowing water is the carrying agent of finer materials. A long tom is usually crudely constructed from rough lumber and consists of two sections: the sluice box and the riffle box. The slope is generally one inch per foot of length, but may vary according to conditions.

The sluice box section should be about 12 feet long. At its head, or upper end, the box is from 15 to 24 inches wide; at the tail, or lower end, it is from 24 to 36 inches wide. The side planks should be from one to 2 inches thick and at least 8 inches high at the tail. At the head end is a flume or iron pipe from which a steady flow of water is fed into the sluice box. A screen or piece of perforated sheet metal placed across the tail prevents the coarse material from passing on into the riffle box.

The riffle box is usually shorter than the sluice box and slightly wider at its tail end. This catchment box is set just below the first opening in the screen and is sometimes set at a more gentle slope. Here the riffles (ridges set crosswise to the flow of water, see page 40) are attached to catch any gold coming over them. The box may be lined with canvas, and it is best to use removable riffles. The sluice box needs heavy planks to withstand gravel abrasion, but the riffle box can be made of lighter lumber. The capacity of a long tom is from 4 to 6 yards of gravel per man in a ten-hour day, two to four men working at shoveling in the gold-bearing gravel and picking out the larger rocks by hand.

The ground to be worked is shoveled into the sluice box and washed by the water coming from the head end. One man works the material in the trough with a heavy iron rake or a fork, taking the coarser gravel out after it has been washed clean and keeping the screen at the tail from clogging. Clean-ups are made whenever necessary, usually at the end of each day or oftener, if the riffles become loaded. As in panning, the concentrates are stockpiled.

THE ROCKER

The rocker, a kind of mechanical panning device, has the advantage of requiring much less water than a long tom—hence the rocker's use in areas where the water supply is short. By using 150 to 300 gallons of water in a ten-hour day, two men can wash from 3 to 5 yards of gravel and save any contained gold during the concentration process.

Rockers vary greatly in size, shape, and general construction, depending on available construction materials, the size of the gold particles to be recovered, and the builder's mining experience. A variety of commercially built steel rockers are available from mining supply houses. Rockers range in length from 24 to 60 inches, in width from 12 to 15 inches, and in height from 6 to 24 inches. Some have a single apron, others two aprons, and screens with holes as much as one-half inch in diameter. Among the devices used to recover gold particles are riffles of all kinds, blankets, carpeting, gunny sack, burlap, cocoa mat, rubber mat, canvas, and occasionally an amalgamated copper plate.

A fairly efficient and easy method of constructing riffles for a rocker consists of clamping a three-eighths-inch metal lath over a double thickness of blanket, the clamps allowing the blanket to be removed easily for cleaning.

A rocker is built in three parts: a body or sluice box, a screen, and an apron. The floor of the body holds the riffles for catching gold. The screen catches the coarser materials and is placed where clay can be broken up to remove all contained particles of gold. The apron is used to carry all material to the head of the rocker, and it is made of canvas stretched loosely over a frame. The apron sags into a low place, or pocket, in which coarse gold and black sands can be collected.

The accompanying illustration shows the general outlines of the placer equipment described so briefly above. Detailed directions for assembling and making them may be obtained from almost any library, state department of mines and minerals, or from the publication referred to in the the footnote on page 31.

OPERATING A ROCKER

When the ground to be worked has been found, the placer operator selects a place near a water supply. The bed plates should be set so that the holding spikes fit into the holes in the plates to provide a proper slope, determined according to the type of ground to be worked. Where most gold is coarse and there is no clay, the head bed plate should be 2 to 4 inches higher than the tail bed plate; where most of the gold is fine, or clay is present, this slope must be lessened somewhat, perhaps to only one inch. It is difficult to save very fine gold if muddy water is used, since the operation does not allow enough time for the fine gold to settle but rather floats it off.

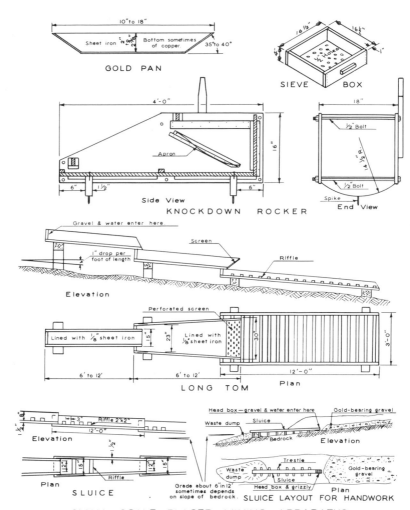

SMALL SCALE PLACER MINING APPARATUS
(Taken from U.S.B.M. I.C. 6611R)

After the rocker is in position, the box is filled with gravel, which is washed by pouring water over it with a dipper. The larger gravel, when clean, is picked out, and all clay lumps are pulverized by hand. The machine is then rocked vigorously for several minutes while more water is added.

It is important to use only the right amount of water. Too much water will carry the material through the rocker too quickly and with it much fine gold; too little water will make a mud that will not allow fine gold to settle.

When all material that will pass through the screen has done so, the box is dumped and the operation is repeated with a new load of gravel, until it is necessary to clean the apron. Because all coarse gold is caught in the apron, the latter should be cleaned several times a day. The concentrates are put in a container or stockpiled for further cleaning. The riffles are cleaned whenever necessary, but not as frequently as the apron. When a blanket is used, it should be washed out carefully in a tub of water, as here a good percentage of any fine gold is caught. All concentrates must finally be cleaned further by ordinary panning.

THE SLUICE BOX

Sluicing is a method of working auriferous gravel in a flume or ditch. The sluice box is a crude sloping trough, having riffles on the bottom to catch the gold. Dimensions vary greatly and are governed by the amount of material to be washed through the sluice. The slope ranges from 5 to 18 inches per 12 feet. The riffles also vary, and a sluice may contain several kinds of riffles, some of which can be quite elaborate and require considerable work in laying.

In the long tom and rocker, the coarse materials are removed directly, but in the sluice everything is allowed to pass through, except in boxes that have a grizzly (a heavy steel-bar screening arrangement) at the head. The coarse material that is allowed to pass through the box serves to grind and clean the gold, thereby making the gold easier to amalgamate or pan, and freeing some material otherwise mechanically held.

In sluicing, much of the manual labor done in the preceding methods is eliminated, as the water does all of the carrying. The mining itself may be done by hydraulicking, or a stream of water may be allowed to fall over a bank, washing the material into the sluice. Sluicing requires more water than the previously described methods, since it takes 20 to 80 cubic feet of water to move one cubic foot of gravel.

Coarse gravel requires more water than fine gravel mixed with sand, but as the slope or grade is increased, the amount of water needed is lessened. The capacity of a sluice box is governed by its grade, by the amount of water available, and by the dimensions of the box. In ground sluicing, a ditch is dug along bedrock containing natural irregularities that furnish pockets that catch the gold.

The Dip Box

The dip box is a modification of the sluice box, 6 to 8 feet long, which may be used where water is scarce and the grade too low for an ordinary sluice. A dip box is portable, being readily carried in an automobile. A dip box will permit handling about as much dirt in a day as a rocker, if the larger stones are thrown out by hand.

The Puddling Box

If a muddy or clayey material is to be sluiced, the first box of the string is made into a "puddling" box. This box can be 6 feet long by 3 feet wide, or any convenient dimensions, with 6-to-8-inch sides and no riffles. The clayey material is shoveled into this box and broken up with a hoe or rake before it passes on into the main sluice, since unbroken lumps of clay may pick up and carry away gold particles.

Dry Placer Operations

There are many widespread auriferous gravel deposits scattered over the dry deserts of the western and southwestern states. In general, no large-scale effort to obtain gold from these placer areas has proven financially successful. Nevertheless, early Mexicans and Indians, working alone or in pairs, produced considerable placer gold by "dry wash" methods from Nevada to northern Mexico and from eastern California to New Mexico.

Ideal ground for dry concentration of gold is difficult to find. Once the dry surface is penetrated, most ground will be found to contain enough moisture to prevent separation of the light and heavy particles. The moisture content of "dry placers" is the principal obstacle to successful operation, although given time and ingenuity, moist ground can be dried out in stages before it is worked.

WINNOWING

Winnowing is the fundamental dry-wash method. It involves screening out all the coarse gravel, then placing the "fines" in a blanket and tossing them into a good breeze, one man at each end of the blanket. The lighter particles are blown away by the wind, while the heavier particles fall back onto the blanket. The weave of the blanket tends to catch and hold onto fine gold.

THE DRY WASHER

A dry washer is usually a home-built piece of equipment that works on the winnowing principle to separate heavy from lightweight particles after hand-picking has removed any larger rocks. Commercial models are also available. Gravel is shoveled into a box having a slanted screen; the coarse material passes off the lower end, and the fines go into a hopper. From the lower end of the hopper, the sand falls onto another sloping plate lined with riffles. A bellows arrangement, which may be either worked by a hand crank or from a gasoline engine operating an eccentric shaft (often used on larger machines) blows air across the riffles, lifting the lighter particles of sand over the riffles and off the lower end of the washer. At best this is a dusty process. The heavier particles of gold and black sand remain on the upslope sides of the riffles; these sides must be periodically cleaned and stockpiled for further concentration and cleaning of the gold.

The crankshaft is equipped with a cam to vibrate the screen, and with a pulley wheel. The wheel transmits power by belt to an eccentric, which operates the bellows.

To insure a flat surface and an even distribution of air in the gold-recovery process, a riffle unit is built up as follows: A well-braced heavy screen is covered with several layers of burlap, overlain by a piece of window screen or fly screen, and covered with handkerchief linen. Above this the riffles are placed from 4 to 6 inches apart. The riffles are made from one-half-inch-to-three-quarter-inch half-round moulding with the flat face on the upper side. If amalgamation of flour gold is desired, pockets to hold mercury are placed in front of the riffles. Some flour gold also passes through the linen and is caught in the burlap.

A motor-driven dry washer can handle up to three-quarters cubic yard of dirt per hour. Hand washers operated by two men have a capacity of one or more cubic yards of dirt during an eight-hour stint, depending upon the size and nature of the material to be handled.

Riffles

Every method of placer mining, except winnowing, depends upon riffles to catch gold particles. Riffles are obstacles placed along the bottom of a sluice or rocker that form pockets to catch gold by concentrating the heavier materials. There are many forms of riffles.

COMMON OR SLAT RIFFLES

These are simple strips of wood, iron, or steel extending across the bottom of the catchment plate and spaced a few inches apart. Normal abrasion is so great on wooden riffles that frequent replacement is required. Other types of riffles are therefore preferred in larger-scale operations.

POLE RIFFLES

Two-to-4-inch peeled poles, frequently used to make riffles, are easily obtained in timbered country. Sections of debarked saplings with the knots smoothed off may be placed either across the sluice box at intervals, or lengthwise along the bottom. This type is used with coarse material and is efficient in concentrating both coarse and fine gold.

BLOCK RIFFLES

Block riffles are made by paving the floor of the sluice box with crosscut wooden blocks about 4 or more inches high, their height depending on the depth and width of the sluice. A narrow slat is nailed to each row of blocks, and the slots are then set across the bottom of the trough. In this manner a space is left between the rows of blocks, forming riffles. The blocks may be made either square or round, and the use of this type of riffle is efficient for both coarse and fine materials.

STONE RIFFLES

Rock or stone riffles are made by paving the sluice box floor with rock, either stream pebbles or flat stones quarried for the purpose. They are held in place by strips of wood nailed at intervals across the bottom. This method works nicely for fine and coarse material and is especially good for cemented material.

ZIGZAG RIFFLES

When slats are placed part way across the floor of the sluice box alternately from each side to form a zigzag pattern, fine gold is concentrated, very much as in panning.

Undercurrents

An undercurrent is a wide, flat sluice box placed beneath the main sluice box and is used to save fine gold. It is usually 5 to 20 times as wide as the main sluice and from 10 to 50 feet long. The undercurrent receives its feed from a grizzly or screen placed in the floor of the main sluice, through which the fine material drops into a trough that distributes the feed evenly across the width. Undercurrents usually have a greater slope than the main sluice, because the shallow water flow is more retarded by friction.

5

Lode-Gold Mining

Lode mining for gold is not a weekend entertainment nor a hobby to be undertaken upon retirement. It is a serious business requiring considerable metallurgical experience to determine whether the auriferous rock one has found is actually "ore." Ore, for the miner's purposes, is any rock from which a valuable mineral can be extracted *profitably*. If sampling reveals that you have found gold ore, then and only then can you plan a positive program for getting it out; if you can arrange for adequate capital, you can then put your plan into operation.

The minimum cost of a semipermanent installation that can be used to produce gold from a hard-rock mine will run several thousand dollars. And with inadequate supervision or under adverse mining conditions, the minimum costs can be exorbitant. Many good small mines fail because there simply is not enough capital available for development, payroll, equipment, and supplies. A working capital of $5,000 to $10,000 is necessary to permit a small lode mine to carry through the early stages to the period when the obtained gold concentrates will allow a worthwhile return. This means that adequate financing should be available before exploration of the mine begins.

Preparation and Costs*

Once you have filed a claim on a promising outcrop, mineralized zone, or geologic structure indicating the possibilities of an ore deposit, the

*Abstracted from Harry E. Krumlauf, *Exploration and Development of Small Mines,* rev. ed. (Tucson: University of Arizona, 1966), Arizona Bureau of Mines Bull. 164, and

first step before committing your capital to development is to consider general costs. Unless you take into account the value of the ore body, cost of beneficiation (treatment), availability of market, transportation costs in getting your ore or concentrates to the refinery, availability of cheap electric power, and many other factors, disaster can be the end result of even the hardest work.

SAMPLING A DEPOSIT

Before any kind of development is started, it will be necessary to obtain a complete picture of the quantity of ore presumed to be beneath your claim. This is termed "blocking out" an ore body, and it usually requires the expert services of a geologist. His sampling may include surface trenching, pit sampling, wagon or diamond drilling, a detailed geologic mapping of the boundaries of the ore body as determined from the drilling, and perhaps geophysical or geochemical prospecting—all merely to determine the outline and boundaries and probable total tonnage of ore. This procedure is usually financed by a partnership, syndicate, or mining company, or by sale of stock in a company you may wish to form to develop the mine. There is no inexpensive way to get around adequate sampling today.

Each scientifically gathered sample, properly representative of its part of the presumed body of ore, must be assayed, and this cost alone may be considerable. But once an actual ore body has been blocked out, the formal steps in developing a mine follow in natural order.

GETTING EXPERT ASSISTANCE

At this point the next sure step, even for an experienced prospector, is to hire a competent, state-licensed consulting mining engineer either to direct all subsequent operations or to advise on the best procedure for developing your particular type of lode deposit. The minimal consulting fee will run $100 a day and up, depending on the inflationary spiral, plus

from *The Making of a Mine* (Socorro, N. Mex.: New Mexico Institute of Mining and Technology, 1956), Circular 41.

all expenses. On a full-time salary basis, a mining engineer's annual salary may run several thousand dollars a month.

Unlike placer mining, the development of a lode mine requires the services of technically trained personnel and a considerable payroll of mine employees before the first carload or truckload of ore can be shipped to a smelter and the first cash returns come in toward paying back the investment. Too often, if the mine happens to be in a remote area, the bullion returns are absorbed entirely by the shipping cost, leaving little to nothing for mine development itself.

EXPENSIVE STEPS IN MINE DEVELOPMENT

It is impossible in this field book to describe in detail the technical aspects of opening a new mine or renovating an old mine for renewed operations. Every large library includes technical books on the subject, including the works cited here.

There are, certainly, very many one-man gold mines that, in the past, have produced a modest profit for the owner-operator, and one- and two-man mines may occasionally still come into production (on a very small scale, of course). In these instances, rest assured that the prospectors involved are thoroughly experienced in professional mining. There is no substitute for mining experience that has been learned through prolonged practical application. Inexperienced people generally "mine" simply by gophering or gouging out a surface hole or digging a short tunnel or shallow shaft. Unless the ore is of high grade, such operations generally fail.

Preliminary mine development consists of excavating pits and shafts, sometimes blasting out long tunnels, and closely spaced drilling to determine and appraise more definitely the ore reserves. An ore body can be classified in three ways: ore measured, ore indicated, and ore inferred. The value of the mine must be sufficient in the first and second categories to amortize the total investment. The possible profit lies in the third category.

The follow-on step, also elaborate and expensive, involves the installation of a milling and treatment plant. The plant, which is designed primarily to concentrate the ore values for shipment to a smelter, includes the further development of the mine; the power and water supplies; pumping facilities to dewater the lower levels if necessary; transportation facilities; and, in the case of large, isolated deposits, a campsite or townsite. Generally speaking, only a large, well-financed company or corporation can afford and engineer such a plant.

The transition into full production status is much less costly. If each of the preceding steps has proved out, then in time you can expect some measure of profit to show up on your balance sheets. All but the last step involve investments of increasing amounts of capital. During and after each of the preliminary steps, you may abandon the enterprise with only minimal loss to pay for your own increasing mining experience, should developments reveal that further investment would not be profitable.

Factors Governing Choice of Method

From information obtained during the exploration program, a mining method may be selected for removing the ore. If the exploration consisted only of drilling and test-pitting, a method cannot be chosen until a shaft has been put down, or an adit driven, and enough drifting and crosscutting has been done to determine the physical characteristics of the ore and its wall rocks.

Of the many factors or conditions that may influence the choice of a mining method, the more important are as follows:

STRENGTH. The strength of the ore and wall rock, including such physical characteristics as texture, hardness, jointing, and faulting, indicate the ability of the ground to stand unsupported over a given span. A strong ore or wall rock is, in general, one that will stand unsupported over a width of 10 to 15 feet and a length of 100 feet. A weak ore or wall rock is one that will not stand unsupported over an area about 6 by 6 feet for more than a short period of time. The possible length of an unsupported span is generally the principal factor in determining the most suitable mining method to be used for a given body of ore.

VALUE OF THE ORE. High-grade—that is, rich—ores should be mined by a method that permits a high percentage of ore recovery. Such methods tend to be relatively expensive, but the value of the additional ore recovered should more than pay for the additional cost. Conversely, low-grade ores must be mined by low-cost methods, even though the percentage of recovered ore may be lower.

DIP. The "dip" that a vein of ore makes with reference to the surface of the land is an angle in a vertical plane and is always measured downward from the horizontal plane, or horizon. On a geologic map dip is always shown with a "strike," which is the intersection of the surface with any horizontal plane, and is always perpendicular to the latter.

The ore from flat-dipping veins—those that are nearly horizontal—must be removed mechanically or by hand. Where the flat-dipping ore body lies almost parallel to the surface beneath a broad spread of sediments, the "overburden" is usually scraped off or removed by power shovels and dumped elsewhere, as in strip coal mining. Ore from steeply dipping veins will move to the haulage levels by the force of gravity, usually into a hopper, which loads directly into an ore car or truck directly below. Certain mining methods do not lend themselves to successful operations on flat-dipping veins.

LENGTH AND WIDTH OF ORE BODIES. A vein of ore underground may vary greatly in length and width. As a tunnel is driven along a vein, the roof may require supports to keep it from caving in. Mining textbooks describe various methods of support, especially for ore bodies of substantial length. Where the wall rock is strong, tunnels shorter than 50 feet may not require the regular use of supporting timbers.

A short ore body, especially where the ore is close to the surface, may be simply mined by "gophering"—that is, literally scooping out the pay dirt by hand or with a power shovel, leaving behind a big hole in the ground. In the case of small, rich ore bodies, no regular method of mining is used.

Before undertaking to mine an ore body more than 5 feet in width and well beneath the surface, one should refer to a technical mining textbook and, from the several methods described in detail, select the one that appears to provide the greatest profit for the labor and equipment to be used. There are far too many variables in the characteristics of the ore and wall rock to permit detailed consideration in this field book.

CAPITAL AVAILABLE. If the amount of capital available is small, only those mining methods that require a small amount of development work need be considered. Sublevel stoping (steplike excavation) is a low-cost method of mining, but, unfortunately, it requires a substantial amount of capital for preliminary development. Hence a small amount of available preliminary capital will prevent use of this method, necessitating a less economical technique.

Why Many Small Mines Fail

There are many reasons why small mines prove unprofitable, chief of which are lack of ore, poor management, shortage of capital, and premature construction of a mill.

The exploration program may have been inadequate or had indicated only a small quantity of ore; when this ore was mined out, the mine simply had to be closed. There are very many instances where attempts have been made to mine material that was too low in grade to return production costs.

Failures of this kind are almost always due to poor management. Gold mining, no matter how small in scale, is strictly a business proposition. Good management may provide a profit from low-grade ore, but poor management requires a bonanza mine to return any kind of profit.

There are many other reasons why small mines may fail. The most common fault is misunderstanding among those who are interested in the property. Another important cause is changes in the market price, not only of gold but also of associated metals that, normally, might have paid operating costs, leaving the gold to profit. Other factors include excessive quantities of water in the mine workings and disasters, such as fires and cave-ins, which can turn even a good small mine into a losing venture.

Mills for Small Mines

Hard-rock ore has to be crushed in some type of mill so that the contained gold can be extracted, either mechanically or through treatment with mercury for amalgamation, or by cyaniding. Comparatively few small mines have sufficient ore reserves to justify construction of a mill for concentrating the ore. The amount of this reserve will depend upon the cost of custom milling or smelting plus the cost of transporting the raw ore, as compared with the cost of concentrating in a small mill on the mine property. The reserves, obviously, must be large enough so that the savings on their treatment will pay for the mill.

Mills are expensive to build, and their operating costs are high. If the mill is not located near a public-utility power line, the cost of a power-generating unit must also be figured in. A mill using selective flotation will cost substantially more than one operating on a gravity feed system, and a fine-grind cyanide plant will cost much more than a flotation mill.

A considerable knowledge of ore concentration is necessary to determine the proper flow sheet (engineering design for an orderly series of step-by-step procedures) for the various types of gold ore. Unless the operator is thoroughly trained in this field, he should certainly obtain the services of a consulting metallurgist to test the ore and to design an appropriate mill. A correctly designed mill may mean the difference between a profitable operation and a financial fiasco.

6

Tips for Gold Hunters

The saying that "gold is where you find it" is a truism. There are, nevertheless, certain conditions that are so unfavorable to the occurrence of gold in any profitable quantity that gold hunters can safely avoid areas exhibiting them. There are also favorable areas in which prospecting can be more profitably spent.

Unfavorable Gold-Hunting Areas

1. Areas where large masses of granite and related coarse-grained, crystalline igneous rocks outcrop, especially if the outcrops are not cut by dikes or other intrusions of finer-grained, usually light-colored igneous rocks such as porphyry, rhyolite, or andesite.

2. Areas where large masses of gneisses or other crystalline schists outcrop, unless they are cut by or in the vicinity of dikes or other intrusions of igneous rocks.

3. Areas where large masses of sedimentary rocks such as limestone, sandstone, and shale outcrop, unless they are cut by dikes or other intrusions of relatively fine-grained, light-colored igneous rocks. Even where such intrusions do occur, unless the sediments have been metamorphosed by heat and pressure to marble, quartzite, or slate, sedimentary areas rarely contain workable quantities of gold.

4. Areas where large masses of dark-colored, relatively heavy igneous rocks outcrop, such as basalt, diabase, and peridotite.

5. Areas in which nothing but the unconsolidated or loosely consolidated material that fills valleys between mountain ranges occurs.

It is not true that valuable deposits of gold never occur in the areas described, but a prospector will usually save time and money by avoiding them. Advanced prospecting techniques, such as geophysical exploration, which produced Nevada's Carlin Mine, might locate a body of worthwhile gold ore, but these techniques are normally beyond the capabilities of the ordinary pick-and-pan gold hunter.

Areas Favorable for Hunting Gold

1. Areas in which some gold has already been found or mined are probably among the most favorable places to look for heretofore undiscovered gold. Such places are described in Parts II and III of this book.

2. The second most likely area in which to prospect for gold is one where the country rock is made up of surface flows, sills, dikes, and other intrusions of relatively fine-grained, light-colored, Tertiary igneous rocks such as rhyolite, trachyte, latite, phonolite, and andesite.

3. As suggested in noting the unfavorable areas, prospecting around outcrops of relatively fine-grained, light-colored igneous rocks may reveal the presence of gold.

4. Areas in which the country rock is some type of porphyry (and especially if several varieties formed at different times occur) may contain deposits of gold that can be worked successfully.

5. Gold lodes that may be worked profitably are sometimes formed around the borders of great masses of granitic igneous rocks, both in the granite and in the surrounding rocks, but more commonly in the latter.

Identifying Gold in Place

Sometimes specks, grains, or thin flakes of gold are visible in the outcrop of a lode. In this case, the gold may be recovered simply by pulverizing and panning. It is, therefore, important that you be able to identify such material and, especially, train yourself to distinguish gold from such gold-colored minerals as marcasite, pyrite, chalcopyrite, and yellow flakes of iron-stained biotite mica.

In most instances, gold occurs as tiny grains that are distributed through the gangue or are included within other minerals and are quite visible. These tiny grains may sometimes be recovered by panning, but frequently they are so small or so firmly locked up in the associated minerals that only fire assaying will reveal their presence. In rare instances, gold is combined with tellurium to form the tellurides, and these can be recognized by any mineralogist who is familiar with their characteristics. It is fortunate that tellurides are not common, because special methods of assaying and treating them must be used.

Despite the fact that visible gold is not common in lodes, it is important to be able to recognize it when it does occur. Gold is the only soft, yellow, metallic substance found in nature that may be easily flattened without breaking and readily cut with a knife blade or indented with a needle. Gold may be superficially confused with the "fool's gold" minerals, but these are all hard, brittle, and crush into a black powder instead of cutting cleanly. Mica plates, on the other hand, while much softer than gold, yield a white powder when scratched with the point of a knife or a needle.

Sometimes there is so much silver present as an impurity in gold that the metallic finds are almost silver-white. However, it is rare that there is not some yellow tint to the natural alloy. Chemical tests are necessary to distinguish gold containing 20 per cent or more silver from metallic silver. In any case, finding a lead-silver-gold lode can be much more profitable than discovering a vein carrying only gold. Silver and lead are much more abundant metals than gold and often defray the entire costs of developing and operating a mine, leaving the "by-product," gold, as pure profit.

Skin Diving for Gold

In recent years skin diving for gold has become increasingly popular, especially with the young, vigorous generation. Much publicity has been given to a few discoveries of gold by divers using self-contained underwater breathing apparatus (scuba) equipment.

As in the case with small-scale placer mining, skin diving for gold offers little opportunity for earning a living. Virtually all of those who prospect and dive for gold do so strictly as a hobby with a special summertime appeal.

Because scuba diving is such a popular pastime, nearly every gold-producing state through its department of mines and minerals or water resources has prepared directions on how to go about diving for gold in their auriferous streams. Most states place strict limitations on the amount of disturbance they permit in gold-bearing streams that also contain trout. Perhaps one of the best sources of information, and the basis for many other state brochures, is *Diving for Gold*, by William B. Clark, Geologist, California Division of Mines and Geology, P.O. Box 2980, Sacramento, California 95812, periodically revised and brought up to date.

Regardless of whether any money is to be made, the lure of gold on the bottom of a stream and the challenge to find it will always attract adventuresome gold hunters. Skin diving for gold involves little expense and can prove to be an interesting experience for those who wish to pursue it. There is some danger attached to skin diving, and any such venture should always be undertaken with another diver.

7

Selling the Gold You Find

The principal purpose behind prospecting for gold is to discover a raw commodity that can be turned into cash. By itself the possession of gold is more of a burden than a blessing, except for specimens in a gem and mineral collection. In the gold-rush era, dust and nuggets could be brought into a trading post, weighed out on a delicately balanced pair of gold scales on the proprietor's counter, and used as a direct medium of exchange for supplies of all kinds. Not so today. In order to obtain any concrete benefits from finding gold, it becomes necessary to sell it for its dollar value and then to use those dollars, or credits, to purchase the goods or services available in our more complicated world.

The Marketing Process

The first step in the marketing procedure is to subject your gold to an assayer to determine the metallic content of the dust or nuggets of raw gold, or the amount of metallic gold in a batch of ore or ore concentrates. Assaying is a necessary and rather costly procedure—and is often disappointing, because of the "shrinkage" that appears.

Shrinkage is the difference between gross weight of an assay lot and the melted or bullion weight that is the end product of the assay or refining process. Nongold impurities are invariably present as mineral particles within or attached to gold grains, or as mineral coatings.

Melting the batch separates the metallic content, the nonmetal con-

53

taminants floating off as slag. In some cases, quartz or mineral inclusions can be clearly seen in larger particles of raw gold. This type of impurity, as well as surface coatings, become increasingly important (as a proportional part of the particle weight) as the gold particle size decreases.

Prospectors often speak of their day's "clean-up," meaning that after a few hours of panning, sluicing, or operating a long tom or rocker, they have scraped up and sacked all the apparent gold particles caught in their riffles, along with worthless quantities of black sand and other heavy minerals. Before submitting such clean-ups for assay, further cleaning needs to be done by washing, blowing, and magnetic separation to remove the many particles of foreign material from the dust. Then when the residue is offered for sale or to a refinery for further treatment, some advance estimate can be made of the true amount of recoverable gold. The differences in thoroughness of preliminary cleaning cause banks or other purchasers to make different advance payments for different lots of gold, even if ultimately assays would prove that the fineness of the gold was the same in each lot submitted.

Always, the costs of assay and cleaning for a batch of raw gold must be deducted from the final settlement.

Sampling necessarily precedes assaying in the marketing process. A small portion of the material is selected for assay to determine the basis for cash settlement. Well-cleaned concentrates will generally be more uniform in value, and the sampling will generally be more nearly representative of the uniform material.

Buying Practice and Processing

Coarse-grained gold and nuggets will generally command a premium over the world market price of gold for specimens and jewelry. An added advantage to the gold hunter for this type of sale lies in avoiding assay costs. The price may be determined by size and condition of the gold without reference to its fineness. Also, the cleaning process in the coarse size ranges will be less time-consuming.

The market for specimen gold and jewelry nuggets varies according to supply and demand. Increased prospecting and mining activity may very well saturate this somewhat limited outlet.

Fine-grained gold is both more difficult to recover and to clean for shipment. In the small-size particle ranges, mercury may become part of the recovery process, necessitating treating the amalgam by retorting off the mercury (dangerous in inexperienced hands because of the very poisonous nature of mercury vapor). In recent years less mercury is being used where the loss of gold is relatively small, because of the financial advantage of dealing in "bright" gold.

Generally, time spent on cleaning your gold will pay off in reduced processing charges and less variation in the settlement returns. Cash advances allowed by a bank or other agent may also improve as your cleaning-up procedure improves with experience.

When your gold reaches the ultimate buyer, his procedure may vary with the size and condition of your shipment. It may be more convenient for him to melt all the bullion and put it in a small shipment as a total sample; a larger shipment may require removing only a representative sample. In each case, the basis for payment will depend upon the weight and fineness of the melted gold. Treatment charges will absorb proportionately greater amounts of the net return for small shipments than for larger ones. Treatment charges vary among dealers, from a fixed percentage of the total gold value to a detailed list of charges for handling each separate lot.

Smelter charges may include a fixed-percentage deduction of metallic content for loss in handling, fixed charges per troy ounce of recovered gold, a minimum refining charge per shipment, and a fixed sampling and assaying charge per lot.

Markets for Raw Gold

Most pay streaks contain a greater amount of gold in small particles (flour gold, dust, flakes, tiny grains) than coarse gold, and many deposits contain no coarse gold at all. Fine-grained gold recovery, cleaning, and marketing are therefore a proportionately large part of the mining process, particularly as it relates to placer gold recovery.

Private buyers are one source of outlet for both nugget and dust gold.

The possibility of finding private buyers has sharply reduced the once-common practice of using mercury as an amalgamating agent in the clean-up process. Sales are negotiated on the basis of the gold at hand, and there is no middleman commission or delay between the offering of gold for sale and the returns in immediate cash or check. As in any other marketplace, the word of caution is for the finder of raw gold to know the buyer well enough to put his faith in an honest deal. While most gold buyers are reputable, there are unscrupulous operators looking for a fast profit for themselves.

Until 1968 gold found in the United States could only be sold to a buyer licensed by the U.S. Treasury to process and refine gold, or to bankers who acted as agents for gold producers. Since January 1, 1975, all Americans have been permitted to own as much gold as they can obtain and to sell it to anybody who wishes to buy it, for jewelry making or for any other purpose, including further processing and refining. However, banks have provided a needed service for a small fee that still has advantages outweighing the cost of the service for most small-scale and new producers of gold.

The advantage of selling your gold through a bank includes a record of the amount and quality of the gold deposited; experienced, careful mailing and handling procedures (since in any case the gold is usually forwarded to a refinery); experience and familiarity with the market outlets and refinery procedures; and a cash advance to cover a large percentage of the appraised value, in view of the obvious fact that returns from a refiner may take a month or more from the date of shipment.

Not all banks are willing to handle raw gold, but almost any bank manager can provide you with the names and addresses of bankers who will act in your behalf as agent to the refiners. The small fee charged will more than compensate you in time and assurance of honesty.

Licensed Refiners

Where direct shipment to buyers and refiners seems desirable, it is always advisable to contact the possible dealer before shipping the concentrate or samples. Each gold processor has his own methods for handling purchases.

Both capacity and requirements regarding the form of material shipped may vary from time to time within each company. Moreover, the addresses given on dealer lists may be office addresses rather than receiving points. Shipment without prior confirmation of terms and company policy can result in delay in returns as well as dissatisfaction with the amount of the net proceeds.

Many commercial buyers of minerals do not deal in raw gold. Others may deal in already refined gold only, as scrap gold, and still others only handle refined bullion. A direct shipment of raw gold concentrate to those addresses listed monthly in the mining industry trade journals (such as *Ore Buyers' Guide*) would only produce delay, frustration, and additional costs of return and new shipment.

Small-lot shipments are especially difficult, because fixed, minimum charges may consume too large a part of the actual gold value. The following is a list of known purchasers of raw gold as well as of retort sponge (the residue from the amalgamation process after the mercury has been driven off):

> American Smelting & Refining Co., Inc.
> Southwestern Ore Purchasing Department
> P.O. Box 5747
> Tucson, Arizona 85703
>
> Eastern Smelting & Refining Corp.
> 37–39 Bubier Street
> Lynn, Massachusetts 01903
>
> Englehard Minerals & Chemicals Corp.
> 430 Mountain Avenue
> Murray Hill, New Jersey 07974
>
> Handy & Harman
> California Branch
> P.O. Box 150
> El Monte, California 91734

The following markets are taken from a list of possible gold buyers supplied by the Arizona Department of Mineral Resources:

American Chemical & Refining Co., Inc.
Sheffield Street
Waterbury, Connecticut 06714

Associated Metals & Minerals Corp.
11944 Mayfield Avenue
Los Angeles, California 90049

Homestake Mining Company
650 California Street
San Francisco, California 94108
(does not buy amalgam)

P.E.P. Industries, Inc.
13429 Alondra Avenue
Santa Fe Springs, California 90670

Sabin Metal Corporation
316–34 Meserole Street
Brooklyn, New York 11206

Western Alloy Refining Co.
366 East 58th Street
Los Angeles, California 90011

Wildberg Bros. Smelting & Refining Co.
349 Oyster Point Blvd.
South San Francisco, California 95080

Choosing a Dealer

In all cases, it is important to realize that the processor has not only to refine your gold and maintain his business, but also to earn a profit. A special problem he faces is that in the case of wide and abrupt market fluctuations in gold prices he may have to pay more for your concentrates than the market will return him and thus may be left holding considerable inventory. So if his deductions seem heavy, remember he can only pay for the contained gold in any case, and his costs have to be covered within a reasonable period of time. You will be expected to pay for the service in one way or another.

Business reputation is an important consideration in choosing a dealer. It is to the honest businessman's advantage to deal fairly with his customers. He has the option of stating his operating and purchasing policies, and you have the option of choosing which set of policies you believe will give you the best returns.

Meanwhile, the dealer's assayers will determine the fineness of your gold and determine the basis for payment. You may not agree with the results in all cases, but those who have undertaken the added expense of sampling and assaying their own product, or have resorted to the more expensive and time-consuming procedure of obtaining an "umpire" assay, have found that the long-term averages run very close to the dealer's figures.

A small shipment hardly warrants the expense of elaborate sampling and assaying procedures. Your results may show up better if you simply stockpile gold concentrate and ship in larger batches.

You should receive from whichever dealer or refiner you select a statement of the basis of settlement showing gross weight, melted weight of the gold content, fineness, credits and treatment charges, and the settlement check for your shipment. Your ultimate choice of dealer may eventually be determined by the ease of understanding the statement.

PART II

Where to Find Gold in the United States

INTRODUCTION

Of the fifty states comprising the United States of America, thirty-two have reported the existence of gold in sufficient showings to interest the casual collector. A few states show gold in only one or a few areas, either as a hypothetical accident of geology or because the Pleistocene glaciers of the Ice Age dropped gold from unknown northern sources in morainal debris.

In 1968 the United States Geological Survey published a comprehensive paper* listing every major gold-producing district and mine in twenty-one states where 10,000 or more troy ounces of placer and lode gold had been mined through 1959, often embracing more than a century of production. A total of 508 districts are described, together with their basic geology, named mines and prospects, and geographic locations. According to the Survey, more than 75 per cent of America's total recorded production of 307,182,000 gold ozs. came from only five western states: Alaska, California, Colorado, Nevada, and South Dakota.

This recorded total, however, is not the whole story, inasmuch as countless early-day prospectors and miners and later amateur and semi-professional gold seekers never recorded their finds. Thousands of ounces of gold are still being gleaned every year from the old districts by amateurs, scuba divers, and knowledgeable prospectors who do not

*A. H. Koschmann and M. H. Bergendahl, *Principal Gold-Producing Districts of the United States* (Washington, D.C.: U.S. Geological Survey), Professional Paper 610.

make official reports. More often than not, these modern-day descendants of the Old West (and not a few in the Atlantic States) treat their finds as gem-worthy specimen materials carrying a far higher value per ounce of gold than even the current high world price of gold in the commercial or national marketplace.

The purpose of this section is not only to highlight the major localities described in Paper 610 and make that information more readily available to all gold hunters, but also to include a comprehensive listing of all other gold localities included in individual state and county surveys where less than 10,000 ozs. of gold have been found. Even minor showings of auriferous minerals will interest the casual prospector as clues to yet possible undiscovered and more valuable deposits overlooked by the "high grade"-minded prospectors of the past.

Wherever possible, every potential gold-collecting locality has been pinpointed with reference to the nearest road-map community, taken in alphabetical order by county and the counties listed in alphabetical order by state. The route and highway numbers, compass orientations, and mileages are also listed in ascending order to permit following any standard oil company road map. Directions are abridged and coded somewhat to save space. Thus: "SANTA RITA, ESE, on U.S. 93 and Rtes. 78 and 46, to Bear Paw Mts., at W base, in watercourse and slope-wash gravels—placer gold" should be translated as "Go east-southeast from the town of Santa Rita, first on U.S. Highway 93 and branching successively on state Routes 78 and 46 to an area encompassing the western base of the Bear Paw Mountains, there prospect all the ravines, draws, and slopes for gravel deposits where early reports show that they contain placer gold."

Because so very many gold localities exist far from any significant population center, the gold seeker must get a topographic or quadrangle map of an area, either from the nearest county, state, or federal geological survey office for a nominal charge or from a nearby U.S. Forest Service ranger station. These maps carry a grid system of township and range, with townships further subdivided into 36-mile-square sections. In describing locations, two systems of coding information are used. Normally a mine might be described as being "in sec. 5, T. 13N, R.

28W," to be translated as "in section 5, township 13 north, range 28 west." In the shorthand that will be used in this book, this locality information is consistently abbreviated to "sec. 5, (13N–28W)." In either case, the gold hunter will find township lines running east and west across the map and range lines running north and south, usually also with degrees and minutes of longitude (indicated at top and bottom of the map margins) and latitude (at the sides of the map), since township and range information depends also on base lines of longitude and latitude. On such maps the section boundaries and their designated numbers are clearly shown for each township-range square. Since these surveyors' directions are not included on oil company road maps, the quadrangle map is a necessary acquisition.

The majority of old gold mines (whose dumps often contain gold-rich ore specimens along with other collectible minerals and gemstones) cannot be easily reached, except with a four-wheel-drive vehicle over very rough roads. Access to many mines may be solely by old-time mule trail that has not been maintained for decades. The gold hunter must be left to his own devices, initiative, and the acquisition of appropriate maps in order for him to reach the more remote or inaccessible localities, but his reward may be the greater because fewer seekers get into such difficult areas.

The gold hunter should exercise the utmost caution in uninhabited country and, always, bear in mind that long-abandoned mines and prospects are dangerous because of decay and cave-ins. Where mining operations are relatively recent or currently in progress, access roads will be better. Permission to look over the mine dumps, where waste gangue may contain specimens of ore showing gold, or to pan nearby stream gravels should be obtained from either the mine owner or the foreman of operations.

In the western states and Alaska, much of the land is federally or state owned (for example, 99.7 per cent in Alaska, 87 per cent in Nevada, and 55 per cent in Arizona). All such land, except where restricted by the military, is wide open to prospecting. Even on military reservations and bombing ranges, prospecting and mineral collecting may be permitted on weekends and holidays, with permission necessarily being sought from the cognizant commanding officer.

Almost all of the land lying east of the 100th Meridian—that is, east of the eastern foothills of the Rocky Mountains—is either privately owned or under strict control of county or state agencies as public parks or Forest Service preserves. Permission to trespass is mandatory from the owners or agencies involved. Private land owners may grant free permission or charge a small fee for prospecting and mineral collecting on their properties. In any case, whether on public or private land, entrance through a closed gate in a fence requires that the gate be immediately closed after passage. Farmers and ranchers take a dim view of gates left open to allow their livestock to escape.

There are literally thousands more auriferous localities than can be listed individually here. However, every mining and placer area of any consequence at all in the thirty-two gold-producing states is listed here, with individual mines named, and usually with indications of nearby places to investigate. From these primary districts or mines, the gold seeker can use his imagination to look farther. He may even discover a gold deposit not found by his predecessors and, perhaps, even one rich enough to warrant claiming and developing into his own small mine. Moreover, many of the old mine dumps today are commercially worth reworking on a lease arrangement from the heirs of the original owners. The lease will, of course, allow for a percentage of any profits to go to the real owners, who may live in a far-distant city.

The county boundaries for each city and community named herein were determined from the collective maps of the Automobile Association of America. Road maps in general show county lines so dimly printed as to be almost invisible. Therefore, it is sometimes necessary to look up a town or city name first in the map index and locate it in the map's number-letter grid system. The name of the county will invariably be shown somewhere nearby, and the county outlines can be discovered and traced out. This method will reveal in what section of a state any unfamiliar county lies. Knowing the county boundaries is especially important where a locality is entered only as "in the far NE corner" where no communities at all exist in an area of several hundred square miles. The western states and Alaska contain many such uninhabited regions, but all such localities are further identified by either section, township, and range, or by latitude and longitude, the latter especially in Alaska.

Before attempting to investigate any remote area, the visitor should always identify himself to the nearest authority (a county or town official, Forest Service ranger, or area rancher) and indicate his probable route and probable time of return. Then he should positively report back on his way out. In case he is afflicted with a mechanical breakdown or suffers an accident and fails to show up on a reasonable schedule, rescue operations can be put into action with a minimum of delay. Every year, somebody fails to take this elementary precaution and perishes, when a little forethought would have insured his safe return home.

ALABAMA

Alabama is scarcely remembered today as a gold-rush state, even by its own citizens. Nevertheless, the third sizable gold rush in American history took place in Alabama following the initial discovery of gold in 1830. The first historic gold rush occurred in North Carolina after 1803, with the second big rush coming in Georgia in 1828. The Alabama rush created enormous excitement between 1835 and 1850, at which time most of the prospectors and miners departed for the California Mother Lode. There, the prospecting and mining skills they had developed at home contributed materially to the development of the far western gold fields.

The study of a geologic map reveals that a belt of gold-bearing gneisses and schists extends from gold-rich Georgia into east-central Alabama. The richest concentrations of gold occur in the Talladega Slate, the Hillabee Schist, the Wedowee Formation, and the Ashland Mica Schist —formations that range in age from the Precambrian era to the Carboniferous period.

Along this belt embracing ancient mountains that merge northeastward into the Appalachian Plateau, the gold rush gave rise to many new communities. Some of the early boom camps became cities; others faded away like their successors in the Far West. In the decade following 1836, Arbacoochee, for instance, reached a population of five thousand inhabitants, with its lode mines giving employment to six hundred miners. Now relatively insignificant, this community is still occasionally referred to nostalgically as the "largest town in Alabama." The discovery of gold at Goldville in 1842 brought in three thousand residents. Where there were once fourteen stores, saloons, a school, and many homes in town, today there is only a crossroads without a single business establishment.

Even after 140 years, there is enough gold remaining within Alabama's auriferous districts that some mining activity continues to the present. Today's weekend and vacationing gold hunters still find ample opportunity to pan out colors, grains, and nuggets of gold. Here and there, one may even earn a "day's wages" or enough in a week or so to make a substantial down payment on a summer vacation.

For further information, write: State Geologist, Geological Survey of Alabama, P.O. Drawer O, University, Alabama 35486.

Chilton Co.

The southwestern part of the Alabama gold belt includes portions of Chilton and Coosa counties, in a region of crystalline rock exposures that are limited on the southwest by the overlap of the Upper Cretaceous (Tuscaloosa) Formation.

CLANTON, W 13 mi. (in SE part of co.): (1) on small tributary of Mulberry Cr. in NW¼SW¼ sec. 8, (22N–13E), the Franklin (Jemison) Mine, small pits in schists and quartz, site of 10-stamp mill operated until 1923—lode gold; (2) in NW¼SE¼ sec. 15, on S bank of a creek exposing the Hillabee Schist, the B. T. Childers Prospect—lode gold, with pyrite and copper minerals; (3) in sec. 17: (a) along Mulberry Cr. from about 2 mi. below Honeycutts Mill for some distance above, the Mulberry Creek placers, productive in early days; (b) all tributary streambed and bench gravel deposits—placer gold.

VERBENA: (1) area, all regional streams and creeks surrounding the intercornering of Chilton, Coosa, and Elmore cos., in bed and bench gravels—placer gold; (2) on Blue Cr. (short tributary of the Coosa R.), in sec. 17 (21N–16E): (a) the Rippatoe placers, discovered in 1835 to become a famous property with mining continuing vigorously until 1855; gold gravels extend about 1 mi. in a valley not more than 200 yds. wide—fine gold, nuggets; (b) adjoining mt. side outcrops of quartz veins in pyritiferous schists and slate, area prospects (especially toward head of cr.)—lode gold; (3) NE 2 mi., on Rocky Cr.: (a) in sec. 30, (21N–16E), and (b) nearby in SW¼ sec. 29, stream placers (rich panning but often reworked since the 1850s).

Clay Co.

ERIN, S, in gravel bars of Gold Mines Cr.—placer gold, with sillimanite, pyrope garnets, talc.

IDAHO (district, in W part of co., SW of the pyrite mines and in Hillabee Schist): (1) in (19S–7E): (a) in NW¼SE¼ sec. 27, the Eley

Mine, operated in 1899—lode gold; (b) in sec. 33, the Chinca-Pina Property (open cut, inclined shaft, several prospect holes, with best panning in surface debris)—placer gold; (c) NW of NE cor. sec. 33, the Haraldson Mine (old); (d) in SE¼SW¼ sec. 36, the Alabama Gold and Mica Co. Mine, quartz vein in mica schists, site of 5-stamp mill—lode gold (obtained by amalgamation); (2) SW 7 mi., in (20S–6E): (a) in sec. 34, the Harall Gold Mine, once-rich producer—lode gold; (b) 1–2 mi. away, the Shinker Mine—minor lode gold; (3) in (19, 20S–7E), several area notable placer deposits, well worked in early days; (4) in (20S–7E): (a) in sec. 3, the Idaho (Franklin) pits, open cuts to 60 ft. deep along hillside exposing graphitic and garnet schists, quartz stringers—free-panning gold, with abundant garnets—and the Hobbs Pit (shallow excavations, with promising panning); (b) in sec. 4, the Laurel pits (in decomposed ground, easy panning); (c) nearby, the Horn's Peak Mine, well developed and site of old 5-stamp mill—lode gold; (d) in sec. 15, the California Property, site of 10-stamp mill—free gold (obtainable by crushing and panning); (5) in (21S–7E), in SW¼ sec. 23, the Prospect Tunnel (opened for nonexistent copper, but 1930 assays rich)—lode gold.

LINEVILLE, S, in area streams emptying into Crooked Cr., placers.

Clay–Randolph Cos.

CRAGFORD (district, along far E side of Clay Co. and far W side of Randolph Co.): (1) in (20S–9E): (a) in sec. 24, the Grizzel Property, veins in quartz to 30 ft. deep (rich specimen ore near surface)—free gold; (b) in sec. 30, on the H. S. Bradley land, prospects—free gold; (c) in secs. 25, 36, the Manning Placer, old diggings along tributaries of Crooked Cr. (landmarks of 1830–40), thin quartz veins nearby—placer and lode gold; (d) in NW¼SW¼ sec. 36, the Farrar Property, deep shaft and crosscut tunnel, worked before 1860—lode gold; (2) in (20S–10E): (a) in sec. 21, the Morris Property, ore in formation traceable ¼ mi. NE to the Tallapoosa R. (shoal sands show placer gold)—free gold; (b) in sec. 29, at Wildcat Hollow and in NW¼ the Teakle Property (deep shaft known as the Orum Pit, ore worked in an arrastra)—lode gold; (c) in sec. 30, adjacent to Wildcat Hollow, the Bradford Fraction

—lode gold; (d) in SW¼ sec. 30, the Goldberg Mine, open cut, inclined pit, veins 6–15 in. thick—lode gold, with antimony, copper, traces of arsenic; (e) in SE¼NE¼ sec. 30, the W. D. Mitchell's Pine Hill Prospect, 80-ft.-deep inclined shaft and pits along strike in ore body 14 ft. thick—lode gold; (f) in NW¼SE¼ sec. 30, the Bradford Ridge Mine (most extensively prospected in district), site of 10-stamp mill—lode gold, with some arsenopyrite; (3) in (21S–9E), between forks of White Oak and Wesobulga crs., in SE¼ sec. 2, the Dawkins Property, site of old stamp mill—lode gold (saved by amalgamation).

Cleburn Co.

ARBACOOCHEE (district about 10 mi. S of HEFLIN via Rte. 9 and Co. Rd. 19 to within 8–10 mi. of the Randolph Co. line), major gold district overlapping into Randolph Co.: (1) in sec. 34, (16S–11E), the Anna Howe Mines (first gold-bearing quartz discovery in Alabama)—lode gold; (2) NE of town, (16S–12E): (a) in sec. 6, the Marion White Property (once a major source of rich specimen ore and float), lenticular veins in slate—free gold; (b) in sec. 27, the Bennefield Property, open cuts, low grade—lode gold; (c) in sec. 34, the Sutherland Property, extensively worked, site of stamp mill (wooden rods shod with iron shoes)—lode gold; (3) in (17S–10E), in sec. 23, the Eckles Property, 40 acres with 100-ft. shaft, quartz vein in decomposed schist—lode gold; (4) E of town in (17S–12E), in sec. 3, the Middlebrook Property, rich panning—free gold.

CHULAFINNEE (district adjoining the Arbacoochee district on the S and SW): (1) in (17S–9E): (a) along Chulafinnee Cr. and tributaries, in secs. 14, 15, 16, the Chulafinnee placers, auriferous gravels under 5–6 ft. of overburden; (b) in sec. 16, the King Mine, a pit of 2,500 sq. ft. in schist laced with quartz veinlets, site of old stamp mill—lode gold; (c) in sec. 22, the Striplin Property (shallow cuts showing quartz stringers in schist which panned well), and adjoining on the NE, the Higginbottom Property, similar, rich panning at surface; (d) in E½NW¼ sec. 22 (3 mi. W of CHULAFINNEE), Rev. Mr. King's Property, in decomposed quartz, site of stamp mill—free gold; (e) in sec. 23, the Carr Cr. Placer, 240 acres of clay and gravel—placer gold; (f) in sec. 24, area watercourse gravels and sands, placers; (2) in (17S–11E): (a) in sec. 2, the Hicks-Wise Mine (at 110 ft., deepest shaft in state), and the Lee Mine, bedded vein in slate, rich—lode gold; (b) in sec. 3, the Valdor Property (with openings in secs. 5 and 6), comparatively rich—lode gold; (c) in secs. 5, 6, the Arbacoochee Placer (most extraordinary gold placer deposit in Alabama, covering 600 acres on top and sides of Gold Hill, once giving employment to 600 men); (d) in sec. 7, the Clear Cr. Placer, long famed for its rich production; (e) in sec. 17, the Golden Eagle (Prince) Mine, 75-ft. shaft, site of stamp mill—lode gold; (f) in sec. 25, the Crown Pt. Property and the Lucky Joe Property (considerable development and site of 10-stamp mill in 1893, pay ore in "chimneys")—lode gold; (g) in SW¼SW¼ sec. 33, the Ayers Prospect (just E of Blake Cemetery), no mining done but good showings in some pits; (h) in sec. 35, the Mossback Property and the Wood's Hole Copper Mine (first paying copper discovery in Alabama, under an iron gossan)—by-product gold; (i) in sec. 36, the Pritchet Property—panning gold.

Coosa Co.

Area: (1) along Weogufka Cr., the Weogufka Cr. Placer (pans run 4–20 gold particles); (2) in (21N–16E): in secs. 1, 2, the Gold Ridge Mine (originally prospected in 1835 for copper—lode gold; (2) in (22N–16E): (a) in sec. 17, at Flint Hill, heavy quartz veins show gold traces; (b) in sec. 35, at Àlum Bluff, near mouth of Hatchett Cr., the Hatchett Cr. Placer, gravels rich enough to have kept 50 men working in 1840, source of gold probably in nearby quartz vein carrying decomposed pyrite.

Goldbranch, NE 1 mi., in sec. 4, (23N–17E), the Stewart (Parsons) Mine, open cuts and shafts in 200-ft.-wide ridge for ½ mi., site of old stamp mill—lode gold.

Rockford: (1) SE town limits, old prospects—gold showings; (2) along Gin-house Branch and Carrol and Pole branches, the Rockford Placer, productive in early years.

Randolph Co.

The auriferous deposits of Randolph Co. border along the boundary of Cleburne Co., q.v., and are in similar formations.

AREA: (1) regional watercourse and bench sands and gravels, placers; (2) in sec. 4, (17S–12E), the Gold Ridge Property, with the Eckhert vein in a highly garnetiferous mica schist carrying iron-alumina garnets to 3-in. dia., dumps provide good panning—free gold; (3) in sec. 12, (18S–10E), the Pinetucky Gold Mine, discovered in 1845 and extensively worked, quartz veins in garnet-bearing mica schist, site of 20-stamp mill (mine among earliest discoveries of lode veins in Alabama and termed a "rich specimen mine")—lode gold, in pyrite.

OMAHA, in sec. 32, (19S–13E), various prospects—gold colors.

WEDOWEE: (1) area cr. sands and gravels along the Tallapoosa R., placers; (2) NW, a mine on Wedowee Cr.: (a) mine itself—lode gold; (b) nearby stream gravels—placer gold.

Talladega Co.

RIDDLE'S MILL (district, in SE part of co. in (19S–6E) and most important locality where gold mining was carried on in the Talladega series): (1) the Riddle's Mine, 100 ft. deep in quartz lenses and kidneys, free-milling—lode gold, with silver and pyrite in depth; (2) in a lead extending from Riddle's Mine, the Woodward Tract—gold, with silver.

TALLADEGA, S 7½ mi., on E side of the Talladega Cr. Valley, the Story (Warwick and Cogburn, Gold Log) Mine, quartz vein in decomposed slates, crushing mill on cr. nearby—lode and free gold.

Tallapoosa Co.

Tallapoosa County contains four major gold districts: Devil's Backbone, Eagle Creek, Goldville, and Hog Mountain. The Goldville district is about 14 miles long and showed great activity in the early days, when the population of the town reached 3,000. Hog Mountain is unique in Alabama in that gold veins are in granite, and here the cyanide process was first introduced into the state. The Devil's Backbone district lies in a belt of the Wedowee Formation, with its southern end in Elmore County, west of the Tallapoosa River, while east of the river the belt runs northeast into Chamber County. Placer gold is found in all streams that cross the Devil's Backbone.

ALEXANDER CITY, 3 mi. out on the Hillabee bridge rd., the Duncan Property, quartz vein, good showings—lode gold.

DADEVILLE: (1) SW, in (21N–22E); (a) in sec. 10, left of the old Dadeville–Young's Ferry rd., the Holly Prospect, active in 1911—lode gold; (b) in sec. 33, the Gregory Hill Mine, quartz seams in graphitic schist—panning gold; (c) nearby, as continuation of the Gregory deposit, the Blue Hill Property, in surface debris—lode gold. The lower part of the property, along with much of the Gregory Hill deposit, now lies under the waters of Martin L.; (2) NW, in sec. 19, (22N–23E), the Bonner-Terrell Property, caved-in, site of old stamp mill—gold.

GOLDVILLE (district, with crossroads remnant about 17 mi. NE of ALEXANDER CITY, extending SW 14 mi. to vicinity of Hillabee Cr. bridge, very many auriferous prospects, placers, lode mines throughout district): (1) in (23N–22E): (a) in sec. 4, the Mahan pits, heavy sulfides, rich—lode gold; (b) in sec. 8, on E bank of Hillabee Cr., the Ulrich pits and Dutch Bend (Romanoff) Mine (originally opened for copper, but gold discovered during digging a wine cellar), 6 quartz veins in 300-ft.-wide slate belt, site of 20-stamp mill and cyanide plant—lode gold; (c), in sec. 9, the Chisholm Property, a 6-ft.-wide vein—lode gold; (2) in (24N–22E): (a) in SW¼SW¼ sec. 26, the Tallapoosa Mine, 185-ft.-incline shaft and drifts, quartz in slate, site of modern mill—free-milling gold; (b) in sec. 34, the Stone (Croft) pits, long abandoned—lode-gold showings; (c) in SW¼ sec. 36, the Early pits, source of rich ore—lode gold; (3) in (24N–23E): (a) in SW¼NW¼ sec. 4, the Birdsong pits, first worked mine in district (by black slave labor, 1840–50)–lode gold; (b) in sec. 5, the Jones pits, well developed, mill of iron-shod wooden stamps and arrastra—free-milling gold, with pyrite; (c) in sec. 8, the Germany pits (among oldest in co.)—lode gold; (d) in sec. 18, the Houston pits, much early development—lode gold; (e) in sec. 24, the Log pits, 2–4-ft. quartz vein with rich "pocket type" production ($30,-000 at old $20 an oz. price).

JACKSONS GAP: (1) S, in (22N–22E): (a) 1½ mi., in sec. 26, the Alabama King Mine (reached from rd. to the Preacher Gunn Prospect by turning E on first timber access rd. S of U.S. 280), open cuts, incline shaft to 300 ft., surface workings in chlorite schist, site of stamp mill—lode gold; (b) 2 mi., in same sec., the Preacher Gunn Prospect, quartz stringers in chlorite schist, 2 125-ft. adits—lode gold; (2) NE, in (23N–

23E): (a) in SE¼ sec. 24, the Greer Property (a quartz vein traceable
to the Hammock workings), and (b) in SW¼, the Hammock Property,
hard quartz, site of stamp mill—panning gold on dumps; (c) in SE¼
sec. 26, the Tapley Property, numerous caved-in openings to 50-ft. level
—panning gold on dumps; (d) in SW¼, the Jennings Property (continuation of the Tapley and in SW part of the Devil's Backbone),
decomposed quartz veins—free gold; (3) in W½SW¼ sec. 17, (23N–
24E), the Johnson Property, old tunnels and shafts in 1½-mi.-long
quartz outcrop—lode gold; (4) 3 mi. W of old site of GOLDVILLE, q.v.,
in secs. 10, 15 (24N–22E), the Hog Mt. (Hillabee) Mines, quartz veins
in granite intruded into the Wedowee Formation, site of 10-stamp mill;
total production of $250,000 before closed in 1916 by World War I
(modern chance of reopening)—lode gold.

MARTIN DAM, in (20N–22E), the Devil's Backbone district (accessible by graded rd. or by dirt rds. from Union Church): (1) all regional
watercourse gravels, long known for panning and sluicing—placer gold;
(2) in sec. 9, along E shore of Lake Martin: (a) the Dent Hill Prospect
(½ mi. NE of the Silver Hill Mine, reached 1 mi. NE of Union Church
by dirt rd. NW from Hwy. 50), quartz in schist—lode gold; (b) ¼ mi.
NE of the Dent Hill, the Farrar Prospect, several old pits in alluvium
—free gold (possibly from saprolite in the Wedowee Formation); (3) in
secs. 16, 17, the Silver Hill Mine, gold in dark talcose slates between
hornblende slates and in quartz stringers mined to 80 ft. deep, site of
6-stamp mill (part of property under lake waters now)—free-milling
gold; (4) in sec. 19, the Mass Prospect—gold in schist.

ALASKA

Gold was known in Alaska as early as 1848, when a Russian mining
engineer reported finding placer gold in gravels of the Kenai River on
the peninsula of the same name. Within two years after the American
purchase of "Seward's Folly" in 1867, miners who had been disap-
pointed in the Cassiar gold fields of British Columbia found rich placer
gold southeast of Juneau at Windham and Sumdum bays. Three years
after major lode-gold veins were found near Juneau in 1880, the camp

boomed into life to become the center for territorial gold mining and to take on the responsibilities of territorial capital.

U.S. Geological Survey Professional Paper 610 divides Alaska into nine geographic regions: Cook Inlet-Susitna, Copper River, Kuskokwim, Northeastern, Seward Peninsula, Southeastern, Southwestern, Yukon, and Prince William Sound, covering 586,400 square miles. Within this immense territory the paper describes 43 mining districts that produced more than 10,000 ozs. of gold each out of a total gold production between 1880 and 1959 of 29,225,071 ozs.

There are so many gold localities in Alaska in addition to the 43 named districts that the reader must necessarily be referred to the U.S. Geological Survey Mineral Investigations Resource Map MR-32, "Lode Gold and Silver Occurrences in Alaska," available from any office of the Survey, and MR-38, "Placer Gold Occurrences in Alaska" (permanently out of print but available in libraries and at the Survey offices). Map MR-32 subdivides the entire state into 153 quadrangles, each of which details separate gold and silver lodes keyed by number to mapped locations. Since MR-32 measures 39 by 57 inches, it is recommended that the inquirer obtain a copy from which to select the particular quadrangle maps that interest him most, and which can be purchased for a nominal cost in any principal Alaskan city.

It is difficult for a newcomer to Alaska who is not familiar with its vastness and climatic differences to decide on a specific area or district in which to hunt gold. Much will depend on whether he is interested in placer or lode gold, his available finances (nearly all Alaskan prospectors must travel by air, often in specially chartered small planes), his experience in surviving in wilderness terrain perhaps hundreds of miles from the nearest community, and his preferences for prospecting in a loosely settled region or in one that is wild and virtually unexplored. The least expensive and most practical procedure is to spend at least one full summer, or better yet a full round of the seasons, in the state, meeting and obtaining suggestions from as many mining men and prospectors as possible, before setting out to investigate any specific locality. Anchorage and Fairbanks are the most logical cities to use for informational headquarters.

Many large areas in Alaska have been withdrawn from mineral entry by federal agencies or the military for various reasons. Preparations for a prospecting trip should include a check on the status of the intended area. Information can be obtained from the Division of Geological Survey offices, the U.S. Bureau of Land Management, or the University of Alaska. Unlike all the other national parks and monuments in the United States, prospecting is permitted in Mount McKinley National Park and in Glacier Bay National Monument, provided that the superintendents are consulted before entering for that purpose.

A person who cannot or will not accept very real hardships, a lonely wilderness existence for considerable periods of time, and the inevitable disappointments should not try prospecting in Alaska. Most experienced Alaskan prospectors live in good part off the land itself, by fishhook and rifle, and a good rifle is often necessary as a protection against marauding bears. Nevertheless, there are many recurring instances where good discoveries of gold or other valuable minerals have made their finders financially independent for life. The odds are long, but the stakes are high. Perhaps, in the long run, an experience at real wilderness living is beyond price.

For additional information, write: (1) Department of Mines, University of Alaska, College, Alaska 99701, or (2) Mining Information Specialist, State of Alaska Department of Natural Resources, Division of Geological and Geophysical Surveys, 323 East 4th Avenue, Anchorage, Alaska 99501.

Cook Inlet-Susitna Region

This sizable region is bounded by the Alaska Peninsula on the southwest, the Alaska Range on the west and north, and by the Talkeetna Mts. on the east.

ANCHORAGE (district, including Cook Inlet; Anchorage Quad.): (1) head of Archangel Cr., many mines—lode gold; (2) head of Fishhook Cr. (most productive area), many mines—lode gold; (3) on Eagle R., Peters Cr., and other regional streams, many placer- and lode-gold mines. The quadrangle shows the precise location of all important mines, including the Golden Light, Kempf, Lucky Shot, Panhandle, Thorpe, War Baby,

Willow Creek, Arch, Archangle, Bluebird, Fern, Gold Quartz, Mogul, Moose Creek, Northwestern, and many others (cf. also under WASILLA).

KENAI PENINSULA (district, covered by the Seldovia, Kenai, and Seward quads.; an area near the center of the S coast of Alaska, immediately N of the Alaska Peninsula):

GIRDWOOD (minor district opened in 1896), many hard-rock mines— lode gold.

HOPE, MOOSE PASS, SUNRISE (districts primarily of lode-gold mines (for example, the Hirshey Mine, which operated from 1911–45).

NUKA BAY (extreme S part of the peninsula): (1) the Nuka Bay Mines Co., Nukalaska Mining Co., Rosness & Larsen, and Skinner mines— lode gold; (2) the Sather, Goyne, Sonny Fox mines, and many others— lode gold.

SELDOVIA (district at S end of the peninsula; Seldovia Quad.), many lode mines: (1) Alley and the Mills & Trimble mines; (2) Alaska Hills, Little Cr., and other mines—lode gold.

TURNAGAIN ARM (a seaway district S of ANCHORAGE in central and N parts of the peninsula): (1) along Turnagain Arm: (a) many small placer operations; (b) many hard-rock mines, especially around head of Crow Cr.—lode gold; (2) along Mills, Canyon, Falls, and Cooper crs. (opened in 1896), many rich placers; (3) along Crow, Resurrection, Palmer, Sixmile, and other crs. debouching into Turnagain Arm from the N part of the peninsula, rich placers.

VALDEZ CR. (district included in the Healy and Mt. Hayes quads., approximately 125 air mi. S of FAIRBANKS, on the S slopes of the Alaska Range at lat. 63°12′ N, long. 147°20′ W, including the drainage area of Valdez and Clearwater crs.): (1) the "Tammany Channel" (buried ancestral channel of Valdez Cr.), discovered in 1904, placers, extensively hydraulicked; (2) several area lode mines (long dormant)—lode gold; (3) mines of the Alaska Exploration & Mining Co., Campbell & Boedecker, the Accident, Lucky Top, Wagner and Associates, Yellowhorn, etc.— lode gold.

WASILLA, NE by rd. and 21 mi. NW of PALMER, the Willow Cr. district (a 50-sq.-mi. area; Anchorage Quad.): (1) area above MATANUSKA in the upper part of Willow Cr. at the head of Cook Inlet (second most productive lode-gold area in Alaska), very many rich mines—lode gold; (2) in NW corner of quad., such mines as the Galena (as well as those listed under ANCHORAGE, q.v.—lode gold; (3) mines of the LeRoi Mining Co., the Lonesome (Gold Mint), Mary Ann, and Moose Cr.—lode gold.

YENTNA-CACHE CR. (district of 2,000 sq. mi., on SE slopes of the Alaska Range between lat. 61°55' and 62°45' N, long. 150°25' and 151° 05' W; Talkeetna and Tyonek quads.): (1) area: (a) extreme NE cor. of the Talkeetna Quad., the Boedecker mines—lode gold; (b) along Dollar, Thunder, and Willow crs., rich placers; (2) Suisitna region, near head of Nugget Cr., many mines—lode gold; (3) the Yentna R., upper drainage and tributaries, especially along Cache, Mills, Peters, and Long crs., rich placers.

Copper River Region

Elliptically shaped, the Copper River Region includes much of the drainage of the Copper River in southern Alaska. It is bounded on the north by the Alaska Range, on the southwest by the Chugach Mountains, and on the northeast by the Wrangell Mts., roughly between lat. 61°00' and 63°10' N and long. 142°00' and 146°00' W.

COPPER CENTER E to McCARTHY, N of the Copper and Chitina rivers, the Copper R. district (embracing the watershed of the S slopes of the Wrangell Mts., especially along streams feeding into the Copper R. below the mouth of the Chitina R.; Valdez and McCarthy quads.): (1) on the McCarthy Quad.: (a) along Bremmer R., such mines as the Golconda Creek, LeTendre, Nelson, Yellow Band, etc.—lode gold; (b) the Davy, Eleanor, Silver Star, and Mineral Cr. mines—lode gold; (c) the Berg Cr., Nugget Cr., Porcupine Cr., and Pearson mines—lode gold; (2) along the Nelchina R., numerous area placers, discovered in 1914; (3) on the Valdez Quad., in the Tiekel area: (a) the Townsend & Holland Mine—lode gold; (b) the Eagle, Knowles & Backman, Portland, Reis, Ross mines; and (c) the Quartz Cr. and the Wetzler mines —lode gold.

GULKANA, in NW part of the Copper R. basin, near intersection of lat. 63°00' N and long. 145°00' W, including the drainage area of the

Chistochina R. and the S foothills of the Alaska Range, the Chistochina district: (1) area: (a) along the Chisna R., placers, found in 1898; (b) along Slate Cr. and in Miller Gulch, many rich placers; (2) NE from GULKANA: (a) 30 mi., along the Chistochina R., many rich placer operations; (b) 45 mi., the Indian Mine, and (c) 10 mi. SE of the Indian, the Silver Cr. Mine—lode gold.

NIZINA (district, along the Nizina R. tributary of the Chitina R., between lat. 61°12′ and 61°37′ N, long. 142°22′ and 143°00′ W; McCarthy Quad.): (1) area lode mines: (a) the Green Butte and Kennecott mines, and (b) the Williams Peak Mine—lode gold; (2) Chitina Valley: (a) watercourse gravel deposits, rich placers, discovered and worked in 1898–99; (b) all regional bench and terrace gravel deposits, placers; (3) along Chititu Cr., rich placers, discovered in 1902; (4) along Dan Cr., lower-grade placers but productive to the present time, with total production of placer gold through 1959 of 143,500 troy ozs.

Interior Region (Yukon River Drainage)

Larger than many midwestern states, this immense territory contains the most important known placer gold deposits in Alaska. More than $504 million in gold was recovered from its placer mines between 1880 and 1957, with the region between EAGLE (on the border of Canada's Yukon Territory) downriver nearly 500 miles to TANANA being the most productive. USGS bulletins 872 and 907, covering the Yukon–Tanana region, are especially complete in the investigations of the mineral resources of this subarctic part of Alaska. Most of the region's total production of 12,282,250 ozs. of gold were derived from placer mining.

BONNIEFIELD (district about 60 air mi. S of FAIRBANKS, between lat. 63°30′ and 64°50′ N, long. 145°40′ and 149°20′ W, including the Kantishna and Valdez Cr., long an important mining center extending from the Tanana Flats on the N southward to the N slopes of the Alaska Range, on the E by the Delta R. and on the W by the Nenana R.; Healy, Mt. Hayes, Fairbanks, and Kantishna quads.): (1) Liberty Bell, Moose Cr., and Spruce Cr. placer mines, rich; (2) California Cr. and the Prospect Mining Co. placer mines; (3) the Fourth of July Cr. Mine, productive placer; (4) area: (a) along Gold King Cr., placers, begun in

1903; (b) along Eva Cr., placers (especially near the Liberty Bell Mine —lode gold), and near Caribou Cr.—float gold; (c) along Margueritc, Moose, Platte, and Portage crs., many mines and prospects—lode gold, with lead, silver, and stibnite.

CHANDALAR (district embracing the upper drainage of the Chandalar R. between lat. 67°00′ and 68°10′ N, long. 147°00′ and 150°00′ W, discovered in 1906, placer production through 1959 of 30,708 ozs.): (1) area along Big, Dictator, and Little Squaw crs., very many rich placers; (2) the Big Cr., Carter, Enveloe, Gold King, Kelty, and Little Squaw mines—lode gold (with an estimated $1 million ore body blocked out in the Little Squaw Mine in 1961); (3) the Mikado and Summit mines —lode gold.

CHENA DISTRICT (70 mi. E of FAIRBANKS; Big Delta Quad.), along Palmer Cr., the Palmer Cr. placers—abundant gold, with scheelite.

CHULITNA (near CANTWELL on Hwy. 3 just E of Mt. McKinley Nat'l. Park; Talkeetna Mts. Quad.), the upper reaches of the Chulitna R., many mines—lode gold, with lead, silver, zinc.

CIRCLE (district NE of FAIRBANKS between lat. 65°15′ and 66°00′ N, long. 144°00′ and 146°00′ W; Circle Quad.; tot. prod. through 1959 of 705,660 ozs.): (1) area: (a) along Birch Cr., the Birch Cr. placers, most productive of district; (b) all regional crs. and terraces—placer gold; (c) along Mastadon Cr., rich placers; (2) the Porcupine Dome Mine—lode gold.

DONNELLY (district about 35 air mi. S of BIG DELTA; Mt. Hayes Quad.), ½ mi. S of Rapids Roadhouse, along Gunnysack Cr. (extending ½ mi. upstream from rd.)—placer gold, with stibnite.

EAGLE (district immediately W of the Canadian line, between lat. 64°35′ and 65°15′ N, long. 141°00′ and 142°40′ W; Eagle Quad.): (1) (a) along Seventymile, American, and Fourth of July crs., very rich placers; (b) along Alder, Barney, Woodchopper, and Crooked crs., rich placers; (c) all other area crs., benches, terraces, placers (tot. prod. of all placers between 1906–59 of 42,220 ozs.); (2) the Eagle Bluff, Tweedon, and Lilliwig Cr. mines—lode gold; (3) W 65 air mi., on Copper Cr., the Copper Cr. Mine—lode gold.

FAIRBANKS (district of 300 sq. mi. between lat. 64°40′ and 65°20′ N, long. 147°00' and 148°10' W, principal gold placer area in Alaska and third most productive lode gold area; tot. prod., 1902–59, of 7,464,167 ozs.; Fairbanks Quad.): (1) broad area surrounding city, all regional watercourse and gravel deposits—placer gold (but long since well prospected); (2) N 70° E, from Treasure Cr. to lower Fairbanks Cr. (20 mi. long by 1 mi. wide)—placer gold; (3) NE 15–20 mi.: (a) Pedro Dome and Ester Dome areas (most productive mines)—lode gold; (b) along Pedro Cr. (first placer discoveries made in district, in 1902), rich placers; (c) in valleys of Cleary, Gilmore, Goldstream, and Engineer crs., rich placers; (4) along Dome, Ester, Vault, Cleary, and Chatanika crs., huge placer dredge operations, worked to present time but in diminished amounts; (5) in Skoogy Gulch, upper Cleary Cr., and along Fairbanks Cr., many lode mines with production beginning in 1910—lode gold; (6) grouped area lode-gold mines: (a) Barker & McQueen, Billy Sunday, Bluebird, and many others shown on the Fairbanks Quad.; (b) the Bunker Hill, Goodwin, and Scrafford; (c) the Engineer, Freeman & Scharf, Green Mountain, Ridge, Tanana, Tungsten Hill, and Woodpecker mines—lode or by-product gold; (d) the American, American Eagle, Melba Cr., Monte Cristo, and Vought mines—lode gold.

GOODPASTER (district; Big Delta Quad.), many area crs., placer operations.

HEALY (district S of FAIRBANKS, around the headwaters of the Nenana R., primarily lode mines): (1) the Ready Cash Mine, productive; (2) the Copper King, Eagle, Flauier, Golden Zone, Liberty, Lindfors, Lucrata, Mayflower, Riverside, Silver King mines, etc.—lode gold; (3) the North Carolina Mine, productive—lode gold; (4) the Chute Cr., Kansas Cr., and Glory Cr. mines—lode gold.

IDITAROD (district between lat. 62°10' and 63°00' N, long. 157°30' and 158°30' W, along the upper drainage of the Iditarod R. and its tributaries to the lower Yukon R., second most productive gold district in the Yukon Basin, with tot. prod. through 1959 of 1,297,500 ozs., nearly all from placers; Iditarod Quad.): (1) all regional streams, benches, and terraces, in gravel deposits—placer gold, with cinnabar, copper, lead, stibnite, tungsten, and zinc minerals; (2) along Otter Cr. (original placer discovery sites, in 1908); (3) along Flat and Willow crs., immense placers worked by dredges, mechanical scrapers, and hydraulicking; (4) the Donlin Cr., Flat Cr., Garnet, Glen Gulch, Golden Horn, and Malamute Cr., mines—lode gold; (5) NNE 50–60 air mi., the Innoko district (between lat. 62°50' and 63°15' N, long. 156°10' and 156°50' W, in the upper drainage area of the Innoko R. immediately NE of the Iditarod R. along the Beaver Mts.; Ophir Quad.): (a) all regional stream gravel deposits, particularly along Ganes Cr. (where first gold was discovered in 1906), placers; (b) in Ophir, Spruce, Little, and Yankee crs., placer mines; (c) 5 mi. SE of old camp of Ophir, on No. 6 Pup Cr., tributary to Little Cr., rich placers; (d) the Independence Mine—lode gold; (e) in the area of Tolstoi, the Book Cr. placers.

KOYUKUK DISTRICT (considered one of most northerly gold-producing districts in the world, between lat. 67°00' and 68°00' N, long. 149°00' and 150°50' W in drainage of the North, Middle, and South forks of the Koyukuk R. to the NE of BETTLES; Wiseman Quad.): (1) area: (a) all sand- and gravel bars along the Koyukuk R., placers; (b) gravel and sandbars along all regional streams tributary to the Koyukuk R. (and its major forks), placers; (c) all regional bench and terrace gravel deposits above the watercourses—placer gold; (2) along Michigan Cr., numerous mines—lode gold; (3) ghost camp of WISEMAN: (a) area streams, benches, terraces, and placers; (b) area hard-rock mines—lode gold.

LIVENGOOD (district 60 air mi. NW of FAIRBANKS; Livengood Quad.; tot. prod. through 1959 of 380,000 ozs. of placer and lode gold): (1) along ridge between Livengood Cr. and the Tolovana R., probable source for all placer gold in the district, numerous lode mines and prospects—lode gold; (2) on Livengood Cr., near mouth of Ruby Cr., rich placers (discovered in 1914): (a) all area cr. bars and (b) along N side of cr., a buried gravel bench, most productive, dredges operating in 1972—placer gold.

MANLEY HOT SPRINGS to EUREKA, the Hot Springs district, between lat. 65°00' and 65°20' N, long. 149°40' and 151°20' W, including the drainages of Baker, Sullivan, and American crs. as major placer occur-rences: (1) at MANLEY HOT SPRINGS, the Barrett Mine—lode gold; (2) along Baker and Eureka crs., placers discovered in 1898 and still active, tot. prod. through 1959 of 447,850 ozs.; (3) in the area of EUREKA, minor lode mines—lode gold. This district adjoins the Rampart district to the N, over a dividing ridge, both shown on the Tanana Quad.

MARSHALL (district, including RUSSIAN MISSION to the E, lying along the lower Yukon R. between lat. 61°40' and 62°00' N, long. 161°30' and 162°10' W, tot. prod. through 1957 of 113,200 ozs. of placer gold; Marshall and Russian River quads.): (1) from MARSHALL: (a) along Wilson Cr., first placer discoveries in 1913; (b) in Bonasila or Stuyahok Valley, productive placers; (c) SE about 10 mi., the Arnold Mine—lode gold; (d) E about 10 mi., the Edgar Cr. Mine—lode gold; (2) from RUSSIAN MISSION, the Cobalt Cr. and Mission Cr. area, several mines —lode gold.

NABESNA (district, between lat. 62°10' and 62°30' N, long. 142°20' and 143°10' W; Nabesna Quad.): (1) the Nabesna Mine (discovery attributed to a bear digging out a marmot in 1929; tot. prod. through 1940 of 63,300 ozs.)—lode gold; (2) other area mines, e.g., the Mineral Pt., Orange Hill, Bonanza Cr., Chathenda Cr., Erie, Carden Cr., Fourmile Cr., and Eureka Cr.—lode gold.

NORTHWAY JCT. (on the Alcan Hwy., the Chisana district, between lat. 61°55' and 62°20' N, long. 141°40' and 142°35' W, in the drainage area of the Chisana R. and its tributaries, tot. prod. through 1959 of 44,760 ozs. of placer gold; Nabesna Quad.): (1) Bonanza Cr., many placers and 1913 original discovery site, still active in a minor way; (2) along Big Eldorado, Little Eldorado, Bonanza, Beaver, and many other crs.—placer gold.

RAMPART (district just N of the Hot Springs district, q.v., between lat. 65°15' and 65°40' N, long. 149°40' and 150°40' W, tot. prod. through 1959 of 86,800 ozs. of placer gold; Tanana and Livengood Quads.): (1) regional cr., bench, and terrace gravel deposits—placer gold; (2) Hess R. and Minook Cr., first placer gold discovered in 1882, numerous placers; (3) along Minook Cr. Valley, terrace gravel deposits—low-grade placers; (4) on Little Minook Cr., major placers (mining began here in 1896, peaking in 1910; most of district's gold production came from here); (5) along Slate and Hoosier crs., placers.

RICHARDSON (TENDERFOOT) DISTRICT, about 60 mi. SE of FAIRBANKS between lat. 64°15' and 64°25' N, long. 146°00' and 146°40' W (a

little-known district lying along the Tanana R.; Big Delta Quad.): (1)
along Tenderfoot Cr., gravel deposits, placers discovered in 1905; (2)
along Buckeye and Democrat crs., placers; (3) area mines, e.g., the
Democrat, Blue Lead, and Grizzly Bear—lode gold.

RUBY (district, ill defined but extending from RUBY, on the lower
Yukon R., for 50–60 mi. S, to include POORMAN and adjacent camps
between lat. 63°40' and 64°45' N, long. 154°40' and 156°20' W, tot.
prod. of 389,100 ozs. of placer gold; Ruby Quad.): (1) Ruby Cr., discovered in 1907, rich placers; (2) Long Cr., opened in 1910, a once
flourishing mining center, placers; (3) Poorman Cr., discovered in 1912,
placers still in production; (4) on Beaver Cr., a mine—lode gold.

STEELE CR., the Fortymile district (from about 40 mi. S of EAGLE
along the Canadian border, including the upper drainage of the Fortymile R. between lat. 64°00' and 64°30' N, long. 141°00' and 142°20'
W, oldest gold-producing area in the Yukon Basin with uninterrupted
output since 1883, with tot. prod. of 400,000 ozs. of placer gold through
1959; Eagle and Big Delta quads.): (1) all area cr. and river gravels,
placers; (2) 10 mi. NW of CHICKEN, along Fortyfive Pup Cr. (tributary
of Buckskin Cr.), abundant placer workings; (3) Jack Wade, Walker
Fork, and such lesser camps as Stonehouse, Ingle, Lost Chicken, Napoleon, Franklin, Davis, and Poker Cr. (all on the Eagle Quad.), abundant
placer workings surrounding each center; (4) Dome, Wade, and
Chicken crs., most abundantly productive placers in district; (5) the
Tweedon and Lilliwig mines (near CHICKEN)—lode gold.

STEVENS VILLAGE (100 mi. NW of FAIRBANKS, on N bank of the
Yukon R. near end of the Yukon Flats region; Beaver and Bettles quads.
(The author's daughter was born here in 1938, when he was an Indian
Service teacher and community worker.): (1) area Yukon R. sandbars—
pannable colors; (2) S, to mouth of the Dall R., thence up the Dall into
a region approx. 100 mi. in dia., extending well N of the Arctic Circle,
many old prospects—placer gold; (3) S 9 mi., abandoned World War
I military post of Ft. Hamlin, area gravel bars and in crs. tributary to
the Yukon R., placer showings.

TANANA (World War I military post approx. 125 air mi. WNW from
FAIRBANKS, on N bank of the Yukon R.), W about 50 air mi., on

Morelock and Grant crs., the Gold Hill district, placers.

TOKLAT, the Kantishna district (about 4,500 sq. mi. between lat. 63° 25' and 65°00' N, long. 149°00' and 151°10' W, including that part of the Alaska Range foothills on the S and part of the Tanana Flats on the N, bounded on the E by the Nenana R. and on the W by tributaries of the Kantishna R., tot. placer-gold production, 1905–57, of 45,925 ozs.; Kantishna R., Mt. McKinley, and Healy quads.): (1) all regional stream gravels, shallow placers locally rich; (2) along the Toklat R. and along its tributaries, rich placers discovered in 1905; (3) along the S boundary of the Fairbanks Quad., such mines as: (a) Liberty Bell, Moose Cr., and Spruce Cr.; (b) California Cr. and the Prospect Mining Co.; (c) the Fourth of July Cr. Mine—all lode gold.

TOLOVANA DISTRICT (NW of FAIRBANKS, adjoining the Livengood district, between lat. 65°20' and 65°45' N, long. 147°50' and 149°00' W, in the upper drainage of the Tolovana R., tributary of the Yukon R., tot. prod. through 1959 of 375,000 ozs. of placer gold and still in production; Livengood Quad.): (1) along Livengood Cr., rich placers discovered in 1915; (2) along Gertrude, Ruth, Lillian, and Olive crs., rich placers (still operating); (3) very many area lode mines, e.g., the Discovery Claim and, above it, Claim No. 16, and scores of others shown on the quad. —lode gold.

Kuskokwim River Region

This immense region, 400 miles long by 75–100 miles wide, of generally sea-level land, comprises the drainage basin of the Kuskokwim River and its tributaries lying south of the Yukon River. Although placer gold was discovered along the Kuskokwim River in 1898, mining did not begin until 10 years later, with a total production through 1959 of 640,084 ozs.

EEK, KANEKTOK, TOGIAK (rivers feeding into the E side of Kuskokwim Bay on the Bering Sea; Bethel and Goodnews quads.), all regional river, tributary, bench, and terrace gravels—placer gold.

GEORGETOWN (district, at mouth of the George R., a tributary of the Kuskokwim R., between lat. 62°00' and 62°15' N, long. 157°15' and 158° 15' W, tot. prod. through 1959 of about 14,500 ozs. of placer and lode gold; Sleetmute Quad.): (1) area: (a) all regional stream gravels, bench deposits, and buried channels—placer gold; (b) along the George R., especially its tributaries of Donlin and Julian crs. (chief placer area, known since 1909), rich placers; (c) upper reaches of the George R., placers; (2) central part of the Kuskokwim Valley (about 45 air mi. S of IDITAROD, q.v., in the Yukon R. Valley, very many placer mines.

KOLIGANEK, NE, and about halfway between the Nushagak and Mulchatna Rivers, at Slcitat Mt. (Taylor Mts. quad.), area mines—lode gold.

MCGRATH, MEDFRA, NIXON FORKS, STIRLING LANDING, OPHIR, TAKOTNA (all close together in the E part of the Kuskokwim Valley, the McKinley district; Mt. McKinley, McGrath, Ophir quads.): (1) many productive placers surrounding each mining camp center; (2) area of Nixon Forks (Medfra Quad.), numerous mines—lode gold

PLATINUM (richest area for platinum mining in North America), the Goodnews Bay district, along SW coast of Alaska between lat. 59°00' and 59°40' N, long. 160°40' and 162°00' W, tot. prod. of gold from 1900 through 1947 of 29,700 ozs of placer gold): (1) all regional stream, bench, and terrace gravel deposits—placer gold; (2) along the Arolic R, placers discovered in 1900 and intensively worked until 1947; (3) Goodnews Bay (small indentation on Bering Sea coast on E side of Kuskokwim Bay, about 125 mi. S of BETHEL): (a) all regional tributary streams, high-grade placers—gold, with platinum group native metal nuggets; (b) Goodnews R., along its full length, little prospected for placer gold but promising.

STONY R., S, toward Goodnews Bay, across the Hoholitna, Holitna, and upper Aniak rivers (Russian Mission and Bethel quads.), a barely explored mineralized region in need of extensive prospecting, uncounted placer-gold areas in stream and bench gravels, terrace and buried channel deposits, with considerable potential also for platinum.

TULUKSAK-ANIAK (district named from its two main rivers and comprising their drainage basins, between lat. 60°30' and 61°30' N, long. 159° 00' and 161°00' W; Russian R. Quad.): (1) Innoko and Holitna rivers, discovered in 1900, large-scale placer operations; (2) Bear Cr. area of the Tuluksak watershed, and (3) gravel deposits of the Aniak R.—placer gold; (4) all regional flood-plain and bench gravels have proved profitable —placer gold. Tot. prod. of district: 230,555 ozs.

Northwestern Alaska

This vast, sparsely populated region lies north of the Yukon drainage basin and the Seward Peninsula and includes the drainage systems of the Kobuk, Noatak, and Alatna rivers, where most of the gold districts are found. Because almost the only access is by air, interested prospectors must be well financed and thoroughly familiar with techniques of living off the land and surviving in one of the most remote regions of Alaska. The total recorded production of 23,000 ozs. is presumably all from placers.

KIANA, the Kobuk R. Valley, and all its tributaries (an immense region; Ambler, Kiana, and Shungnak quads.): (1) all regional stream sand- and gravel bars, benchlands, and terraces, placer showings but little prospected; (2) gravel deposits of the Squirrel R., especially its Klery Cr. tributary, numerous placers and most profitable for today's gold hunter.

SHUNGNAK (district in the Kobuk R. Valley, between lat. 66°50' and 67°10' N, long. 156°50' and 157°25' W; Shungnak Quad.; tot. prod., 1898–1955, est. 10,000–15,000 ozs.): (1) many area placer workings in and about community; (2) along Wesley Cr., 6 mi. W of the Dahl Cr. tremolite mine, near head of cr., placers; (3) N 10 mi., the Aurora Mountain, Riley Cr., and Ruby Cr. mines—lode gold; (4) NE 40 mi., the Shishakshinovik Pass Mine—lode gold; (5) W 50 mi., in valley of the Ambler R., many rich placers.

Prince William Sound

Prince William Sound lies along the southern coast of Alaska east of the Kenai Peninsula and is mapped on the Valdez and Cordova quadrangles. Within the restricted area between the Chugach Mountains on the north and the waters of the sound lie the notable gold and copper mining centers of PORT WELLS, PORT VALDEZ, and ELLAMAR. Between 1894, when placer gold was discovered near VALDEZ, through 1956, a total 137,600 ozs. of gold was produced, either from lode mines or as a by-product of copper refining.

CORDOVA (Cordova Quad.): (1) N 5 mi., the Wilson Point Mine; (2)

ESE 15–20 mi., lode-gold mines of the Bear Creek Mining Co., Lucky
Strike Mining Co., and McKinley Lake Mining Co.

ELLAMAR (Cordova Quad.): (1) the Bligh Island, Cloudman Bay, and
Ellamar mines—lode gold; (2) the Alaska Commercial Co., Banzer,
Galena Bay, Hemple Copper Co., Hoodoo, Montezuma, Standard Cop-
per Mines Co., and Threeman—lode gold or by-product gold; (3) the
Dickey Copper Co. Mine—by-product gold; (4) the Fidalgo Mine—
lode gold.

LATOUCHE (island, Blying Sound Quad.): (1) the Seattle Alaska Copper Co. Mine—by-product gold; (2) just N (in Seward Quad.), at
Horseshoe Bay, the RADCO Mine—by-product gold.

VALDEZ (coastal terminus of the Richardson Hwy., i.e., State Hwy.
4, to FAIRBANKS and shipping port terminus of the oil pipeline from
Prudhoe Bay, Valdez Quad.): (1) several mi. down the fjord, the Cliff
Mine (staked in 1906 and largest gold producer, still operating) lode
gold, (2) the Alaska Gold Hill, Alice, Bluebird, Bunkerhill, and many
others—lode gold; (3) a group of lode-gold mines: (a) the Mayfield and
the National; (b) the Cameron-Johnson, Gold King, Minnie, Oleson,
Rambler, Rough & Tough—lode gold; (4) a great many other area
lode-gold mines grouped in various locations within the district, e.g., (a)
the Curley Kidney Mine; (b) the Alaskan, Big Four, Blue Ribbon,
Chesna, etc.; (c) the Ibex, Pinochle, Ramsay-Rutherford, Rose Johnson,
Valdez, Valdez Bonanza, and others—lode gold.

PASSAGE CANAL waterway, numerous mines—lode gold.

UNAKWIK Inlet waterway, numerous mines—lode gold.

Seward Peninsula

The Seward Peninsula is the westernmost extremity of North America and second only to the Yukon-Tanana region in the production of
placer gold, largely from the rich concentrations in the sands of the
Bering Sea beaches at NOME. Although gold was discovered as early as
1855, nothing was done about it until 1898, when the Nome district
was organized. The total production through 1959 was recorded as
6,060,000 gold ozs., all but about 10,000 ozs. being from placer operations.

BLUFF (Solomon Quad.): (1) W 7 mi., on California and Coca-Cola crs., rich placers; (2) Daniels Cr.: (a) along course, many placers; (b) beach sands at mouth, placers; (3) N, on Eldorado and Sweede crs., placers.

COARSE GOLD (about 70 air. mi. NNE of NOME; Teller Quad.), an old placer district with many operations.

COUNCIL (district in S part of the peninsula, including all drainage area of Golovnin Bay extending E almost to the Tubutulik R., discovered in 1865, with tot. prod. through 1959 of about 588,000 ozs. of placer gold; Solomon Quad.): (1) (a) along Aggie Cr. tributary of the Fish R., placers; (b) along Crooked Cr. and in Benson Gulch (tributary of Melsing Cr.), placers; (c) along Ophir Cr., discovered in 1896 and most productive in district, rich placers; (2) all cr. gravels and bench deposits in the drainage basin of the Niukluk R., including Ophir, Melsing, Goldbottom, Mystery, and Elkhorn crs.—placer gold; (3) along the Fish R. and Slate, Iron, Wheeler, West, Flynn, Spruce, Post, Daniels, and Koyana crs.—lode mines (gold); (4) The Big Hurrah Mine and on Bunker Hill, mines—lode gold; (5) NE, in the Bendeleben Mts., the Bendeleben district (50–75 mi. NE of NOME; Bendeleben Quad.); (a) along Crooked Cr. and the Inmachuk R., mines—lode gold; (b) the Omilak, Perkeypile and Ford, and Timber Cr. mines—lode gold.

DEERING, the Fairhaven district (including the Candle and Inmachuck districts, 40 mi. long immediately S of Kotzebue Sound, between lat. 65°40' and 66°10' N, long. 161°40' and 163°20' W; Candle Quad.): (1) Old Glory and Hannum crs., initial placers discovered in 1900; (2) along Candle Cr., richest placer in district, discovered in 1901 and producing 379,200 ozs. of gold through 1959; (3) along Bear Cr. and the Inmachuk R., many productive placers; (4) up the Kiwalik R.: (a) area gravels, placers (b) along Quartz Cr. and on Gold Run, placers; (c) a few mi. below mouth of Quartz Cr., rich placers.

KOYUK (district in SE corner of peninsula, between lat. 64°55' and 65° 40' N, long. 160°20' and 162°00' W, including the drainage of the Koyuk R.; Candle Quad.): (1) along Alameda and Knowles crs., placers operated since 1900 to the present; (2) at DIME LANDING and HAYCOCK, area crs., many rich placers; (3) all other regional crs. and in bench gravels, especially along Monument Cr., open-cut operations—placer gold; (4) between Little and Dry crs., extending a considerable distance with extensive dredging operations—placer gold.

NOME (district in S-central part of peninsula, between lat. 64°25' and 64°57' N, long. 165°00' and 165°30' W; Nome Quad. More than half the gold mined on the Seward Peninsula has come from this district, with production between 1897 and 1959 pegged at 3,606,000 ozs. of placer gold.): (1) the Nome R. placers, discovered in 1897; (2) Anvil Cr. and many other area crs.—placer gold; (3) sands of the Bering Sea beaches (most profitable of entire district)—placer gold; (4) many area lode mines along various streams and gulches.

PORT CLARENCE (district of about 2,000 sq. mi. at W end of the peninsula; Teller Quad.): (1) Bluestone and Agiapuk R. Basin, many placers, (2) Grantly Harbor, all area stream gravels, placers; (3) area lode mines, e.g., the Bessie & Maple, Brooks Mountain, Lost River, and Yankee Girl—lode gold.

SOLOMON, gold area lying along S side of the peninsula between lat. 64°30' and 65°00' N, long. 163°30' and 164°30' W (Solomon Quad.), all area stream and bench gravel deposits and numerous lode mines operated from 1900 through 1937.

TAYLOR (district in central part of peninsula, between lat. 65°10' and 65°45' N, long. 164°20' and 165°20' W; Teller, Bendeleben quads.): (1) along the Kougarok R.: (a) near mouth of Henry Cr. and (b) near head of r., especially on Macklin Cr. and tributaries, many rich placers, (2) in S part of district: (a) the Coffee Dome area, (b) along Iron and American crs., many placers; (3) many lode mines in general area.

UNALAKLEET, the Bonanza district forming the E border of Norton Sound; Norton Bay and Unalakleet quads., all regional streams—placer gold.

YORK, area placers, productive.

Southeastern Alaska

The total gold production of the Panhandle region between 1880 and 1959 is recorded as 7,788,514 ozs., of which all but 173,723 ozs. came from lode deposits. There are far too many mines to list here, but they are shown on the 13 quadrangle maps that cover the area.

CRAIG (district and Quad., on Prince of Wales Is. about 63 air mi. W of KETCHIKAN): (1) around town, many lode mines; (2) at head of the Kasan Peninsula, around SALT CHUCK, many lode-gold mines; (3) on Dall Is.: (a) N part, and (b) S part, lode-gold mines; (4) at S tip of Prince of Wales Is. (Prince Rupert Quad.), the Nelson & Tift Mine—lode gold; (5) E of HYDABURG, in large area, many lode-gold mines.

CHICHAGOF (district of some 4,500 sq. mi., including Baranof, Chichagof, Kruzof, and Sitka islands): (1) around town, many lode mines, of which the Alaska Chichagof Mine was the principal producer; (2) 6 mi. NW of PELICAN, numerous lode-gold mines; (3) around SITKA, area old mines and prospects where early attempts to mine gold (1871) failed commercially; (4) SE arm of Sitka Sound, many mines—lode gold.

JUNEAU (district, including Douglas and Admiralty islands, with tot. prod. between 1882 and 1959 of 6,883,556 ozs. of lode gold; Douglas and Admiralty quads.): (1) in town, the Alaska-Juneau Gold Mine, opened in 1893 and closed in 1944, now a tourist attraction, with tot. prod. of 2,874,361 ozs. of lode gold; (2) along Lemon Cr., placer deposits discovered in the 1880s; (3) at DOUGLAS (across bridge on Douglas Is.): (a) the great Alaska-Treadwell Mine, now flooded out; (b) many surrounding lode-gold mines; (4) along E side of Lynn Canal, numerous lode mines; (5) NW 45 mi., on W side of Lynn Canal, big lode-gold mines; (6) NNW 50 mi., on E side of Lynn Canal, a group of lode-gold mines; (7) Hawk Inlet and Funter Bay on Admiralty Is., many lode mines; (8) Sumdum and Windham bays, area where first placers were discovered in 1869; (9) NW 20 mi. from SUMDUM, in two contiguous areas, many lode-gold mines; (10) SE 50 mi. from SUMDUM, the Colp & Lee Mine—lode gold.

KETCHIKAN-HYDER (district, including the S end of the Alaska Panhandle), many lode-gold mines.

MT. FAIRWEATHER DISTRICT (N of Cross Sound to the NW of the Chichagof district; Mt. Fairweather Quad.): (1) Lituya Bay: (a) area beach sands, placers; (b) the Lituya Bay Mine—lode gold: (2) halfway up Glacier Bay and E of Mt. Fairweather, many area lode-gold mines; (3) at the mouth of Glacier Bay, the Francis I Mine—lode gold.

PETERSBURG (district embracing Kupreanof and Kuiu islands; Petersburg Quad.): (1) area lode-gold mines, productive; (2) S and SE 15–22 mi., several important mines, especially on Wodwaski Is.; (3) SE 50 mi., the Thomas Bay Mine—lode gold; (4) WNW 30 mi., on N tip of Kuiu Is. (Port Alexander Quad.), the Keku Islet Mine—lode gold.

SKAGWAY (Porcupine) district, along the Porcupine Cr. tributary of the Kleheni R.; Skagway Quad.: (1) N 10–12 mi., along the White Pass & Yukon Route RR, the Inspiration Point Mine—lode gold; (2) along the Porcupine R.: (a) the Summit Cr. Mine—lode gold; (b) river and tributary cr. gravels—placer gold; (c) all area side benches, both low and high, placer gravels.

WRANGLE (district, Petersburg Quad.): (1) E about 10 mi., the Berg Basin, Groundhog, and Lake mines—lode gold; (2) on Zarembo Is. (about 20 mi. SW of town), the Exchange Mine—lode gold.

YAKATAGA (district of 1,000 sq. mi. along the coast just W of the N end of the Panhandle, between lat. 60°00′ and 60°30′ N, long. 141°20′ and 141°10′ W; Bering Glacier Quad.): (1) area beach sands, placers, discovered in 1897; (2) along the White R., bench gravels worked hydraulically to the present, tot. prod. through 1959 of 15,709 ozs. of placer gold.

Southwestern Alaska

ALEUTIAN Is., on Unga Is. in the Shumagin Group, the Apollo Consolidated Mine, operated steadily from 1891 to the present, with tot. prod. of 107,900 ozs. of lode gold, with lead, copper, zinc.

BRISTOL BAY (including Nushagak Bay, Dillingham and Iliamna quads.), along upper reaches of the Mulchatna R. and its tributaries, placer deposits.

KATMAI NAT'L. MONUMENT (on the Alaska Peninsula, Iliamna Quad.), all area outside Monument boundaries, in gravels of streams flowing into Iliamna Lake—placer gold.

KODIAK (Is., Karluk Quad.), W coast beaches and tributary lode deposits—gold, with copper, lead, silver, tin, and zinc.

Topknot of North America

The Arctic Slope of Alaska lies within the Polar Zone, some 600 miles east and west by 150 miles north and south, sloping from the foothills of the Brooks Range to the Arctic Ocean. Known primarily for recent great discoveries of petroleum, especially around Prudhoe Bay, this immense region of tundras has been very little prospected for gold.

ANAKTUVIK PASS (area of the N slopes of the Endicott Mts., Chandlar Lake Quad.), numerous mineralized outcrops—lode-gold possibilities, with likelihood of placer deposits in the regional watercourse and bench gravel deposits.

OPILAK R. (S of Barter Is., the Mt. Mitchelson and Demarkation Point quads.), along main r. and presumably in gravels of its tributary streams—placer gold likely.

ARIZONA

The first gold found in Arizona was located near the old Mexican settlements of Tucson and Tubac, and in the Dome (Gila City) district, where rich placers were found, in 1859. The first lode deposits were found in 1863 in the Bradshaw Mountains of Yavapai County. From 1860 through 1965, Arizona produced 13,321,000 ozs. of placer and lode gold, ranking eighth among America's gold-producing states. Forty-two districts in 10 counties produced more than 10,000 ozs. apiece, with very many smaller districts producing less amounts. Nearly 80 per cent of Arizona's lode gold, and much of its placer gold, has come from a northeast-trending belt of mountains about 65 miles wide bordering the southwest margin of the Colorado Plateau.

Apache, Coconino, and Navajo Cos.

PAINTED DESERT, a region embracing a large geographic area of the Triassic Chinle Formation in which microscopic particles of gold are disseminated throughout the residual clays.

Cochise Co.

Third among Arizona's gold-producing counties, Cochise County produced approximately 2,723,000 ozs. of gold between 1879 and 1959, more than half coming as a by-product of the Bisbee copper ores. Only 950 ozs. of placer gold are included in the total.

Bisbee (Warren) district: (1) the Copper Queen, Calumet, and Denn copper mines—by-product gold; (2) SE 4 mi., in the upper part of Gold Gulch, small placer deposits.

Dos Cabezas (district, 18 mi. SE of Wilcox in the Dos Cabezas Mts.): (1) area copper-lead-silver mines—by-product gold; (2) vicinity: (a) the Dos Cabezas placers, 1901 to present; (b) all area arroyos and gulches, benches and terraces—placer gold (flat, ragged, rather coarse); (3) NW 1½ mi., the Le Roy Property—lode gold (with lead, silver); (4) N 2½ mi.: (a) the Dives (Bear Cave) Mine, site of 50-ton flotation mill built in 1934—lode gold; (b) E of the Dives, the Gold Ridge (Casey) Mine—lode gold; (5) NE 2¾ mi., the Gold Prince (Murphy) Mine— lode gold; (6) at N foot of the Dos Cabezas Mts., the Teviston placers, including all area gulches—nugget gold (often to large size).

Gleeson: (1) area (all privately owned), the Gleeson placers (dry-wash)—fine gold to medium-size nuggets; (2) in gulch W of the Copper Belle Mine, placers worked in 1930s—coarse gold.

Hereford, SW 12 mi. and about 3 mi. N of Mexico, in Ash Canyon in SE part of the Huachuca Mts., the Huachuca placers (along canyon bottom for 3 mi.

Manzoro (RR siding in N part of co.), S ¾ mi., at NE foot of the Dragoon Mts., the Golden Rule (Old Terrible) Mine, rich—lode gold.

Pearce, at E and W margins of Pearce Hill, area gravel deposits known as the Pearce placers.

Tombstone (district in the Tombstone Hills, primarily silver mines that produced 271,000 ozs. of by-product gold between 1877 and 1932), S ½ mi., erratically distributed oxidized lead-silver deposits, locally rich —free gold.

Turquoise (Courtland, Gleeson) district, 14 mi. E of Tomb-stone, on E side of the Dragoon Mts. 18 mi. N of Bisbee, numerous mines with gold finely divided, tot. prod., 1908–55, of 70,000 ozs.

Gila Co.

Eighth among Arizona's gold-producing counties, Gila County produced 240,500 ozs. of gold through 1959, about two thirds of it as a by-product of copper refining.

BANNER (Christmas) district, in extreme S tip of co. at SE end of the Dripping Springs Mts., tot. prod., 26,000 ozs.; at the Christmas Mine, major producer—lode gold.

GLOBE (usually combined with adjacent MIAMI as the "Globe-Miami district"): (1) up Pinal Cr. from town, placers; (2) along Lost Gulch and Pinto Cr., placers; (3) N, in the Apache Mts., and W, in small gulches draining into Richmond Basin, placer deposits.

MIAMI: (1) area copper mines, e.g., Miami-Inspiration, Castle Dome, Copper Cities, and Cactus—by-product gold; (2) gulches peripheral to the Castle Dome Mine and, NE, the Golden Eagle Mine, placer gravels.

PAYSON, area mines to the SSE, S, and SW, at varying distances, e.g., the Gowan, Single Standard, Golden Wonder, Ox Bow, Silver Butte, Zulu, and Bishop's Knoll, all quartz—free gold (with local concentrations of copper minerals carrying free gold).

Graham Co.

ASHURST, SW on the Klondyke rd., in broad pass between the Santa Teresa and Pinaleno Mts., the Clark district (19 mi. from RR at Cork siding): (1) many area prospects on quartz outcrops—free gold; (2) the Chance claims, small production—lode gold.

SAFFORD: (1) E 14 mi. on U.S. 70 and 7 mi. N on dirt rd. to Bonita Cr., area upstream from mouth along the Gila R., the Gila River placers (richest deposits within arcs of some curves): (a) all stream, bench, and terrace gravels, placers (flake to wire gold); (b) on N side of the Gila R., between Bonita and Spring crs., numerous placers (e.g., the Neel, Smith-Boyls, Hammond-Serna, and Colvin); (2) NE, in SW part of the Gila Mts., the Lone Star district: (a) the San Juan and Lone Star mines; (b) in mts. to N, the Roper, West, Wickersham, Merrill, and other mines, in quartz veins—free gold.

Greenlee Co.

CLIFTON: (1) area mines—gold, copper; (2) above town, along the San Francisco R., Quaternary bench gravels, placers; (3) along the lower part of r. as far as mouth of Eagle Cr., placers; (4) out 12 mi., the Smuggler claims, extensive sluicing—placer gold; (5) between town and the Old

Rock House, along Chase Cr.: (a) in creekbed, thin gravel placers; (b) in all tributary gulches—placer gold.

DUNCAN, W 12 mi., the Ash Peak district, silver with by-product gold prod. of 11,296 ozs., 1936–54.

METCALF, E 2 mi., near Copper King Mt.: (1) the Copper King Mine —lode gold; (2) NE of the Copper King, vein outcrops (granite) following porphyry dikes—panning gold.

MORENCI (area of large-scale copper mines with by-product gold prod. of 203,000 ozs., 1873–1959): (1) ESE 1 mi., the Hormeyer Mine— by-product gold; (2) W 3 mi., in Gold Gulch, placers; (3) SW 4½ mi., in Gold Gulch (area of many auriferous veins worked intermittently for years), the Lakeman Mine—lode gold.

OROVILLE, area gravels, once hydraulicked—placer gold.

Maricopa Co.

As the fifth gold-producing Arizona county, Maricopa County produced 428,000 ozs. of gold, primarily from the Vulture Mine, with 3,000 ozs. attributed additionally to placer deposits.

AGUILA, S 15 mi., in NW part of the Big Horn Mts., the El Tigre Mine (12 claims)—lode gold.

CAVE CR. (district, 25–55 mi. N of PHOENIX): (1) S 1 mi., the Mormon Girl Mine—minor lode gold; (2) SW 2½ mi., the Copper Top Mine—lode gold; (3) Cave Creek-New River turnoff: (a) SW 4 mi., both sides of rd.—panning gold; (b) W 2½ mi. to N trending rd., along this rd. 1½ mi., area around the old Go John Mine—panning gold; (4) N a few mi., on slope E of the main watercourse, the Phoenix and Maricopa mines, site of 100-stamp mill and cyanide plant—lode gold (17,000 ozs. prod., 1890–1959), with molybdenum and vanadium; (5) N 13 mi., dirt rd. to W: (a) the Rackensack Mine and (b) S to old mine —lode gold; (c) other area mines: the Mex Hill, Lucky Gus, A. B. Bell, Dallas-Ft. Worth, and Gold Reef—all lode gold.

PHOENIX: (1) S 9 mi., in N part of the Salt River Mts., the Max Delta Mine—lode gold; (2) out 18 mi. in N foothills of the Phoenix Mts., the Winifred district, the Jack White Mine—lode gold; (3) NW 45 mi.: (a) along San Domingo Wash for 6–7 mi., the San Domingo placers; (b) along Old Woman Gulch (a S tributary), large placers—large, coarse gold nuggets; (c) all area arroyos and gulches, in black-sand deposits—placer gold; (d) lower country gravels, often in black sands, placers—finely distributed gold.

WICKENBURG: (1) area of entire region shows scores of old mine dumps, which occasionally yield large gold nuggets due to carelessness of early miners; (2) SE 7 mi., sands and gravels of entire length of the Hassayampa R.: (a) notably abundant for several mi. below mouth of San Domingo Wash—placer gold; (b) the Hassayampa placers, very productive 1934–49; (3) SW 14 mi. and 9 mi. W of the Hassayampa R., at S margin of the Vulture Mts.: (a) the Vulture Mine, producer of 366,-000 ozs. of gold, 1863–1959, site of big mills and 100-ton cyanide plant (used to leach old dump tailings in 1934); (b) in pediment of Red Top Basin NW of the Vulture Mine, area of about 3 sq. mi., the Vulture placers (continuing down Vulture Wash for 2 mi. SW of the mine), many dry-wash workings, pits, etc.—placer gold; (4) W 18 mi. and 2½ mi. S of U.S. 60, the Sunrise Mine—lode gold; (5) NE, a large mining area extending into Yavapai Co.—lode gold.

Mohave Co.

Ranking second among Arizona's gold-producing counties, Mohave County produced 2,461,000 ozs. of lode and placer gold through 1959, with about 50 per cent coming from the San Francisco district at Oatman.

AREA: (1) in NE part of the Cottonwood Cliffs, in upper reaches and tributaries of Wright Cr., small placers; (2) along the Colorado R. downstream from mouth of the Grand Canyon: (a) all river bars, placers; (b) all elevated bars and benches formed by tributary streams, placers; (c) at Willow Beach (65 mi. NW of KINGMAN near Hoover Dam Hwy.), an ancient river bar near outer bow of the Colorado R. (250 acres), the Sandy Harris Placer; (d) 2½ mi. N of Pyramid Rock, Colorado R. benches, placers.

ALAMO CROSSING (of the Bill Williams R. about 40 mi. N of U.S. 60 at WENDON, in Yuma Co.): (1) many regional old mines, e.g., the Little Kimball (5 mi. back of the crossing with its Jim Rogers Mill abandoned at the crossing)—lode gold; (2) N 16 mi., at ghost camp of Rawhide straddling the "Owl Hoot" outlaw trail and temporary stopover where desperadoes could dig raw gold from adjacent Rawhide Butte—free gold.

CHLORIDE (60 mi. NW of KINGMAN): (1) in E part of the White Hills, the Gold Basin district: (a) many area mines—lode gold; (2) W 9 mi., the Pilgrim Mine—lode gold; (3) NW: (a) 25 mi., the Mockingbird Mine and (b) on E side of the Black Mts., 30 mi. from town, the Gold Bug Mine—lode gold; (4) in the Cerbat Mts., the Cerbat Mountains district (gold, lead, silver, zinc): (a) E 1½ mi from CHLORIDE, near middle of W slope of mts., the Pay Roll Mine, and (b) W 1½ mi., on the pediment, the Tintic Mine—lode gold; (c) E 2½ mi., the Rainbow Mine, and (d) E 3½ mi., the Samoa Mine (most active and constant producer through 1908)—lode gold, silver.

HACKBERRY: (1) E 9 mi., in N part of the Cottonwood Cliffs plateau, the Walkover Mine—lode gold and gold sulfides; (2) N 25 mi., in foothills of the Grand Wash Cliffs, the Music Mountain district, productive mines such as the Ellen Jane, Mary E, and Southwick—lode gold.

HOOVER DAM, area downstream from mouth of the Grand Canyon, river bars, benches, and terraces—placer gold (finely disseminated).

KINGMAN: (1) SW 3 mi. and ½ mi. NE of McConnico, on property of the old Bi-Metal Mine, in area draws and gullies, the Lewis Placer—coarse gold, (2) SE 6 mi., in the Maynard district, the Lookout Placer. (a) along bed of a shallow gulch, and (b) hillside gravels—coarse wire gold; (3) S, in the Lost Basin Range, the Lost Basin district, numerous mines—lode gold; (4) NW 15–18 mi. on U.S. 93, the Wallapai district (near center of the Cerbat Mts.), including the old camps of Cerbat, CHLORIDE, q.v., Mineral Park, and Stockton, very many old lead-silver-zinc mines that from 1904–56 produced 125,063 ozs. of by-product gold; (5) 10–25 mi. W and NW from CHLORIDE, the Weaver district in N part of the Black Mts., divided into 3 sections: (a) on W slope, the Virginia camp; (b) on E slope, the camps of Mockingbird, Pyramid, and Pilgrim; (c) a few mi. N of Pilgrim, the Gold Bug camp, tot. prod., 1900–59, of 63,200 ozs. of lode and by-product gold; (6) NW 56 mi., at N end of Red Lake playa (9 mi. by dirt rd. branching N from U.S. 93), in T. 28 and 29 N, R. 17 and 18 W, the Gold Basin placers, rather extensively worked by dry-wash equipment; (7) NW 72 mi. via U.S. 93, in T. 29, 30 N, R. 17 W and 8 mi. from the Colorado R., the King Tut placers (an area 8 mi. long).

OATMAN (the San Francisco district, including the Vivian, Gold Road, and Boundary Cone localities in an area about 10 mi. long by 7 mi. wide on W slopes of S part of the Black Mts., with Union Pass sometimes included, discovered in 1864, with tot. prod. through 1931 of 2,045,400 ozs. of lode gold): (1) area: (a) the Tom Reed Property (extremely rich), including the Pasadena, Aztec Center, and Big Jim-Aztec mines—lode gold; (b) beneath town, the Tip Top ore body; (c) 1½ mi. NE of the Tom Reed, the Gold Road Mine, discovered in 1900, very rich—lode gold; (2) very many other rich lode mines, e.g., the Gold Dust, Ben Harrison, Leland, Midnight, Sunnyside, Iowa, Lazy Boy, etc.; (3) the Moss Vein (reputedly produced $240,000 from a pit of 10 cu. ft.); (4) N 1 mi., at Goldroad (Katherine), rich mines—lode gold; (5) SW 1½ mi., the Pioneer (German-American) Mine—lode gold; (6) NW 5 mi. (6 mi. downstream from U.S. 66 in valley of Silver Cr.), placers; (7) NW 7 mi. and 2 mi. N of Silver Cr., the Moss Mine (probably first discovered in district)—lode gold; (8) NNW 12 mi., extending from Union Pass NW in the Black Mts. (summit of Rte. 68 to BULLHEAD CITY), W to the Colorado R., the Union Pass (Katherine) district: (a) 2 mi. E of the Colorado R., in sec. 5, (21N–20W), the Katherine Mine, discovered in 1900, rich producer—85 per cent lode gold, 15 per cent silver; (b) 4 mi. E of the Gold Standard mill, the Roadside Mine—lode gold, silver; (c) 3½ mi. SW of the Roadside, in sec. 20, the Arabian Mine—lode gold, with twice as much silver; (d) other rich area lode mines are: the Tyro, Sheeptail-Boulevard, Frisco, Black Dyke Group, Pyramid, and Golden Cycle—gold, silver.

MINERAL PARK (6¾ mi. SE of CHLORIDE, q.v.): (1) the Tyler Mine —lode gold, with silver and lead; (2) SE 1¼ mi., near W foot of the Cerbat Range, numerous area mines—lode gold, silver; (3) the Golden Gem, Flores, Cerbat, and Oro Plata mines—lode gold, lead, silver, zinc.

PIERCE FERRY (accessible from U.S. 93 N of turnoff to CHLORIDE 12

mi., or 54 mi. SW of BOULDER CITY, Nev.): (1) S 8 mi. from the
Colorado R., in T. 28 and 29 N, R. 17 and 18 W, placer deposits; (2)
S 9 mi., in same T. and R., placers.

TOPOCK: (1) SE 18 mi.: (a) in SW part of co., in foothills of the
Chemehuevis or Mohave Mts., the Chemehuevis placers; (b) at SW
foot of the mts., in area of the Red Hills, the Mexican or Spanish
diggings (most productive)—placer gold; (c) NE side of mts., in Prin-
ter's and Dutch gulches, placers; (2) SE, in the Mohave Mts., area placer
and lode-gold mines; (3) SE 55 mi.: (a) the Best Bet (Kempf) Mine—
lode gold; (b) a few mi. N of the Best Bet, the Gold Wing Mine (and
district), with other area prospects—lode gold; (c) on opposite side of
mts., the Dutch Flat Mine—lode gold.

Pima Co.

Pima County ranks seventh among Arizona's gold-producing counties, with a total production of $3,212,000 through 1931, of which two
thirds came as a by-product of copper refining.

AJO (major copper center with enormous open-pit mine, which produced 990,000 ozs. of by-product gold, 1924–59): (1) the New Cornelia
Mine, open-pit, spectacular; (2) S 30 mi. via branch from old Sonoita
rd. and about 6 mi. W of Dripping Springs in the Puerto Blanco Mts.,
the Golden Bell Mine—lode gold.

ARIVACA: (1) in area of the Las Guijas Mts.: (a) the Las Guijas
(Arivaca) placers, very productive; (b) along Arivaca Cr., in large channel along SW slope of mts., and (c) along Las Guijas Cr., along the NE
slope (much smaller, drier), placers worked intermittently to the present;
(2) regional pediment slopes, mesas, and watercourse beds, placers
(mostly dry-wash); (3) pediment at N foot of mts.: (a) from pediment
itself, and (b) from area arroyos and gulches, large-scale placers (operated
in 1933).

CONTINENTAL, W 6 mi., along upper course of Amargosa Arroyo: (1)
numerous dry-wash placers; (2) all tributaries, watercourse beds and
banks, dry placers; (3) in secs. 20, 21, 28, and 29, (18S–12E), in thin
soil and hillside detritus—placer gold with magnetic sand.

COVERED WELLS, S 6 mi., the Morgan Mine—lode gold.

GREATERVILLE (district, 34 air mi. SE of TUCSON, producer of 4,140 gold ozs., 1903–59): (1) area of about 8 sq. mi. on lower slope of the Santa Rita Mts., all watercourses, benches, etc.—placer gold; (2) productive gulches of Boston, Kentucky, Harshaw, Sucker, Graham, Lousiana, Hughes, and Ophir (below its jct. with Hughes Gulch)—placer gold; (3) upper parts of Los Pozos and Colorado gulches, Chisapa Gulch (on rd. from town to Enzenberg camp), and Empire Gulch (below its jct. with Chispa Gulch)—abundant placer gold; (4) in Hughes Gulch, area lead-silver mines—lode gold.

QUIJOTOA (Pima village about 70 mi. W of TUCSON on Rte. 86), in area of the Quijotoa Mts. (15 mi. long by 5 mi. wide encompassing about 100 sq. mi.): (1) the Quijotoa placers; (2) many regional unmined and poorly prospected quartz outcrops and hematitic brecciated zones—lode gold (some exposures have provided spectacular specimens): (3) 3 mi. S of Pozo Blanco and 1 mi. W of the foot of the Quijotoa Mts., several area placer diggings, rich—coarse gold; (4) 6 mi. N of the mts. (area of several ghost mining camps from 1880s boom), placer operations.

REDINGTON, SW, on N slope of Alder Canyon in the Santa Catalina Mts., from near USFS boundary to within a few mi. of the San Pedro R., placers—coarse, flat, ragged gold.

ROBLES JCT. (25 mi. W of TUCSON on Rte. 86), SW, in the Baboquivari Mts.: (1) the Gold Bullion Mine—lode gold; (2) 5–6 mi. SE of Baboquivari Peak, at E foot of mts., in benches and bars along a large E-trending wash, the Baboquivari placers.

SELLS: (1) in the South Comobabi Mts. (on the Papago Indian Reservation), the Cababi (Comobabi) district: (a) at S base of the mts. and 4 mi. W of Hwy 86, the Akron Mine (20 claims)—lode gold; (b) NW 1½ mi. from the mine, the Corona Group (7½ claims)—minor lode gold; (c) other area mines include the Hawkview, Faro Bank Group (4 mi. N of town), and minor prospects—lode gold; (2) SE 21 mi., on W side of the Baboquivari Mts., the Allison (Chance) Mine, major district producer from 1898 to mid-1930s—lode gold.

TUCSON: (1) SW 30 mi., at Papago (Aquajito), the Papago district, along Ash Cr., the Sunshine-Sunrise Group of claims in Pescola Canyon, placers; (2) SSE 30 mi., at NW base of the Santa Rita Mts., in area of Madera Canyon, the Old Baldy placers (richest below the rd. forks in deposits trenched 40–50 ft.), rather extensively worked.

Pinal Co.

Pinal County ranks sixth among Arizona's gold-producing counties, with 893,350 ozs. of lode and placer gold mined between 1858 and 1959.

APACHE JCT., NE on Hwy. 88 (the Apache Trail) to the Goldfields district between the Superstition and Goldfield Mts. 36 mi. E of PHOE-NIX in Maricopa Co., the Young (Mammoth) Mine, 1,000 ft. deep, mill and cyanide plant—lode gold.

CASA GRANDE, out 32 mi. on the hwy. to COVERED WELLS, then 1 mi. by branch rd. to W foot of the Slate Mts., the Mammon Mine and mill site—free gold (erratically distributed).

MAMMOTH (Old Hat district 50 mi. NE of TUCSON, in SE part of co., on E flank of the Black Hills; of tot. prod. of 403,000 gold ozs., about 40,000 came as by-product of the San Manuel copper ores): (1) the Mammoth Mine—lode gold primarily, with molybdenum, (2) the San Manuel Mine, primarily copper—by-product gold; (3) SE 3 mi., in sec. 26, (8S–16E), the Collins Property and the Mohawk Mine (both close to the Mammoth)—lode gold, with lead, copper, zinc.

ORACLE, S 4–10 mi. (extending into Pima Co.), near NW base of the Santa Catalina Mts., the Canada del Oro (Old Hat) placers, most noted gold-producers in co.: (1) many old pits, trenches, and tunnels—placer gold; (2) area arroyos and gulches, especially on N side of main watercourse, many placers worked in 1930s.

RAY (in NE part of co. about 17 mi. S of MIAMI), between the Dripping Springs Range to the E and the Tortilla Range to the W, the Mineral Creek district and the Ray Consolidated Copper Co. Mine—by-product gold (tot. prod., 35,250 ozs.).

SUPERIOR (the Pioneer district, with tot. prod. of 397,700 ozs. of gold, 1875–1959): (1) at E edge of town, between Queen Cr. and the Magma Mine, the Lake Superior and Arizona Mine (copper)—by-product gold; (2) the Magma (Silver Queen) Mine—by-product gold; (3) S of Queen Cr. a short distance, the Queen Cr. Mine (S extension of the Superior and Arizona vein)—minor lode gold; (4) SE 4 mi., the Belmont Mine (Smith Lease)—lode gold.

Santa Cruz Co.

HARSHAW (10 mi. S of PATAGONIA), between Sonoita Cr. on NW and Alum Canyon on SW, placers about 1 mi. sq.—gold, native lead.

NOGALES, N 6 mi., in Guebabi Canyon, in Quaternary gravels along stream, the Nogales placers, well worked in early 1900s.

ORO BLANCO-RUBY area, the Oro Blanco placers, especially: (1) in Oro Blanco and Viejo gulches, placers: (a) area hillsides, bench and terrace gravels—placer gold, silver; (b) near mouth of Warsaw Cr. (2½ mi. N of Mexico), productive placers; (2) in Alamo Gulch, productive placers.

PATAGONIA: (1) S 9 mi. (6 mi. N of Mexico), on E slopes of the Patagonia Mts., the Patagonia (Mowry) placers, productive: (a) along Mowry Wash and tributaries, in Quaternary gravels; (b) along main wash at E border of Quajolote Flat (½ mi. N of the Mowry Mine); (c) 1½ mi. SE of the Mowry Mine, in two N-side parallel tributary gulches, intermittently worked by Mexicans to the present; and (d) downstream from the old Mowry smelter, in Quajolote Wash, gravel deposits—all produce placer gold (angular, nuggets to 2 ozs., coarse); (2) SW 6 mi. and 2½ mi. NW of the Three R Mine, in 320 acres of Quaternary gravels, the Palmetto placers, worked by sluice and dragline in 1927.

RUBY (with ARIVACA in Pima Co.), the Oro Blanco district, productive since Spanish times with tot. prod. of 100,200 ozs. of lode and placer gold: (1) W 2 mi. and W of the rd. to WARSAW: (a) the Old Glory Mine —lode gold; (b) N, the Margarita Mine—minor lode gold: (2) NW 2¾ mi., on W side of the ARIVACA rd., the Austerlitz Mine—lode gold; (3) SW 4 mi. by rough rd., the Tres Amigos, Dos Amigos, and Oro Blanco mines—free gold (medium-fine to coarse).

SALERO, SW 2¼ mi. and 1 mi. S of Mt. Allen at SW base of the Grosvenor Hills, on each side of the twp. line in SW¼ sec. 35, the Tyndall placers (and district).

WASHINGTON (district, about 17 mi. E of NOGALES via graded dirt rd.): (1) NW 3 mi., the Four Metals Mine, and (2) 2 mi. W, the Proto Mine—lode-gold traces in base-metal ores.

Yavapai Co.

In central Arizona, Yavapai County ranks first among the gold-producing counties, having produced 3,476,150 ozs. of lode gold and 266,-804 ozs. of placer gold through 1959.

BAGDAD (42 mi. W of PRESCOTT), many area mines in SW part of the Eureka (Bagdad) district, tot. prod., 59,787 ozs., 1887–1951, principally the Hillside Mine—lode gold.

BLACK CANYON CITY (district, in SE part of co. between E foothills of the Bradshaw Mts. and the Agua Fria R., tot. prod., 1914–59, of 46,700 ozs.): (1) area mines, especially (a) the Howard, and (b) 1 mi. below, productive placers; (2) all regional watercourse and slope-wash gravels (district is 18 mi. long by 8 mi. wide)—placer gold; (3) along Black Canyon at several places: (a) area small gravel deposits, placers; (b) near Turkey Cr. Station, small placers worked annually; (c) between mouth of Arrastra Cr. and CLEATOR and (d) in American and Mexican gulches, productive placers; (4) for 20 mi. along Humbug, French, and Cow crs., the Humbug placers: (a) upper gravels to 20 ft. deep—flaky to floury gold; (b) at or near bedrock—coarse gold.

BUMBLEBEE (N of BLACK CANYON CITY, q.v.): (1) the Bumblebee and Bland Hill mines—lode gold; (2) S, the Nigger Brown, Blanchiana, and Gillespie mines—lode gold; (3) S 3 mi., a gravel bar in Black Canyon Cr., placer; (4) E 4 mi. (9 mi. from CORDES, q.v.), on W brink of Agua Fria Canyon, the Richinbar Mine—lode gold, silver.

CASTLE HOT SPRINGS (in S part of co.), near upper Castle Cr., the Castle Creek district: (1) area mines such as the Golden Aster (Lehman), Swallow, Whipsaw, Jones, and Copperopolis—lode gold; (2) in the Wickenburg Mts. 1 mi. N of the Castle Hot Springs rd. and 10 mi. E of MORRISTOWN in Maricopa Co., the Golden Slipper Mine—lode gold.

CHERRY: (1) N 1½ mi.: (a) the Bunker (Wheatly) Mine, 8 claims—finely divided gold; (b) the Federal and Leghorn mines, intermittently worked—lode gold; (2) WNW 2 mi., the Gold Bullion (Copper Bullion) Mine—lode gold, minor copper; (3) SW 2 mi., the Logan Mine—lode gold.

CLEATOR, W on rd. along old RR grade, in area of several sq. mi. N of Crown King and E of Towers Mt., the Pine Grove district: (1) the Crown King Group (of mines), very rich, site of 300-ton flotation mill operated in 1934—lode gold, silver (minor copper, zinc); (a) NE of the Crown King, the Philadelphia Mine and, just E, the Nelson Mine—lode gold, silver; (b) N 2 mi., the Fairview Tunnel (elev., 7,200 ft.; extension of the Nelson vein)—lode gold; (c) many other area lode-gold mines; (2) S of the Pine Grove district, on the Humbug side of the divide, the Tiger district, on S slopes of Wasson Peak, the Oro Belle and Gray Eagle mines and mill, principal producers 1907–9—lode (gold, 5,662 ozs.; silver, 16,301 ozs.; copper, 23,830 lbs.).

CONGRESS: (1) E 2½ mi., old mining camp of Stanton, many area mines—lode gold; (2) NW 3 mi., in SE part of the Date Creek Mts.: (a) Congress Mine (tot. prod., 1887–1959, of 396,300 ozs.)—lode gold; (b) ½ mi. W, the Congress Extension—lode gold; (3) NW several mi., at SW margin of the Date Creek Mts., the Martinez district, numerous old mines—lode gold; (4) SE 4½ mi., old camp of Octave: (a) just above camp and on flat-topped mt. above adjoining Stanton, the Weaver-Rich Hill district, area mines and placers (nuggets in surface debris of Rich Hill); (b) along SW front of the Weaver Mts., lode mines produced 308,000 ozs. and placers 104,000 ozs. of gold; (5) NE 5 mi., the Alvarado (Planet Saturn) Mine—lode gold;

CONSTELLATION (15 mi. NE of WICKENBURG in Maricopa Co.): (1) area: (a) Gold Bar (O'Brien) Mine—lode gold; (b) other area mines of the Black Rock district (tot. prod., 1900–59, of 12,190 ozs.)—lode gold; (2) 4½ mi. N of WICKENBURG, the Oro Grande Mine, major producer —lode gold; (3) E 7 mi., near head of Amazon Gulch (16 mi. NE of WICKENBURG), the Groom Mine and 30-ton ball mill—lode gold; (4) NW, the Arizona Copper Belt Mining Co. Mine—lode gold, copper.

CORDES (jct. 61 mi. N of PHOENIX): (1) Poland Jct., W past power station, take left or lower dirt rd., in stream gravels along rd.—placer gold; (2) SW, to S foothills of the Bradshaw Mts. (about 45 mi. NNW of PHOENIX), the Tiptop Mine (and district), producer of about 10,000 ozs. lode gold, 1875–1959.

DEWEY, out 16 mi., in S part of the Black Hills on headwaters of Cherry Cr., the Cherry Cr. district (22 mi. from CLEMENCEAU): (1) at

E foot of the Black Hills, the Monarch Mine (and several others nearby —lode gold; (2) S of the Monarch, the Etta, Gold Ring, and Conger mines—lode gold.

HILLSIDE STA.: (1) N 8½ mi. and within ¾ mi. of the Santa Maria R., the Mammoth (Hubbard) Mine, 8 claims—lode gold; (2) out 13 mi. and 3½ mi. W of the BAGDAD hwy., in secs. 4 and 9, (13N–8W), the Crosby Mine—lode gold; (3) NW 18 mi., the Eureka placers: (a) along Burro Cr., principal producers; (b) all area gulches, bench gravels— placer gold; (4) out BAGDAD hwy. 20 mi., take rd. W 1 mi., the Cowboy Mine (4 claims)—lode gold, with lead; (5) out the KINGMAN hwy. 28 mi., then 4½ mi. S to the S part of Grayback Mt., the Southern Cross Mine, shallow workings—lode gold, minor copper; (6) out 32 mi.; (a) in a deep canyon on E side of Boulder Cr., in secs. 16 and 21, (15N–9W), the Hillside Mine (2 mi. underground workings)—lode gold (most abundant in galena), wire silver; (b) 1½ mi. S of the Hillside Mine, on a tributary of Boulder Cr., the Comstock and Dexter Property—lode gold.

HUMBUG (reached 22¼ mi. NE of MORRISTOWN in Maricopa Co.) in SW part of the Bradshaw Mts., the Humbug district (about 100 old mines and claims), including the Fogarty, Queen, Little Annie, Heinie, Lind, and Columbia groups of mines—lode gold.

JEROME (Verde district, on E slope of the Black Hills just W of the Verde R., mammoth copper district abandoned about 1950), primarily the United Verde and United Verde Extension mines, tot. prod. of by-product gold through 1959 of 1,571,000 ozs.

KIRKLAND: (1) SSE 9 mi., in Placerita, French, and Cherry gulches, many productive placers (extensively worked in 1930s); (2) Blind Indian Cr. and Mill Cr. drainages—placer gold.

MAYER, on pediment at NE foot of the Bradshaw Mts., along Big Bug Cr., large general area of Big Bug district centered about: (1) the Big Bug Placers W of Big Bug Mesa in Agua Fria Valley; (2) NW 3 mi., W of Big Bug Cr., the Pantle Bros. Lease, extensively worked in early 1930s—placer gold.

McCABE: (1) S a short distance, in Galena Gulch, the McCabe-Gladstone Mine—lode gold, silver, copper, lead, zinc; (2) SW 1¾ mi., in upper part of Chaparral Gulch: (a) the Union Mine—lode gold; (b)

1,700 ft. S of the Union, the Little Jesse Mine—lode gold; (c) SW of the Jesse, on a ridge, the Leland-Dividend Mine and 100-ton flotation mill—lode gold and lead in sulfide ores.

PEEPLES VALLEY: (1) W, along Model Cr., placers; (2) N ¾ mi., take rd. through gate 2–3 mi. to W side of valley, in area of Model Cr.: (a) small gravel deposits, placers; (b) area lode-vein prospects—free gold.

POLAND (at N foot of Big Bug Mesa, accessible from the Black Canyon hwy.): (1) area: (a) the Poland-Walker Tunnel, on dumps—gold showings; (b) 1¼ mi. W of tunnel and mesa (from rd. branching NE from the SENATOR hwy. ⅛ mi. S of the Hassayampa bridge, the Money Metals Mine—lode gold; (2) W 1 mi. and about ½ mi. N of Big Bug Cr., the Henrietta (Big Bug) Mine and mill—lode gold, with silver, lead, copper; (3) S of mesa, in area of Turkey Cr., the Turkey Cr. district (noted early-day producer of gold and silver): (a) 1 mi. N of Pine Flat, the Cumberland Mine and mill (1908)—lode gold; (b) watercourse and slope-wash gravels of area—placer gold.

PRESCOTT (originally settled by placer miners, becoming territorial capital in 1864): (1) area: (a) in the Columbia district, area creeks— placer gold; (b) S, along upper reaches and main course of Granite Cr., the Granite Cr. placers (discovered in 1860s); (c) S, to headwaters of the Hassayampa R., along entire course of r. to co. line 2 mi. N of WICKEN-BURG (Maricopa Co.), many lode mines and placers (all regional water-course beds, benchs, terraces, and hillsides); (2) E 4 mi. and just S of the DEWEY rd., the Bullwhacker Mine—high-grade, coarse gold; (3) S 4–6 mi.: (a) in New England Gulch (tributary of Granite Cr.), rich early placers; (b) all regional cr. and watercourse beds, benches, and slope-wash gravels—placer gold; (4) S 5 mi., area of upper Groom Cr., the Groom Cr. district, many mines, e.g., Midnight Test (National Gold), Empire, King-Kelly-Monte Cristo, Victor, and Home Run—lode gold; (5) S 6 mi., on W slope of the Bradshaw Mts., the Hassayampa-Groom district: (a) along the Hassayampa R., many placers (prod., 18,500 ozs. of gold through 1959); (b) regional lode mines (tot. prod., 108,300 ozs. of gold); (6) S 6 mi., in area of Hassayampa Cr., the Oro Flame and Sterling mines (reached by rd. branching S from the Wolf Cr. rd. 2⅓ air mi. from U.S. 89), 19 claims, mill—lode gold, some silver; (7) SE and

about 4½ mi. NE of Mayer, in the Agua Fria district, the Stoddard
Mine (one of earliest in state; tot. prod., 12,710 ozs.)—by-product gold
from copper-silver ores; (8) ESE 12 mi., on NE slope of the Bradshaw
Mts.: (a) the Big Bug Mine and Iron King Mine (producers of 627,000
ozs. lode gold, 1867–1959); (b) area gravel deposits in streams, benches,
on slopes, placers (tot. prod., 42,700 ozs. of gold); (c) the McCabe-
Gladstone Mine (southernmost of several rich lode mines along the
7-mi. Silver Belt-McCabe vein—gold in sulfide ores; (9) near head of
Hassayampa Cr., just below Mt. Union Pass (elev., 7,188 ft.): (a) the
Senator Mine—free-milling gold; (b) 1,000 ft. E, the Cash Mine—lode
gold, lead, copper, zinc; (c) 1 mi. W of the Senator Mine, on S side of
Hassayampa Cr., the Big Pine Mine (4 tunnels)—lode gold, silver; (10)
out 14 mi. and 3 mi. W of the Senator hwy., on W slope of the
Bradshaw Mts. near head of Slate Cr.: (a) the Davis-Dunkirk Mine
(6,000-ft. underground workings)—lode gold, silver; (b) ⅞ mi. E, the
Tillie Starbuck Mine—lode and free gold, silver; (11) SE 20 mi., in Peck
Canyon drainage area, the Peck Mine (and district)—by-product gold
(15,500 ozs., 1875–1959); (12) N 22 mi., at Del Rio, area gravels,
intermittently worked—placer gold; (13) S 25 mi., along Minnehaha
Cr.: (a) below 5,000-ft. elev., the Minnehaha placers; (b) on Oak Cr.,
below Fenton's ranch, productive placers; (c) the Minnehaha and Montgomery mines—lode gold; (14) SE 40 mi. by rd., the Pine Grove-Tiger
(Crown King) district in the heart of the Bradshaw Mts., tot. prod.,
130,275 ozs.: (a) the Pine Grove camp, several area mines—lode gold;
(b) immediately S of Pine Grove, the Crown King Mine, major production—lode gold.

Rock Springs (46 mi. N of Phoenix), at Mile Post 246 turnoff to
the Maggie Mine, placers.

Skull Valley (27 air mi. WSW of Prescott), NE, on rd. to Prescott in Copper Basin Wash, the Copper Basin district between Skull
Valley and the Sierra Prieta, many area placers.

Venezia (15 mi. from Prescott via the Senator hwy.), in Crook's
Canyon, the "Trapshooter" Reilly Group (Crook, Venezia, Starlight,
Mt. Union mines), rich—lode gold, silver.

Wagoner, SE (and SW of the Crown King Mine), the Minnehaha

area mines: (1) the Button Shaft Mine, and (2) 1½ mi. SW, the Boaz
Mine and mill—lode gold.

WALKER (7 mi. SE of Prescott): (1) area lode mines produced 43,000
ozs. of lode gold, 1910–59; (2) SW 1 mi.: (a) the Sheldon Mine—lode
gold, with copper and lead; (b) nearby, the Mudhole Mine—lode gold,
silver; (3) along Lynx Cr. 6 mi. to jct. with Agua Fria Cr., the Lynx
Creek placers, one of most productive and extensively worked placers in
Arizona, tot. prod., 1863–1959, of 97,000 ozs. of gold.

Yuma Co.

Yuma County ranks fourth among Arizona's gold-producing counties.
Nine districts produced 771,000 ozs. of gold, while many other areas
produced less than 10,000 ozs. each. Total production of $13,250,000
came largely from the Kofa, Fortuna, and Harquahala mines.

AREA, the Tank Mts.: (1) regional watercourse and bench gravels—
placer gold; (2) at E foot of mts., the Puzzles, Golden Harp, Ramey, and
Regal mines and prospects, all worked by dry-wash methods—rather
coarse placer gold.

BOUSE: (1) area old mines, in dark-red rhyolite, plainly visible—fine
gold flakes (almost impossible to mine successfully); (2) NW 5 mi.: (a)
the Old Maid Mine, intermittently operated—lode gold; (b) 1 mi.
farther NW, in low, narrow N part of the Plomosa Mts., the Dutchman
Mine—lode gold (visible flakes in copper-stained shale); (3) N 28 mi.,
in area of the Bill Williams R., the Planet district, 2 mi. S of r., the
Planet Lease—lode gold (abundant, finely divided when panned).

CASTLE DOME DISTRICT (31 rd. mi. NE of DOME in S-central part
of co. in S part of the Castle Dome Mts. 20–25 mi. N of WELLTON;
tot. prod., 1863–1959, of 9,500–10,000 ozs. of placer and lode gold): (1)
many area old mines, both placer and lode; (2) the Big Eye Mine, major
producer—lode gold; (3) E and S of the Big Eye, very many gravel
deposits—placer gold.

CIBOLA (due S of Blythe, Calif.), on E side of Colorado R., reached
22½ mi. S of QUARTZITE via U.S. 95, turn W on dirt rd. to this almost
totally abandoned camp, area old mine dumps—gold showings.

DOME (district 20 mi. ENE of YUMA, discovered in 1858, tot. prod.

through 1959 of 24,765 ozs. of gold, mostly prior to 1865 from area 2
mi. long by ¼–¾ mi. wide. Altogether, this region embraces about
5,000 sq. mi. between U.S. 60, 70 and U.S. 80, east of the Colorado R.
to the Maricopa Co. line, with hundreds of old mines and prospects
worked primarily for gold): (1) W 1½ mi., in Monitor Gulch (and all
tributary gulches and terraces), rich placers; (2) in S and central parts
of T. 7, 8 S, R. 18, 19, and 20 W, in canyons around Muggins (Klotho,
Coronation) Peak and at base of Long Mt., rich placers; (3) out 10 mi.
by rough rd. to Burro Canyon (trending S from Muggins Peak), rich
early placers that still produce gold; (4) E, in the Muggins Mts., in S
and central parts of T. 8, 9 S, R. 8, 9, and 10 W, many dry-placer gravel
deposits.

KOFA (district about 27 air mi. SE of QUARTZITE, reached via U.S.
95 and E-trending dirt rd.): (1) the King of Arizona and North Star
mines (tot. prod., 1896–1941, of 237,000 ozs.)—lode gold; (2) same
area, the Kofa Queen Mine, big producer; (3) N of the King, in a gulch
draining W from the N part of detached hills, all area outwash gravel
deposits in debris to 70 ft. thick over a 60-acre parcel—placer gold.

LAGUNA (just N of the Gila R and E of the Colorado R in S end of
the Laguna Mts. in R. 21 and 22 W, tot. district prod. to 1959 of 10,500
ozs. of gold): (1) Laguna Dam, area at E end about 10 mi. NE of YUMA,
as well as (a) the Las Flores area, and (b) the McPhaul area, many local
placers; (2) along Colorado R., in potholes and tributary gulches—colors
and nuggets; (3) N of the Gila R. and the Gila Mts., in R. 21 and 22
W, embracing the S, SE, and SW portions of the Laguna or San Pablo
Mts., placer deposits.

PARKER: (1) NE 5 mi.: (a) the Rio Vista Northside Mine (12 claims)
—lode gold; (b) on the N, the Capilano Mine, shallow shaft and open
cuts, rich—lode gold, with copper; (2) NE 5–8 mi., the Cienega district,
tot. prod., 1870–1959, exceeding 10,000 ozs. from numerous mines—
lode gold copper; (3) NE 7 mi.: (a) the Lion Hill Mine and 25–ton
amalgamation mill, and (b) 1 mi. NE, the Billy Mack Mine, good
producer—lode gold.

QUARTZITE: (1) on top of highest peak visible from town to the E
(very arduous hike), a one-man mine—lode gold; (2) SE 5 mi., the

Plomosa district: (a) includes E and W margins of the La Posa Plain and covering 7,500 acres, the Plomosa placers (gravels worked extensively to depths of 70 ft., dry-wash); (2) area of the Dome Rock Mts.: (a) La Cholla, in area 4–5 mi. long S of I-10 bordering the E foot of mts., abundant placer sands at bedrock level; (b) Middle Camp, at E foot of mts., just N of the Oro Fino Placers, q.v., an area 4–5 mi. E–W and 1 mi. N–S, rich seams of gravel at bedrock—placer gold; (c) Oro Fino, at E foot of mts., in vicinity of I-10, the Oro Fino placers, rich; (3) W 8 mi., the La Paz (Weaver) district, the Goodman Mine (producer of $100,000)—free gold; (4) W 9 mi. and 6 mi. E of the Colorado R., along W side of the Dome Mts., extensive placers and several lode mines (tot. prod., 1862–1934, about 100,000 gold ozs. from dry-wash placers and 4,000 ozs. of lode gold from area veins; (a) all regional gulches and tributaries draining W slopes of mts., dry placer gravels; (b) Goodman Arroyo and Arroyo La Paz especially rich placers; (c) Ferrar Gulch, most productive placers in district; (5) out 22 mi., at W base of the Dome Rock Mts., in T. 2 N, R. 21 W, the Trigo placers: (a) area arroyo bottom gravels, and (b) in ancient bar and channel gravels—placer gold (coarse, flat grains); (6) S on U.S. 95 to Stone Cabin, turn E into SW margin of the Kofa Mts., the Kofa district (cf. under KOFA).

SALOME: (1) SE 5 mi., at N base of the Harquahala Mts., the Hercules Mine, productive 1909–34—lode gold; (2) S 8 mi., in Harquahala Gulch, gravel deposits worked 1886–87—placer gold; (3) S 10 mi., the Ellsworth (Harquahala) district in the Little Harquahala Mts., tot. prod. through 1959, about 134,000 ozs. of lode and placer gold; at Harquahala (ghost town), area great mines: (a) the Harquahala Bonanza, Extension, Summit Lode, Narrow Gauge, and Grand View mines; (b) to the NE, the Golden Eagle and subsidiary shafts—lode gold; (4) S 11 mi.: (a) at S base of the Harquahala Mts., the Socorro Mine and mill—lode gold; (b) take rd. branching E from the SALOME-HASSAYAMPA rd. for 5 mi., near S base of the Harquahala Mts., the Hidden Treasure Mine—lode gold (fine to medium-coarse); (c) 8 mi. farther along rd. (toward AGUILA), on the plain S of the Harquahala Mts., the Alaskan Mine, discovered in 1920—lode gold; (5) SE, along rd. across desert to TONO-PAH (Maricopa Co.): (a) very many regional old mine dumps—gold showings; (b) area of the Tank Mts. (cf. under AREA).

VICKSBURG, NE 30 mi., to Alamo Spring (accessible by 13-mi. dirt rd. branching W from the Sheep Tanks rd. at E entrance of New Water Pass), the Alamo Region: (1) within 1 mi. of S margin of the Little Horn Mts. and ¼ mi. S of the SE cor. T. 1 N, R. 15 W: (a) the Sheep Tanks Mine (and district) and 100-ton cyanide mill, active 1909–34—lode gold; (b) nearby, the Resolution and Black Eagle Claims—lode gold; (2) 5 mi. E of Sheep Tank and just N of rd. to PALOMAS, the Davis Prospect, shallow cuts made 1931–32—lode gold.

WELLTON, S 6 mi., the Wellton Hills (La Posa) district, such mines as the Double Eagle, Poorman, Draghi, Donaldson, Wanamaker, Welltonia, Northern, and Shirley May—lode gold.

WENDEN, SE 5 mi., at N base of the Harquahala Mts., the San Marcose Mine, high-grade—lode gold.

YUMA: (1) E 16½ mi. on I-8 (U.S. 80), take dirt rd. S to W base of the Gila Mts. (about 14½ mi. S of the N end), the La Fortuna district (many outcrops of gold-bearing quartz): (a) the Fortuna Mine, major producer ($2.5 million, 1896–1904), intermittently worked—lode gold; (b) many area prospects, cuts, and drifts—gold showings; (c) 3 mi. N of the Fortuna, on crest of the Gila Mts. (rough terrain), gold-quartz outcrops, some traceable for ½ mi., very little prospected—lode and free gold; (2) NE on U.S. 95: (a) to McPhaul bridge, ¾ mi. N take branch rd. 3¼ mi. to SE margin of the Laguna Mts., the Las Flores district, numerous area mines, including the Traeger (Agate), Golden Queen, and Pandino claim—lode gold; (b) to area of the Castle Dome Mts., old camp of Thumb Butte, then 10 mi. E across mts. to E side, the Big Eye Mine—lode gold.

CALIFORNIA

Because of the fascinating story of the great gold rush of 1849 and the subsequent intensive mining of gold ever since, on April 23, 1965, Governor Edmund G. Brown signed legislation designating native gold as California's official state mineral, thus adding luster to the state's principal title: "The Golden State." Even today, despite high mining costs, gold ranks fourth in California's mineral production. Since Spanish times of the late eighteenth century, when placer gold was being mined from many southern California localities, the state's total production is recorded between 1848 and 1965 (with production continuing and beginning to expand again) of about 68,200,000 ozs. of placer gold and 37,900 ozs. of lode gold, with total production pegged at 106,-130,214 ozs.

For more information and many interesting brochures, write: California Division of Mines and Geology, Division Headquarters Resources Building, Room 1341, 1416 Ninth Street, Sacramento, California 95814 (or P.O. Box 2980, Sacramento, California 95812).

Amador Co.

Most productive of the Mother Lode counties, Amador County produced 6,320,000 ozs. of placer gold and 7,675,000 ozs. of lode gold through 1959. Amateur panning and sluicing operations continue every summer.

AREA, the Mother Lode district (about 1 mi. wide across the W-central part of co. from S–N, many mines)—lode gold: (1) the Old Eureka Mine (at 1,350 ft., the deepest gold-mining shaft in America, and largest producer on the Mother Lode in early days); (2) the Kennedy, Argonaut, and Keystone mines—lode gold.

FIDDLETOWN (district in S part of T. 8 N, R. 11 E: (1) area, sizable dredging and drift operations, 1850–1950 (prod. between 10,000 and 100,000 ozs.)—placer gold; (2) along Indian Gulch, area gravels, placer workings.

JACKSON, the Gwin Mine—lode gold (occasional masses of crystallized gold in arsenopyrite).

PLYMOUTH (NW part of co.): (1) along the Cosumes R., the Cosumes River placers, tot. prod., 10,900 ozs.—placer gold; (2) Loafer Hill, area near OLETA, small gravel deposits—placer gold.

VOLCANO (district, in T. 7 N, R. 12 E, in W-central part of co., tot. prod. estimated around 100,000 ozs. of placer gold prior to 1932): (1) area of Jackson Gulch and Rancheria, placers; (2) Mokelumne R., area hydraulic operations—placer gold.

Butte Co.

Total production from Butte County, 1880–1959, is reported as 3,123,115 ozs. of placer gold and 103,800 ozs. of lode gold.

AREA, in T. 21 N, R. 4 E, the Surcease Mine—lode gold.

CHEROKEE CITY: (1) area stream gravels, placers; (2) on Cherokee Flat, alluvial gravel deposits—placer gold.

MAGALIA (in N-central part of co.; prod. unknown prior to 1932, but from then through 1959, a record of 15,976 ozs. of gold): (1) area Tertiary gravel deposits—placer gold; (2) the Perschbaker Mine, major producer prior to 1910—lode gold.

OROVILLE (district, in S part of co. along the Feather R., tot. prod., 1903–59, of 1,964,130 ozs., hence most productive area in co.): (1) area gravel deposits, placers; (2) Thompson's Flat, large placers.

YANKEE HILL (district, in T. 21 N, R. 4 and 5 E, in central part of co., tot. prod. of 5,154 ozs. of placer gold and 34,427 ozs. of lode gold), all along the Feather R., many placer operations and area lode-gold mines.

Calaveras Co.

After placer gold was discovered in 1849, rich lode veins were opened in 1850 above the placer workings. Placer gold production is estimated at 2,415,000 ozs. and lode gold, mined from the East and West belts and the Mother Lode, at 2,045,700 ozs.

AREA: (1) Calaveras R., channel and tributaries, placers; (2) Table Mt., area gravels, placers; (3) in T. 3 N, R. 10 E, along the Calaveras R., the Jenny Lind district, large-scale dredge and dragline operations, est. prod., 100,000 to 1,000,000 ozs.—placer gold.

ANGELES CAMP, many area mines and prospects, e.g., (1) the Keystone, Lancha Plana, and Union mines—gold by-product of copper; (2) the Utica and Gold Cliff mines, major producers—lode gold.

CAMANCHE (district, in NW part of co., tot. prod. estimated between 100,000 and 1,000,000 ozs.), along the Mokelumne R., huge, bucket-type dredge operations—placer gold.

CAMPO SECO (district, in T. 4 and 5 N, R. 10 E, in NW part of co., tot. prod., about 60,000 ozs.): (1) area tributary stream gravels, rich placers; (2) the Pern Mine (copper)—by-product gold.

MELONES (districts along the Mother Lode, East Belt, and West Belt running N–S through W part of co., some 800 lode mines and prospects): (1) Carson Hill, classic discoveries and most productive area, many mines—lode gold; (2) in East Belt district, in NW½NW½ sec. 33, (4N–14E), the Sheepranch Mine, big producer—lode gold; (3) in West Belt district, the Royal Mine, producer, 1870–1938, of $3,000,000 —lode gold.

MOKELUMNE HILL (district, in T. 5 N, R. 11 E): (1) S of the hill 2½ mi., and (2) the Eclipse, Infernal, and other mines 3 mi. S of the hill —lode gold.

MURPHYS, in T. 3 N, R. 14 E, Tertiary gravel deposits in old channels —placer gold.

SAN ANDREAS, in T. 4 N, R. 12 E, area Tertiary channel gravels— placer gold.

Contra Costa Co.

WALNUT CR., E, on Mt. Diablo, in Mitchell Canyon, a prospect in a ravine tributary—gold, with bornite and chalcopyrite.

Del Norte Co.

Between 1880 and 1959 Del Norte County produced 44,000 placer-gold ozs., plus 700 ozs. from lode veins.

AREA: (1) the Diamond Creek district, area copper mines—by-product gold; (2) E part of co., some lode gold mines; (3) the Low Divide district in NW part of co., area mines—lode gold.

ROCKLAND (district), the Keystone Mine—lode gold.

SHELLY CR.: (1) area quartz mines—lode gold; (2) along upper Monkey Cr., lode gold mines.

SMITH R.: (1) along river, placer operations produced about 40,000 ozs. of gold; (2) all tributary stream gravels—placer gold.

El Dorado Co.

James Marshall discovered gold at Coloma in 1848 on the S fork of the American River 8 mi. NW of Placerville to initiate the great gold rush of 1849–50. From 1880 through 1959 the county produced 1,267,700 ozs. of gold, earlier production not being recorded but probably very large.

Area: (1) along the American and Cosumnes rivers and their tributaries, very many placer operations; (2) in the Georgia Slade district, in T. 12 N, R. 10 E, in NW part of co.: (a) area quartz mines, rich—lode gold; (b) area weathered gravel deposits, extensive hydraulic operations—placer gold; (3) throughout the Mother Lode, East Belt, and West Belt districts (comprising a zone 10–20 mi. wide extending N–S in W part of co.; tot. prod., around 1,000,000 ozs.): (a) the Union and the Church mines—lode gold; (b) the Big Canyon, Mt. Pleasant, Pyramid, Sliger, Taylor, and Zantgraf mines, all major producers—lode gold.

Fairplay: (1) area lode mines and placer operations; (2) Indian diggings, placer grounds, and lode mines.

Georgetown, in N part of T. 12 N, R. 10 E, many area placers.

Nashville: (1) area lode mines, opened in 1851—lode gold; (2) area placer mines (reactivated in the 1930s, when large, floating dragline dredges were introduced)—placer gold.

Placerville: (1) many area old placer mines (including all regional streams); (2) NE 5.6 mi., along the American R., placers; (3) gravel deposits in Cedar Ravine, Forest Hill, Smith's Flat, Webber Hill, and White Rock Canyon—placer gold; (4) S 18 mi., in T. 9 N, R. 13 E, the Grizzly Flat Placers; (5) in W-central part of T. 10 N, R. 11 E, mines—lode gold.

Fresno Co.

Placer gravels along the San Joaquin River produced 121,000 ozs. of gold between 1880 and 1959, when it was part of Madera County.

Area, all sand and gravel operations along the San Joaquin R. between Friant and Herndon—placer gold (as by-product).

FRIANT DAM. The gravel excavated for use in building the dam produced $196,977 in placer gold, 1940–42.

Humboldt Co.

Gold is the principal mineral resource of Humboldt County, primarily from placers along the Klamath and Trinity rivers. The total production through 1959 was 131,300 ozs.

AREA: There are six lode-gold mines shown on USFS quadrangles.

ORICK, area ocean beach sands N and S of the mouth of the Klamath R., once worked—placer gold, platinum.

ORLEANS (NE part of co.), very many gravel bars and benches along the Klamath R. to SOMES BAR in Siskiyou Co. along Hwy. 96, many bars with colorful names attesting to gold-placer operations prior to 1900.

Imperial Co.

Gold occurs throughout Imperial County in its arid mountain ranges. Here is where the classic pick-and-pan burro prospector of the nineteenth century crisscrossed the desert between water holes. A minimum estimate of 235,000 ozs. of lode and placer gold have come from this county.

OGILBY (NW of YUMA, Arizona, in SE part of co.), the Cargo Muchacho district, many old mines worked since Mexican times; tot. prod., about 193,000 ozs.: (1) all regional arroyo bottoms, benches, terraces, and bajadas, dry-wash placers abundant; (2) many abandoned area lode mines—gold (fine, grain, wire, nuggets), often with copper.

PICACHO (camp on Colorado R. due N of YUMA, Ariz., in extreme SE cor. of co.), the Chocolate Mts. area: (1) area placer and lode claims—gold; (2) SW, in the Picacho Mts., many gold-bearing veins in gneisses and schists overlain by lavas, tuffs, and conglomerates—lode and free gold; (3) the Paymaster district, minor—lode gold; (4) S 5 mi., the Picacho Mine, Bluejacket, and others—lode gold.

TUMCO (district and ghost camp), once a good producer from several area mines—lode gold.

Inyo Co.

Inyo County produced 496,000 ozs. of gold between 1880 and 1959, primarily from lode mines scattered throughout the county, with a considerable percentage as by-product from lead-silver, copper, and tungsten ores.

AREA, in the Inyo Range, the Russ district (opened 1861): (1) area mines—lode gold; (2) W and E slopes of mts., in Mazourka and Marble canyons, many small-scale placer workings.

BALLARAT (South Park district), lat. 36°00′ N, long. 117°10′ W, in the Panamint Range of the S-central part of co., producer of $300,000 to $1,000,000: (1) area old mine dumps—gold traces; (2) the Ratcliff Mine, chief producer—lode gold; (3) SW 10–15 mi., in T. 23 S, R 42 and 43 E, in the Argus Range, the Sherman district (tot. prod., 1939–41, of 14,184 ozs. of lode gold): (a) area lead-silver mines—by-product gold; (b) the Arondo Mine—free gold; (c) the Ruth Mine—free gold, with pyrite.

BISHOP, W 17 mi., the Willshire-Bishop Creek district (on E front of the Sierra Nevada Range and in the Tungsten Hills; tot. prod., 75,000–100,000 ozs. of by-product gold): (1) the Bishop Cr. Mine and, at head of Bishop Cr., the Willshire Mine—lode gold; (2) the Pine Cr. Mine (largest domestic tungsten mine)—by-product gold; (3) the Cardinal Gold Mining Co. Mine—lode gold only.

DEATH VALLEY (Nat'l. Mon., on slope of the Funeral Range at lat. 36°40′ N, long. 116°55′ W, the Chloride Cliff district; tot. prod., 1931–59, of 60,000 ozs.), the Keane Wonder Mine (now in ruins)—lode gold.

LONE PINE, E, into the Inyo Mts. of N-central part of co., the Union (Inyo Range) district, lat. 36°35′ N, long. 118°00′ to 118°01′ W; tot. prod., 1860s–1959, between 10,000 and 50,000 ozs.: (1) the Reward and Brown Monster mines, major producers—lode gold; (2) area canyon and gulch gravels, slopes and drainage channels, placers.

TECOPA (SE cor. of co.), E 5–10 mi., the Resting Springs district, tot. prod. through 1959 of 15,005 ozs. of lode gold from lead-silver ores, the Shoshone Group of mines.

WILD ROSE (district, ranger sta. in Monument, W, in the Panamint

Range, tot. prod., 1906–59, about 73,000 ozs. of lode gold), the Skidoo Mine.

Kern Co.

A total of 1,777,000 ozs. of gold came from Kern County between 1851 and 1959, primarily from 6 major districts.

AMALIE (district, between S summit of the Piute Mts. and Caliente Cr., in T. 30 S, R. 33 and 34 E, tot. prod. of 30,000 ozs. of lode gold): (1) the Amalie Mine, major producer; (2) several other area mines—lode gold.

BODFISH: (1) SE by dirt rd., the Green Mountain district (between PIUTE, on W slope of the Piute Mts., and edge of Kelsey Valley on E side of mts.), the Bright Star Mine, prod., 33,100 ozs.—lode gold; (2) 7 mi. NW of PIUTE, in T. 29 S, R. 33 E, the Joe Walker Mine, producer of $600,000—lode gold.

KERNVILLE, in T. 25 S, R. 33 E, the Cove district (tot. prod., 262,800 ozs.), the Big Blue Mine—free gold, with arsenopyrite.

LAKE ISABELLA: (1) W, in the Greenhorn Mts.: (a) along Greenhorn Gulch, extensive placers; (b) area quartz mines—lode gold; (2) W, in T. 26 S, R. 32 and 33 E, the Keyes district (tot. prod., 39,600 ozs. through 1959), numerous area mines—lode gold.

MOHAVE: (1) in T. 11 N, R. 15 W, the Pine Tree Mine, produced $250,000—lode gold; (2) SW 4 mi., in T. 10 and 11 N, R. 11, 12, and 13 W, the Rosamond-Mohave district, tot. prod., 278,250 ozs. of gold plus silver: (a) W of ROSAMOND several mi., the Tropico Mine, major producer into 1950s; (b) other area mines around adjacent Wheeler Springs—lode gold; (3) NE 25 mi., in the El Paso Mts.: (a) all regional gravels—placer gold (surprisingly abundant); (b) such major mines as the Cudahy Camp, Owens Camp, Burro Schmidt's Tunnel, Colorado Camp, etc.—lode gold.

RANDSBURG (heart of the Rand district, which lies along the Kern Co.-San Bernardino Co. line, with larger half of district in the latter co. Nearly all of the 836,300 ozs. of gold is from the Kern Co. half, with silver a by-product): (1) the Yellow Aster Mine, largest producer, and other area mines—lode gold; (2) NW 9 mi., in Goler Wash, placer deposits worked 1893–94.

WELDON, S 16 mi., in T. 28 S, R. 35 E, the St. Johns Mine—lode gold.

Lassen Co.

Two main districts, the Diamond Mountain and Hayden Hill, produced a total of 147,500 ozs. of gold in Lassen County, with a considerable amount of unrecorded placer gold produced before 1880.

HAYDEN HILL (in N-central part of co., community destroyed by fire in 1910), the Hayden Hill district in T. 36 and 37 N, R. 10 and 11 E, producer of 116,000 ozs. of lode gold, 1870–1910, from several mines.

SUSANVILLE, S 6 mi., in T. 29 N, R. 11 and 12 E, the Diamond Mountain district (tot. prod., 9,700 ozs. of placer gold and 21,800 ozs. of lode gold): (1) area stream gravels of Tertiary age—placer gold; (2) several area gold-quartz lode veins in three types of igneous rocks, mines.

Los Angeles Co.

Gold placers were worked in Los Angeles County between 1834 and 1838 by Mexican and Spanish miners, and by 1858 more than 6,000 miners were working placer deposits 35 miles northwest of the Los Angeles city hall. Most of the county's total production of gold through 1959—109,200 ozs.—came from lode deposits, but small yields of placer gold are garnered every year by amateur gold hunters from many places, especially from sand and gravel pits and from the streams of the San Gabriel Mountains above AZUSA.

ACTON (in N-central part of co.), the Cedar and Mt. Gleason district, many area mines and prospect pits, notably the Governor Mine, tot. prod. of district, 1880–1959, at least 50,000 ozs.—lode gold.

AZUSA, N, up the San Gabriel Canyon, the San Gabriel district (tot. prod., 1848–1957, about 165,000 ozs. of gold): (1) San Gabriel R. gravels, worked 1848–80 for est. 120,000 ozs. of placer gold and still productive of colors and nuggets to weekend panners and sluice operators; (2) several area lode mines in area gold-quartz veins cutting igneous and metamorphic rocks—lode gold; (3) East Fork of the San Gabriel R., old site of Eldoradoville (gold camp of early 1860s and favorite amateur gold-hunting area today), in area watercourse and bench gravels—placer gold (some sizable nuggets).

LANCASTER, N, along the Kern Co. line and S of NEENACH, the Antelope Valley district (gold discovered in 1934, with tot. prod. through 1946 of 9,700 ozs.): (1) many area claims and prospects, and (2) the Rivera Mining Co. claims (most productive)—lode gold.

NEWHALL, E, in Placerita Canyon (State Park), area of original productive placer operations, in present gravel deposits—placer gold.

SAN FERNANDO, NE 12 mi., in Pacoima Canyon: (1) headwater and area gravels and slopewash deposits—placer gold; (2) 12 mi. up canyon from its mouth, the Denver Mining and Milling Property—lode gold.

TUJUNGA, N, in Tujunga Canyon, area gravel deposits—colors and small nuggets.

Madera Co.

Total gold production of Madera County through 1959 is calculated at 79,281 ozs.

AREA: (1) along the Fresno, Chowchilla, and San Joaquin rivers, many dredging operations from discovery years into the 1950s—placer gold; (2) Fish Cr., N ½ mi., the North Fork district, area lode-gold mines and placer workings.

COARSEGOLD: (1) area lode-gold mines hidden back in oak-covered hills; (2) N 2 mi., old mine dumps—gold showings; (3) NE 5 mi., area gold-quartz mines, on dumps—gold showings.

HILDRETH, area extending to Grub Gulch, quartz mines—lode gold.

RAYMOND, the Mt. Raymond district, the Star Mine—gold, with lead.

Mariposa Co.

Southernmost of the Mother Lode counties, Mariposa County has had a long and productive gold-mining history, with total production of 2,144,500 ozs. of gold recovered 1880–1959, of which 583,500 ozs. came from placer operations. Following the closure of mining during World War II, the annual gold production has been about 1,000 ounces.

AREA: (1) regional Mother Lode, East Belt, and West Belt mines (cf. also HORNITOS, COULTERVILLE, etc.), very many important lode-gold and placer-gold producers; (2) Agua Fria and Mariposa crs. (first placers discovered before 1849), many placer gravels; (3) in sec. 19, (5S–19E), the Copper Queen Mine—by-product gold; (4) in secs. 9 and 10, (4S–15E), the Fitch Mine—lode gold; (5) many placers in regional Tertiary gravels produced about 75,000 ozs. of gold: (a) NW part of co., the Blanchard district; (b) along the Tuolumne Co. line, just S of Jawbone Ridge; and (c) on ridge between Moore and Jordan crs. 4–5 mi. NW of Bower Cave.

BAGBY, W, in Quaternary gravels along the Merced R., placers produced 50,000 ozs. of gold, 1860–70.

BEAR VALLEY, area mines, especially the Josephine—lode gold.

COULTERVILLE (more or less central in the Mother Lode and East Belt areas, a NW-trending zone 3–4 mi. wide extending from MORMON BAR into Tuolumne Co., tot. prod., about 1,009,000 ozs.): (1) major mines include the Princeton, Pine Tree, Mt. Ophir, etc.—lode gold; (2) East Belt mines include the Original and Ferguson, Hite, and Mariposa —lode gold.

HORNITOS (district, lat. 37°30′ N, long. 120°14′ W, more than 500,-000 gold ozs.): (1) along Hornitos Cr., in Quaternary gravels—abundant placer gold; (2) West Belt lode mines, numerous major producers—lode gold (500,000 ozs. through 1959).

MARIPOSA, S 2½ mi., in T. 5 S, R. 18 E, the Mormon Bar district (tot. prod., 75,000 gold ozs.): (1) headwaters of Mariposa Cr. and its tributaries—placer gold; (2) Mormon Bar, area of extensive dragline operations in the 1930s—placer gold; (3) SE 6 mi., the Silver Bar Mine —lode gold.

Mendocino Co.

NAVARRO, area placer gravels along the Navarro R. in Anderson Valley—gold, with platinum.

Merced Co.

MERCED FALLS to SNELLING, the alluvial plain between these towns was extensively operated 1929–43 by several connected bucket dredges of the Yosemite Mining and Dredging Co. Tot. prod., 1880–1959, of 516,346 ozs.—placer gold.

Modoc Co.

Area, in NE part of co., the High Grade district, several quartz mines in volcanics produced about 14,000 ozs.—lode gold.

Mono Co.

Gold mining began in Mono County in 1862, with a total production of 1,176,200 ozs. through 1959, with activity continuing to the present.

Area, the Blind Springs Hill district, the Diana, Comet, Comanche, etc., mines (mainly copper, silver)—by-product gold.

Benton (district), area lode mines—gold, lead, silver.

Bodie (district and State Monument ghost town in T. 4 N, R. 27 E, discovered 1860 with tot. prod. through 1959 of 1,456,300 ozs. of lode gold): (1) many area old mines and dumps—gold showings; (2) on slopes above town, the Standard Mine, principal producer.

Bridgeport: (1) headwaters of the Walker R., Virginia Cr., and Dog Cr., productive placers; (2) N of Mono Lake, the Bodie Diggings— placer gold; (3) NE and E of the East Fork of the Walker R., just N of Masonic Mt., in T. 6 N, R. 26 E, the Masonic district: (a) area small mines, and (b) the Pittsburgh-Liberty Mine, major producer to district's tot. prod. of about 34,000 ozs.—lode gold.

Mammoth Lakes, in the Casa Diablo Mts., scattered small mines and prospects—lode gold.

Patterson (district, in the Sweetwater Range), area silver-quartz mines—lode gold, silver.

Tioga Pass (elev., 9,941 ft., and gateway to Yosemite Nat'l. Park), area quartz veins and stringers in granite—gold showings.

Monterey Co.

Los Burros (district), small area mines in quartz veins and stringers —lode gold.

Napa Co.

Napa County's total gold production of 23,225 ozs. came primarily as a by-product of silver mining.

CALISTOGA: (1) N 2 mi., the Mt. St. Helena Mine; (2) the Palisades Mine, principal gold-producer; (3) the Silverado Mine.

Nevada Co.

Both placer and lode-gold mines of Nevada County, opened in 1850 and continuing active to today, produced a total of 17,016,000 gold ozs. through 1959. Many placer deposits are known rich today but have not been touched because of state laws curtailing hydraulic mining operations.

AREA, many Tertiary gravel deposits that produced at least $90 million in placer gold through 1909 include the San Juan Ridge, North Columbia, Scotts Flat, Quaker Hill, and Red Dog-You Bet mines, all of which became big hydraulic operations with huge untouched reserves that are producing small amounts of placer gold by amateur panning and small-scale sluicing efforts.

FRENCH CORRAL, area stream gravels—placer colors, nuggets.

GRASS VALLEY-NEVADA CITY (district, with very many Mother Lode area camps and mines with such picturesque names as Rough and Ready, etc.; tot. prod. estimated at 2,200,000 ozs. of placer gold and 10,408,000 ozs. of lode gold): (1) GRASS VALLEY: (a) on outskirts of town, the Gold Hill Mine (first lode mine in district); (b) Ophir Hill, Rich Hill, Massachusetts Hill; (c) Empire, North Star, Idaho-Maryland (largest production of any lode-gold mine in state)—all lode gold; (2) NEVADA CITY: (a) the Blue Tent and Sailor Flat operations—placer gold; (b) the Champion, Providence, Canada Hill, Hoge, and Nevada City mines—lode gold.

MEADOW LAKE (35 mi. E of GRASS VALLEY), a relatively minor lode-gold district of several mines, worked 1863–1905, with tot. prod. of about 10,000 ozs.—lode gold.

Placer Co.

Although best known for its great placer mines, Placer County also produced substantial amounts of lode gold, with a total of 2,014,000 ozs. recorded 1880–1959 (gold produced after 1849 to 1880 was not recorded, but probably was very considerable).

AUBURN: (1) along the American R., many rich placers; (2) the Ophir district (most productive, with 145,300 ozs. of lode gold): (a) the Green Emigrant Mine, major producer; (b) the Crater, Bellevue, Oro Fino, and Three Stars mines, big producers—lode gold.

COLFAX: (1) W 1½ mi., the Rising Sun Mine (leading gold mine in co., tot. prod. of $2,000,000 between 1866 and 1932, rich pockets in quartz)—lode gold; (2) E 5 mi., the Iowa Hill district, thick Tertiary gravel deposits worked hydraulically and by drifting: (a) area placer operations produced $8,000,000—placer gold; (b) the Morning Star Mine produced about $2,000,000—lode gold.

DUTCH FLAT-GOLD RUN (district, along N boundary of Placer Co. due E of GRASS VALLEY, q.v. in Nevada Co.), all Tertiary stream channel gravels extending S from Nevada Co., many area workings—placer gold (about 492,000 ozs. through 1959).

FOREST HILL (district, in S-central part of co., extensive operations produced tot. of 344,000 gold ozs.): (1) area Tertiary channel gravels, worked by drift mining—placer gold; (2) E 5 mi., in S part of co., the Michigan Bluff district (tot. prod., 300,000 gold ozs.): (a) the Big Gun Mine, largest hydraulic operation in district—placer gold; (b) the Hidden Treasure Mine, richest drift mine in California—placer gold; (c) several area lode-gold mines, e.g., the Pioneer, Rawhide, etc.

LOOMIS, area gravel deposits along the American R., dredged during the 1930s—placer gold.

PENRYN, E 1 mi., the Alabama Mine—lode gold, with lead and silver.

Plumas Co.

Of the approximately 4,582,000 ozs. of gold produced in Plumas County, 1855–1959, more than half came from Tertiary gravel placer mines worked on a large scale by hydraulic methods.

AREA: (1) the Gopher Hill, Nelson Point, Sawpit Flat, and Upper Spanish Cr. mines, all in alluvial gravels—placer gold; (2) the Engles and Superior mines, primarily copper—by-product gold.

BELDEN, along the North Fork of the Feather R., placers.

CRESCENT MILLS (district, in T. 26 N, R. 9 E, tot. prod. through 1959 of 3,255 ozs. of placer gold and 32,069 ozs. of lode gold): (1) area streams, in Quaternary gravels—placer gold; (2) the Green Mountain

Mine, producer of $2,000,000 by 1890—lode gold.

JOHNSVILLE (district, in S-central part of co., in E½ T. 22 N, R. 11 E, tot. prod., 393,000 gold ozs.): (1) regional stream and bench gravels —placer gold; (2) the Plumas-Eureka Mine, major producer—lode gold.

LA PORTE (district, in SW part of co., in T. 21 N, R. 9 E, hydraulic-mining center since 1850s, with tot. prod., 1855–1959, of 2,910,000 gold ozs.), the Yuba R. channel traced NE for 10 mi., 500–1,500 ft. wide and 14–129 ft. deep, with placer gold concentrated in lowermost 2 ft. above bedrock—placer gold.

MUMFORDS MILL, area copper mines—by-product gold.

RICH BAR, area gravels along Indian Cr. tributary of the Feather R., rich placers.

SPRING GARDEN, NE 9 mi., the Walker Mine, primarily copper— by-product gold.

Riverside Co.

A total of 108,800 gold ozs., mostly lode gold, came from many relatively small mines and a few placers scattered throughout Riverside County, 1893–1959.

BLYTHE, SW 20 mi., the McCoy Mt. district, numerous copper mines —by-product gold.

DESERT CENTER, N, a large area along the N boundary of co., in T. 2 and 3 S, R. 10 and 12 E, the Pinon-Dale district (tot. prod., 1943–59, of 75 gold ozs., with 32,000 ozs. produced prior to 1943): (1) the Lost Horse Mine, producer of $350,000—lode gold; (2) the New Eldorado Mine, minor producer—lode gold.

PACKARDS WELL, S 2 mi., in the Palen Mts., area copper mines— some by-product gold.

PERRIS, W and SW, in T. 4 and 5 S, R. 4 W, the Pinanate district, the Good Hope Mine (worked prior to 1850 by Mexicans), with $2,000,000 produced, 1850–96—lode gold.

Sacramento Co.

Among the leading gold-producing counties, Sacramento County pro-duced 5,005,700 ozs. of placer gold and 5,000 ozs. of lode gold between 1880 and 1959.

AREA: (1) along the Cosumnes R., stream and terrace gravels—placer gold; (2) general area S of the Cosumnes R.: (a) in sand and gravel pits, and (b) regional tributary streams and ancient channels—placer gold.

FOLSUM (center of intensive bucketline and dredgeline operations): (1) area placer operations, and (2) drift mines, mostly opened before 1930, district production about 3,000,000 ozs.—placer gold.

SLOUGHHOUSE (district, along the Cosumnes R. in T. 7 N, R. 7 E, with tot. prod. of 1,700,000 ozs. of placer gold produced by early hydraulic methods and dredging during the 1930s), many area stream gravel deposits—placer gold.

San Bernardino Co.

As the largest county in the United States, San Bernardino County has very many mines scattered throughout its desert and barren mountain ranges. Although some placer deposits were worked in the early 1850s, most of the recorded production of 517,000 gold ozs. through 1959 came from lode mines and as a by-product of base-metal mines.

LUDLOW, S 7 mi., in T. 6 and 7 N, R. 8 E, the Stedman district, the Bagdad-Chase Mine, discovered in 1903, produced $4,500,000 million —lode gold.

NEEDLES, SW 36 mi. on U.S. 95 to W-trending rd. into the Turtle Mts., at Carson Wells, the Lost Arch Mine—lode gold.

ORO GRANDE, W 5 mi.: (1) the Silver Mountain district, area silver mines—lode gold; (2) the Black Diamond Mine—gold, with copper and silver.

SAN BERNARDINO, NE 20–25 mi., in T. 2 and 3 N, R. 1, 2, and 3 E, the Holcomb district (tot. prod., 400,000 ozs. of placer gold and 54,500 ozs. of lode gold): (1) the Ozier Mine, major producer since 1850—lode gold; (2) Holcomb Valley: (a) many area rich placer deposits located in the 1860s, producing about 340,000 ozs. of gold; (b) many valley area mines, mostly abandoned now—lode gold.

TWENTY-NINE PALMS, SE 15 mi., in T. 1 S, R. 11 and 12 E, the Dale district, the Brooklyn Mine (opened in 1893, tot. prod. of 63,500 ozs.) —lode gold.

VICTORVILLE, SE 30 mi., the Wild Rose Group (of claims)—lode gold in tremolite.

San Diego Co.

The total gold production of gold in San Diego County, 1869–1959, was about 219,800 ozs. of lode gold, with very minor placer production.

JULIAN (55 mi. NE of SAN DIEGO, most important gold-producing section of co.): (1) area lode mines; (2) E about 5 mi., in vicinity of BANNER, area lode mines; (3) S 3 mi., a copper mine—by-product gold; (4) S 4 mi., the Friday Mine, nickel ores—by-product gold.

RAMONA (district). (1) NE, the Spaulding Mine—lode gold; (2) in flats below the Little Three Mine—placer gold.

WYNOLA, area small placer gravel deposits (earliest gold discoveries in co.).

San Joaquin Co.

The 1885–1959 production of 126,400 ozs. of gold in San Joaquin County came largely from sporadic placer deposits along the Moke-lumne River.

BELOTA (district, along the Calaveras R., in T. 2 N, R. 9 E, in E-central part of co., a W extension of the Jenny Lind district in Calaveras Co.; tot. prod. estimated between 20,000 and 40,000 gold ozs.), numerous area Quaternary gravel deposits—placer gold.

CAMANCHE, along the Mokelumne R. all the way to CLEMENTS, numerous placer deposits.

CLEMENTS (district in NE cor. of co., along the Mokelumne R., prod. estimated between 50,000 and 100,000 gold ozs.), all area Quaternary gravel deposits, worked small-scale before World War I and dredged during 1930s—placer gold.

LINDEN, numerous area gravel deposits along the Calaveras R.—placer gold.

Shasta Co.

Between 1880 and 1959 Shasta County produced a total of 2,033,000 ozs. of gold, of which 375,472 ozs. came from placers. Each summer finds many amateur and "pocket" gold hunters, as well as scuba divers, working the county's streams.

FRENCH GULCH (the Deadwood district, along the W-central border of the co., tot. prod. of 128,000 ozs., 1848–1959, mostly from lode mines): (1) along Clear Cr., earliest placer discoveries; (2) the Washington Mine, discovered in 1842—lode gold; (3) the Greenhorn Mine, in sec. 33, (37N–5W), copper—by-product gold.

PLATINA (SW cor. of co.), the Harrison Gulch district, the Midas Mine (discovered in 1894 and destroyed by fire in 1914 after producing $3,563,587)—lode gold.

REDDING: (1) W 4 mi., the Silver King Mine—by-product gold; (2) SW 10 mi., in T. 31 N, R. 6 W, the Igo district (along Clear Cr., placers produced 115,022 ozs. of gold, primarily 1933–42, with minor amounts since); (3) NW 20 mi., the West Shasta district (copper-zinc, gold prod., 520,000 ozs.)—by-product gold: (a) S part of district, the Iron Mt. Mine —by-product gold, with silver, copper; (b) the Balaklala Mine—by-product gold and silver.

SHASTA (ghost town 6 mi. W of REDDING, now a historical monument), area old mines and site of early-day rush—lode gold.

SOUTH FORK (district near IGO, in sec. 17, (31N–6W), the Big Dike Mine, silver—by-product gold.

WHISKEYTOWN (district 5 mi. SE of FRENCH GULCH): (1) numerous area mines along Clear Cr., e.g.: (a) the Mad Mule Mine, largest producer, and (b) other area mines—free gold in quartz and pyrite.

Sierra Co.

Lode gold mines in Sierra County produced most of the total recorded 2,161,000 ozs. of gold, with an unrecorded production prior to 1880 probably boosting the total well above 3,000,000 oz.

ALLEGHANY, DOWNIEVILLE (districts in SW part of co. about 10 mi. apart, with tot. placer prod. of 194,000 gold ozs. and drift mine tot. from Tertiary gravels of 485,000 ozs.; tot. prod. recorded as 2,173,000 gold ozs.): (1) regional stream gravels and bench deposits—panning colors and nuggets; (2) the Sixteen to One Mine, principal producer of $9,000,000 to 1928 and 17,000 ozs. to 1958 with production continuing —lode gold.

SIERRA CITY, the Sierra Buttes Mine (and district), producer of 825,000 ozs.—lode gold.

Siskiyou Co.

Heart of the famed "Northern Mines" of gold rush history, Siskiyou County contains more than 370 once-active gold mines, which yielded 1,773,000 ozs. of gold, 1880–1959, with a large unrecorded amount produced between 1850 and 1880. Each summer sees scores of scuba divers, pocket hunters, and amateur prospectors working the gold-bearing streams throughout the western half of the county in perhaps the most rugged region in California.

CALLAHAN (44 mi. SW of YREKA at S end of Scott Valley in the poorly defined Scott River district, center of enormous mining activity): (1) bed of the South Fork of Scott R., from town S toward headwaters: (a) great hydraulic operations and Chinese rock piles—placer gold (still found after every winter runoff); (b) all contributary gulches and crs., with many lode mines hidden back in steep mts.—placer and lode gold; (2) N, along the Scott R., several mi. of dredging operations (halted in 1955 by law)—placer gold; (3) SW, by jeep rd., the Martin McKeen Mine, producer of 12,100 ozs.—lode gold; (4) the Porphyry Dike Mine, important producer—lode gold; (5) N 5 mi., Sugar Cr., follow W on USFS rd. to hydraulic workings—placer gold; (6) S a few mi. to Camp Eden: (a) SE, across canyon, the Blue Jay Mine, and (b) the gravels of nearby Jackson Cr. (rich producer of gold nuggets after winter floods); (7) E 10–12 mi.: (a) along Grouse Cr., placers; (b) the Copper King Mine—by-product gold; (8) S 14 mi., Carter Meadows recreation area, along Trail Cr.—placer showings.

CECILVILLE (30 mi. SW of CALLAHAN on S Fork of the Salmon R.): (1) N by steep USFS rd. to summit, the Black Bear Mine, now private property and "hippie" commune, but producer of 150,000 ozs.—lode gold; (2) many area mines in regional canyons—lode gold; (3) E fork of the S fork of the Salmon R., all area stream gravels, benches, and slopewash gravels—placer gold, with platinum nuggets.

FORKS OF SALMON (center of large gold-mining area): (1) all regional stream and bench gravels, placers; (2) immense hydraulic operations on major streams—placer gold; (3) SE about 10 mi., the King Solomon Mine (reached by Matthews Cr. jeep rd.), major producer—lode gold.

FORT JONES: (1) W, along the Scott R. rd., Indian Cr., extensively dredged placers; (2) N and W: (a) all access rds. into the Scott Bar Mts. lead to productive mines—lode gold; (b) all regional gulches, canyons, streambeds, and bench deposits (some extensively worked)—placer gold; (3) N 12 mi., vanished town of Deadwood on old rd. to YREKA, area canyons and gulches, mines—lode gold; (4) W, down Scott R. rd., Cottonwood and Rancheria crs.: (a) large-scale dredging produced $4,000,000 in 1850s; (b) area lode-gold mines, especially the Golden Eagle, produced 48,500 ozs. to 1931—lode gold.

GREENVIEW (in Scott Valley): (1) W 3 mi., Oro Fino: (a) area mines on S and E sides of Quartz Mt., site of gold rush camp—lode gold; (b) the Quartz Hill Mine—lode gold; (2) W 4 mi., in Quartz Valley, MUGGINSVILLE: (a) several old mines on W side of Quartz Mt. and 5-stamp mill remnant near Quartz Valley School—lode gold; (b) area stream gravels—pannable colors.

HAPPY CAMP, area jade mines along Indian Cr.—nephrite jade, gold colors and nuggets, and most prized of all, jade laced with stringers of raw gold (prime collectors' gemstone).

KLAMATH R., N, to Ragged (Ragged Ass) Gulch, cr. bars and bench gravels—placer gold.

SAWYERS BAR (Salmon R. district, roughly 800 sq. mi. of extremely mountainous country between the Marble Mountains and Salmon-Trinity Wilderness areas, tot. prod., 1855–1965, estimated at 15,981 ozs. of placer gold and 18,868 ozs. of lode gold): (1) all regional stream gravels—placer gold and platinum; (2) very many huge hydraulic operations and Chinese diggings, accessible from the ETNA-SAWYERS BAR rd. —placer gold; (3) North Fork of the Salmon R., all gravel and slopewash deposits—placer gold; (4) South Fork gravel bars, especially near mouth of Black Gulch—placer gold; (5) E 3 mi., the Whites Gulch Mine, hydraulic operations worked until 1970—placer gold; (6) E 10 mi. to Idelwild (old sawmill site and USFS campground), gravels of South Russian R.—placer gold.

SCOTT BAR (3 mi. S of jct. of Scott R. with the Klamath R. 3 mi. E of HAMBURG on Hwy. 96), the Scott Bar Mine, operated 1850–1970—lode gold, in hessite.

SOMES BAR: (1) Klamath R. bar and bench gravels, placers; (2) all tributary creeks and bench deposits—placer gold; (3) N, along Klamath R. to HAPPY CAMP, vast bench gravel deposits untouched to today, extremely rich extensions of huge hydraulic operations E of HAPPY CAMP to hwy. turnoff to YREKA, tremendous placer gold potential.

YREKA (co. seat and center of the Northern Mines region): (1) area lode mines, rich (following winter storms, large gold nuggets can be picked up within city limits, especially around the water works where, in December 1964 floods, several nuggets to 4 oz. each were found on the surface; (2) N and W, in T. 46 N, R. 10 and 11 W, the Humbug district (embracing the drainage of Humbug Cr., tot. prod. estimated between 25,000 and 50,000 ozs. of lode and placer gold): (a) area cr. gravels, productive placers to 1901; (b) many area lode mines accessible by logging rds. and shown on USFS quadrangle map; (3) HAWKINSVILLE (suburb of YREKA on the N), rich gold camp of early 1850s. (a) canyon and gulch gravels and terraces—placer gold; (b) numerous area mines in and around townsite—lode gold; (4) Hungry Cr., area gravels and slopewash deposits—placer gold; (5) SE 14 mi., the Peg Leg Mine— lode gold.

Stanislaus Co.

Between 1880 and 1955 the Quaternary gravel deposits of Stanislaus County produced 364,600 ozs. of placer gold.

OAKDALE, gravel bars, benches, and terraces along the Stanislaus R., NE to KNIGHTS FERRY—placer gold.

WATERFORD, all Tertiary gravels and benches along the Tuolumne R., E to LA GRANGE—placer gold.

Trinity Co.

Lode mines and placers produced 2,036,300 ozs. of gold in Trinity County between 1880 and 1959, most of it derived from placer gravels of the Trinity R. basin.

AREA, W edge of co., in gravels of the Trinity R. and the South Fork: (1) stream, bench, and terrace gravels—placer gold; (2) area quartz mines—lode gold.

DEDRICK, DOUGLAS CITY, HAY FORK, all gravel bars, benches, and tributary deposits along the Trinity R.—placer gold, platinum nuggets.

HELENA: (1) area gravel bars and benches along the Trinity R.— placer gold; (2) N, along the East Fork, especially in the Indian Cr. tributary, productive placers.

JUNCTION CITY, numerous area lode and placer mines.

LEWISTON, along a mineralized belt extending E to FRENCH GULCH, q.v. in Shasta Co., numerous mines—lode gold.

TRINITY CENTER (original town now under waters of Clair Engle Lake but central in the Trinity Basin district, tot. prod. of 1,750,000 gold ozs., 1880–1959): (1) area huge hydraulic operations, visible from USFS logging rds.—placer gold; (2) the Golden Jubilee Mine, very productive —lode gold; (3) 3 mi. above old townsite, the Enright Claim—placer gold, with platinum nuggets to several ozs. (not recognized in early days); (4) N 5 mi., at CARRVILLE, near mouth of Coffee Cr.: (a) area gravels—placer colors; (b) area lode mines, producers of $1,000,000 by 1910—lode gold.

WEAVERVILLE: (1) many big productive mines along side rds.—lode gold; (2) the La Grange Mine, largest hydraulic operation in area— placer gold.

Tulare Co.

Of more than 50 lode mines in Tulare County, only those in the White River district, discovered in 1853, produced a significant amount of the total gold production of 20,325 ozs. of lode gold between 1880 and 1959.

MINERAL KING (district), area mine dumps—gold, with arsenopyrite.

PORTERVILLE: (1) S 11 mi., the Deer Creek Mine (silver)—minor associated gold; (2) E 30 mi., on the Middle Fork of Tule R., in sec. 30, (19S–31E), various small mines and prospects (primarily copper)— by-product gold.

WHITE R. (district), area mines, most productive in co.—lode gold.

Tuolumne Co.

One of the Mother Lode counties of central California, Tuolumne County produced 2,580,000 ozs. of lode gold and 7,551,000 ozs. of placer gold between 1850 and 1959, the placers labeled the richest in the state.

COLUMBIA: (1) many area gravel deposits—placer gold; (2) at Sawmill Flat, the Barney Pocket Mine—lode gold.

GROVELAND-MOCCASIN-JACKSONVILLE (district, in S part of co., in T. 1 S, R. 14, 15, and 16 E): (1) area Quaternary stream gravels yielded about $34,000,000 in placer gold prior to 1899; (2) near GROVELAND: (a) the Longfellow Mine, most productive in district, with 24,200 gold ozs. produced before 1899—lode gold; (b) SW, at Big Oak Flat, the Washington Mine—lode gold.

JAMESTOWN-SONORA (the Columbia Basin district, about 2 mi. in dia., in parts of T. 1 and 2 N, R. 14 and 15 E in NW part of co., with tot. prod. about 5,874,000 ozs. of placer gold): (1) all area stream, bench, and terrace gravels—placer gold; (2) 3 mi. S of JAMESTOWN, the Mann Copper Mine—by-product gold; (3) between both towns and the Stanislaus R., the Pocket Belt district (5–6 mi. wide, tot. prod., 267,000 ozs.): (a) many small area mines—lode gold; (b) the Bonanza Mine, so rich that it produced $300,000 in a week—lode gold.

SONORA: (1) area gravel deposits (streambeds, benches), very rich and extensively worked—placer gold; (2) the O'Hara mines—crystallized lode gold.

TUOLUMNE, STANDARD, SOULSBYVILLE (the East Belt district, tot. prod., 965,000 gold ozs.): (1) very many area lode and placer mines; (2) area mines around the settlement of Pooleys Ranch—lode gold.

TUTTLETOWN, W, beginning of the Mother Lode belt of about 40 lode mines extending NW–SE across Tuolumne Co. S to the headwaters of Moccasin Cr., where the belt enters Mariposa Co., tot. prod., minimum of about 1,550,000 ozs. of lode gold: (1) the Harvard Mine and the Dutch Claim (Sweeney and App-Heslep) mines, most productive properties in district with $9 million produced to 1928—lode gold; (2) the Golden Eagle Rule, Dutch-App, and Eagle-Shawmut mines (most productive 1890–1920)—lode gold.

Yuba Co.

Yuba County's placer gravels produced 4,387,100 ozs. of gold, 1880–1959, while its lode mines yielded 907,500 gold ozs.

BROWNS VALLEY-SMARTVILLE (district, in central part of co.), in T. 16 N, R. 5 E: (1) many area mines—lode gold; (2) the Hibbert and Burris mines, most productive—lode gold.

CHALLENGE, numerous area mines—lode gold.

DOBBINS (district): (1) NW 2 mi., in the Indiana Ranch area, the California M Lode—lode gold; (2) in sec. 30, (18N–7E), the Red Ravine Mine—lode gold.

HAMMONTON (district, in S part of co. along the Yuba R., in parts of T. 15 and 16 N, R. 4 and 5 E, tot. prod., $100,000,000): (1) all regional Quaternary gravels, extensive placer operations; (2) all regional buried channel and exposed Tertiary gravel deposits, extensively worked placers.

MARYSVILLE-YUBA CITY, regional gravels of the Feather R., especially upstream toward the mts.—placer gold.

SMARTVILLE, numerous area mines—lode gold.

COLORADO

Known as the Centennial or Silver State, Colorado ranks second among the gold-producing states with an aggregate production of 40,776,000 ozs. of placer and lode gold. Almost all of the gold mines are in the mountainous western half of the state, known as the Colorado Mineral Belt, with mine elevations so high that both altitude and abrupt climatic changes, even in summer, must be taken into account by any gold hunter planning to prospect the gold-producing counties.

Gold occurs in varying degrees in most fissure veins located in every mining district, and for many years gold mining was of prime importance to Colorado's economy. The most important source of gold has been the telluride minerals calaverite and sylvanite, with petzite to a lesser degree. Very many Colorado mines carry significant amounts of gold in pyrite, galena, sphalerite, and arsenopyrite, with pyrite occurrences most productive.

For more information, write: Colorado Division of Mines, Department of Natural Resources, 1845 Sherman Street, Denver, Colorado 80203.

Adams Co.

STRASBURG, all bars and benches along Clear Cr., tributary of the South Platte R., extensive placer operations produced 16,800 gold ozs. between 1922 and 1959.

Alamosa Co.

BLANCA (WEST BLANCA), elev., 10,000–14,000 ft., many area mine dumps—gold showings.

Arapahoe Co.

AREA: (1) along the South Platte R., all gravel bars—placer gold; (2) Cherry Cr. and Dry (Cottonwood) Cr., in T. 5 S, R. 66 and 67 W, many productive placer workings; (3) all regional tributary watercourses, placer gravels.

DENVER, SW, in the Holy Cross Mts., area small mines (some operating 1974–75)—lode gold.

Baca Co.

SPRINGFIELD, SW 45 mi., at Carizzo (Estelene) Cr., in exposures of white sandstone—gold, with copper minerals.

Boulder Co.

Ranking ninth among Colorado's gold-producing counties, Boulder County produced 1,048,200 gold ozs. between 1858 and 1959, all but 3,000 ozs. coming from lode mines.

BOULDER: (1) NW 3–8 mi., the Gold Hill-Sugarloaf district (about 12 sq. mi., tot. prod., minimum of 412,000 ozs., 1859–1959): (a) area cr. beds, terraces, benches, etc.—placer gold (3,000 ozs. produced); (b) Gold Hill Mine, largest producer, including several adjoining mines— lode gold, silver; (c) old camps of Sugarloaf, Rowena, Salina, and Sunshine, many area mines—lode gold, silver; (2) SW 4 mi. to Magnolia (reached by steep grades), numerous high-grade mines—lode gold in tellurides (tot. prod., 130,000 ozs.); (3) W 17 mi. (4 mi. NW of NEDERLAND, q.v.), the Grand Island-Caribou district in SW part of co. (tot.

prod., 1932–59, of 10,006 ozs.): (a) numerous area lead-silver mines—by-product gold; (b) the Cardinal and Eldora mines—lode gold.

JAMESTOWN (Central district 9 mi. NW of BOULDER, tot. prod., 207,000 gold ozs.): (1) many area mines—gold in pyrite and telluride minerals; (2) the JAMESTOWN, GOLD HILL, and WARD area mines—lode gold.

NEDERLAND, N on Rte. 160 toward WARD, turn E onto the Sugarloaf–Sunset rd. to mi. 7, the Oregon Mine—gold in sulfides.

WARD (district 9–13 mi. NW of BOULDER, 12 sq. mi. in headwaters of Lefthand and Fourmile crs.): (1) old camps of Sunset and Copper Rock (more than 50 lode mines in area, tot. prod., 172,000 ozs.)—lode gold; (2) the Niwot and Columbia mines, largest producers—lode gold; (3) in E part of district, many mines—gold tellurides.

Chaffee Co.

Bordering the Continental Divide near central Colorado, Chaffee County produced 370,000 gold ozs., largely from lode mines but with small amounts from placers and base-metal mines.

AREA: (1) gravel bars and benches all along the Arkansas R. from BUENA VISTA SE 25 mi. to the Fremont Co. line, and near GRANITE (close to the co. line 15 mi. NW of SALIDA), many placers; (2) Lost Canyon Gulch, Chalk Cr., Cottonwood Cr., Pine Cr., Bertscheys Gulch, Gold Run Gulch, Gilson Gulch, Oregon Gulch, and Ritchey's Patch, all with abundant placer workings; (3) many small mines scattered throughout co.—lode and by-product gold.

BUENA VISTA: (1) SE 5 mi., at Free Gold on Trout Cr., area mines—lode gold; (2) E, to RIVERSIDE (6 mi. off U.S. 24, last 2 mi. difficult, alt., 12,000–13,000 ft.), area copper-lead-silver mines—by-product gold; (3) NE 13 mi. (just S of Trout Cr. Pass), area mines—lode gold; (4) W 14 mi., near head of Cottonwood Cr., small mines—lode gold; (5) the Chalk Cr. district (W part of co. near headwaters of Chalk Cr. 16 mi. W of NATHROP, tot. prod., 250,000 ozs. of by-product gold): (a) some 20 area mines in T. 15 S, R. 80 and 81 E, but chiefly the Mary Murphy Mine (largest producer, with 220,400 ozs.)—by-product gold; (b) area cr. gravels, minor placer deposits.

GARFIELD, the Monarch district (alt., 10,000–10,500 ft., tot. prod., 15,000–20,000 ozs. of by-product gold): (1) area mines in T. 49 and 50 N, R. 6 E, on old dumps—gold showings; (2) the Madonna Mine, major producer—lode gold.

GRANITE: (1) S 3 mi.: (a) along Clear Cr., area mines—lode gold; (b) 4–10 mi. farther SW, numerous placers; (2) W 15 mi., on Clear Cr. at WINFIELD (La Plata), alt., 9,750–12,000 ft. (mineralized area 1–3 mi W and SW of WINFIELD), small mines—lode gold; (3) T. 11 and 12 S, R. 79 E (with part of district in Lake Co.): (a) area cr. gravels—placer gold; (b) area mines—lode gold; (4) area of N part of co., extending N into Lake Co., along Cache and Clear crs., many placer workings.

NATHROP, S 2–3 mi., along Browns Cr. in Browns Canyon near U.S. 285, placers

SALIDA: (1) W, at Monarch Pass, area mines and dumps—gold showings (2) S 4 mi., the Cleora Mine (copper), near U.S. 50—by-product gold; (2) N 4 mi., in the Trout Cr. Hills via Rte. 291, the abandoned Sedalia Copper Mine, on dumps—by-product gold showings; (3) W 8 mi., on E side of the Arkansas R. Valley, at SEDALIA, area mine dumps —gold in abundant sphalerite; (4) N 11 mi., along Turret Cr., area mines—by-product gold; (5) NNE 16 mi., to CALUMET (Whitehorn) in Fremont Co., alt., 9,500–10,000 ft., area mine dumps—gold showings.

TWIN LAKES, the Red Mt. area, from town W to the Continental Divide, alt., 11,000–12,000 ft. (difficult to reach), many area old mines above valley—lode and by-product gold.

Clear Creek Co.

Located in the north-central part of Colorado in the Front Range west of Denver, Clear Creek County ranks seventh among the state's gold-producing counties, with a total of 2,400,000 ozs. coming mainly from lode mines, 1859–1959, with primary production from 1859 to 1864 from placer deposits.

AREA, in N-central part of co. 7 mi. WNW from CENTRAL CITY in Gilpin Co., the Alice district (extending into Gilpin Co., tot. prod., 23,000 gold ozs., 1883–1959): (1) the Alice Mine (first worked as a placer producer of 2,903 gold ozs., then as a lode mine); (2) the North

Star-Mann Mine, producer of 5,610 ozs. through 1916—lode gold.

DAILEY (ATLANTIC), near head of W fork of Clear Cr. and Butler Gulch, 2 mi. E and SE of Jones Pass, many area mines—lode gold.

GEORGETOWN-SILVER PLUME (GRIFFITH), on U.S. 6 and 40–42 mi. W of DENVER, local steep grades, in W part of co., district of 25 sq. mi., tot. prod., 145,000 gold ozs.: (1) area mines—by-product gold; (2) SW 6–8 mi., the Argentine (West Argentine) district, just E of the Continental Divide, tot. prod., minimum of 25,400 gold ozs.: (a) along Leavenworth Cr., and (b) SE side of Leavenworth Mt., (c) on SE slopes of McClellan Mt. at head of Leavenworth Cr. 6 mi. farther SW, many area mines—lode and by-product gold; (d) on Kelso Mt., area mines—lode gold; (e) the Belmont Silver Lode and the Baker Mine, largest district producers—lode gold.

IDAHO SPRINGS (district, embracing an unbroken succession of gold deposits extending from town to CENTRAL CITY and BLACKHAWK in Gilpin Co., tot. prod. in co. section about 1,805,000 gold ozs.): (1) mines at Cascade, Coral, Jackson Bar, Paynes Bar, Spanish Bar, and Virginia Canyon—lode gold; (2) Chicago Cr., Nevada and Illinois gulches, and Missouri Flats, extensive placers; (3) the Gregory, Russel, Bates, Bobtail, and Mammoth mines—lode gold; (4) NW 2½ mi., at TRAIL (Freeland, Lamartine): (a) along Trail Cr., and (b) the Lamartine Mine (2 mi. SW on the divide between Trail Cr. and Ute Cr. 2–4 mi. off U.S. 6 and U.S. 40)—lode gold; (5) W 6 mi., along Silver Cr., placer workings; (6) NW 10 mi. (2 mi. on U.S. 40 and 8 mi. on Rte. 285), to Alice (Lincoln, Yankee Hill, alt., 10,000–11,000 ft.), numerous area mines—lode gold; (7) camps of Montana (Lawson, Dumont, and Downieville), the W extension of district on U.S. 6 and U.S. 40, area mines—lode gold.

Conejos Co.

PLATORO (W on Rte. 15 from LA JARA), the Axel, Gilmore, Lake Fork, Ute, and Stunner districts, very many regional mines—lode and by-product gold.

Costilla Co.

SAN LUIS, NE 7 mi., to PLOMO (Rito Seco), on Rito Seco Cr., area mines—gold, in pyrite and quartz.

Custer Co.

Ranking seventeenth among Colorado's gold-producing counties, Custer County produced 107,300 gold ozs. between 1872 and 1959, primarily as a by-product of lead-silver mining.

SILVER CLIFF: (1) area mines and dumps—by-product gold; (2) NE, to Oak Cr. (camps of Ilse, Spaulding) on Rte. 143 about 16 mi. SW of FLORENCE in Fremont Co., area mines—by-product gold.

WESTCLIFFE, SE 7 mi., the Rosita Hills (Rosita, Querida) district in the low W foothills of the Wet Mts., tot. prod., 84,660 gold ozs., 1870–1959: (1) area mines and dumps—gold showings; (2) the Bassick Mine, major producer—lode gold.

Dolores Co.

Practically all of the 104,500 gold ozs. produced in Dolores County came as a by-product from lead, silver, and zinc mines.

OPHIR, the San Juan Mts. (a triangle between OURAY in Ouray Co., SILVERTON in San Juan Co., and OPHIR E to the Hinsdale Co. line— steep, rugged access rds., make local inquiry), very many old mines and dumps—gold showings.

RICO (Pioneer) district, near SW end of the Colorado Mineral Belt near headwaters of the Dolores R.: (1) mines along Rte. 145 (36 mi. NE of DOLORES and 27 mi. S of TELLURIDE)—by-product gold; (2) NW 16 mi., at Lone Cone (Dunton), on the West Dolores R. on Rte. 331, area mines—by-product gold.

Douglas Co.

FRANKTOWN: (1) along Cherry Cr. for several mi. N of town, (2) NW 4–5 mi., along Lemon Cr.; (3) S 1 mi., in Russelville Gulch tributary to Cherry Cr., extending 3 mi., many productive placers.

LOUVIERS, along Dry Cr. (tributary of the South Platte R.), in gravel deposits extending NE into Arapahoe Co., placers.

PARKER, NW 1½ mi. on Rte. 83: (1) in Newlin Gulch, and (2) NW of Newlin Gulch, in Happy Canyon, all sand and gravel deposits—placer gold (microscopic grains to pinhead nuggets).

Eagle Co.

Many small gold and silver mines scattered throughout Eagle Co. produced a total of 359,900 gold ozs.

EAGLE: (1) W 1 mi. on U.S. 24, and 6–8 mi. up Brush Cr. on rd. toward FULFORD, area copper mines—by-product gold; (2) SSE 20 mi. to FULFORD, head of Brush Cr., area lead-silver mines—by-product gold.

GILMAN (Battle Mt., Red Cliff district, in SE part of co., on NE flank of the Sawatch Range between town and REDCLIFF (about 20 mi. N of LEADVILLE): (1) the Eagle Mine (fourth largest zinc mine in America with copper, lead, silver)—by-product gold; (2) Battle Mt., area mines operated 1877–78—by-product gold.

McCoy, W, along the Colorado R., bar and bench gravels—placer gold.

MINTURN, SW 10 mi., at head of Cross Cr., at Holy Cross (Eagle River), spotty, high-grade ores—lode gold.

REDCLIFF: (1) along Homestake Cr., several mines—by-product gold; (2) Battle Mt., Belden, and other area mines all way to GILMAN, q.v. —by-product gold.

Elbert Co.

AREA, all Platte R. bars and terrace gravels—placer colors.

ELIZABETH, W and NW 1–1½ mi., along Gold (Ronk) Cr. on Rte. 86 (about 40 mi. SE of DENVER), gravel deposits—placer gold.

El Paso Co.

COLORADO SPRINGS: (1) NW 6 mi., at Blair Athol in the foothills, area gravel deposits—placer gold; (2) SW 7 mi. via Gold Camp rd., the St. Peter's Dome district, numerous mines—lode gold.

Fremont Co.

AREA, gravel bars and terrace gravels along the Arkansas R. from the Chaffee Co. line downstream to FLORENCE, many placers.

BADGER CR. (8 mi. SE of SALIDA in Chaffee Co.), 4 mi. up cr., gravel bars and benches—placer gold, with copper minerals.

CANON CITY, W 13 mi. on U.S. 50, Currant Cr. (Parkdale, Micanite), with mineralized area extending 8 mi. N along cr., area mines—lode and by-product gold.

COTOPAXI, N 9 mi. on U.S. 40 to Red Gulch (24 mi. SE of SALIDA in Chaffee Co.), area copper-silver mines—by-product gold.

WHITEHORN (district, E of and continuous with the Calumet district of Chaffee Co.), many area mines—lode gold.

Garfield Co.

GLENWOOD SPRINGS, NEWCASTLE, N, on dumps of many old mines: (1) along Riffle and Elk crs., and (2) on S flank of the White River Plateau (almost inaccessible today), primarily lead-silver-zinc—by-product gold.

Gilpin Co.

Lying about 30 miles west of Denver, on the east slope of the Front Range, Gilpin County ranks second among the gold-producing counties of Colorado. From 1859 through 1959 a total of 4,207,000 ozs. of lode gold and 47,900 ozs. of placer gold were produced.

AREA: (1) the northern districts of Perigo, Independence, and Pine-Kingston-Apex, covering half a township 20–35 mi. SW of BOULDER in Boulder Co. and 50–60 mi. NW of DENVER, good access rds., very many mines—lode and by-product gold; (2) southern districts of Central, Nevada, Gregory, Russel, and Quartz Mt., 40–50 mi. W of DENVER and SW of BOULDER, very many mines (prospect all dumps)—lode and by-product gold.

BLACKHAWK: (1) rich area mines (mostly base-metal and silver)—by-product gold; (2) the Gregory diggings, rich placers discovered in 1859; (3) at Russel Gulch: (a) area gravel deposits, productive placers; (b) many area lode mines (early production, $400 per day per man)—lode gold.

CENTRAL CITY (district, along S boundary of co., the N segment of the rich chain of ore deposits between town and IDAHO SPRINGS, q.v. in Clear Cr. Co., tot. prod. of 4,170,000 ozs. of lode gold and 30,000 ozs. of placer gold): (1) SW 2.1 mi. on Rte. 279, past ghost town of Russel

Gulch, area old mine dumps—gold showings; (2) turn right onto Rte. 279 (unmarked) to mi. 3.8, the Gloryhole (enormous open pit with dangerous rim)—gold showings in a variety of minerals; (3) take rd. to ghost town of Apex: (a) just N of North Clear Cr., in central part of co., the northern Gilpin district (extending N to the Boulder Co. line), most important mines lie just S of Apex, tot. prod. of 35,000 ozs.—lode gold; (b) Gilpin area, second most important mines, on dumps—gold showings; (c) in Gamble Gulch, placers; (d) the Dirt and Perigo mines, active to 1959—lode gold.

ROLLINSVILLE: (1) all area gulches, placers worked since 1897; (2) along South Boulder Cr., dredging from 1937–39 produced 7,724 ozs. —placer gold.

Grand Co.

GRAND LAKE (Wolverine), E 7 mi. on Rte. 278, area lead-silver mines —by-product gold.

PARSHALL, S, to the head of Williams Fork, the La Plata district (extending a few mi. SE across Jones Pass on the Continental Divide into the headwaters of the West Fork of Clear Cr.)—lode gold, often with pyrite.

Gunnison Co.

Lying west of the Continental Divide, Gunnison County produced 130,000 gold ozs., mostly from lode mines but with unrecorded placer gold production between 1861 and 1880.

AREA: (1) in N part of co., all area gulches, placers; (2) in Washington Gulch, site of first gold discovery, placer workings.

ALMONT (on Rte. 306), NE 7 mi., in Spring Canyon, several lead-silver mines—by-product gold.

CRESTED BUTTE: (1) NW 10 mi. and a few mi. N of Rte. 135, area silver mines—by-product gold; (2) N 22 mi., at Elk Mt. (alt., 9,500–11,000 ft.), including ghost town of Gothic, numerous area copper mines—by-product gold, silver.

GUNNISON, E 13 mi., at Gold Brick (1–4 mi. NE of PITKIN, q.v., on fair rd., alt., 9,000–13,000 ft.), numerous relatively rich mines concentrated just E of Gold Cr. in an area 4 mi. long by 1 mi. wide—lode gold.

PARLIN: (1) Take Rte. 162 to the Quartz Cr. district (cf. also under PITKIN), area mines—lode gold; (2) S 3–4 mi., at Cochetopa (Green Mt., Gold Basin), extending from Cochetopa Cr. 2–4 mi. W, area mines —free-milling gold.

PITKIN, N and NE 1–4 mi., in SE part of co., to include Box Canyon, the Gold Brick-Quartz Creek district (tot. prod., 1879–1959, of 80,000 gold ozs.; includes OHIO CITY, founded 1881): (1) NE 1–4 mi. (near rd. to TINCUP, q.v.), in S end of extensively mineralized zone, many old mines, on dumps—gold showings; (2) S 6 mi., near U.S. 50 about 25 mi. E of GUNNISON, in Box Canyon, the old Independence and Camp Bird mines 3–4 mi. N of Waunito Hot Springs Cr. via steep rds.— free-milling gold.

POWDERHORN, in the Cebolla district (Iola, Domingo, Vulcan, and White Earth), many mines along Cebolla Cr.—lode gold.

SARGENTS, N 10 mi., at Tomichi (Whitepine), a ghost town, area mines—lode and by-product gold.

TINCUP (15 mi. N of PITKIN in N part of co. 25 mi. NE of GUNNISON via Rte. 162, at head of Willow Cr. on extreme SE flank of Taylor Park and on opposite side of 12,000-ft. Cumberland Pass from PITKIN), the Tincup district, tot. prod., about 16,400 ozs. of gold, mostly prior to 1932: (1) along Tincup Gulch, many placers; (2) at headwaters of Willow Cr., area mines (source of placer gold)—lode gold.

Hinsdale Co.

Following silver discoveries in 1871, Hinsdale County produced about 71,365 ozs. of gold through 1959.

LAKE CITY: (1) along Henson Cr. above town and extending for 10 mi. W, many lead-silver-zinc and copper mines—by-product gold in tellurides; (2) S 5 mi., Lake Fork of the Gunnison R., at N end of Lake San Cristobal: (a) many area mines in 5-mi. stretch along Lake Fork, and (b) the Golden Fleece Mine (noted for its tellurides)—lode gold; (3) SW 12 mi., between Lake Fork and its tributary Henson Cr., in several districts, e.g., Burrows Park (Whitecross), on Rte. 351 near head of Lake Fork, alt., 10,500–12,000 ft., many area mines—lode gold; (4) SW 18 mi., at Carson at head of Wagner Gulch, mineralization extending across the divide into head of Lost Trail Cr., many lead-silver mines—by-product gold.

Huerfano Co.

LA VETA, 11 mi. SW of WALNESBURG via U.S. 160 and 5 mi. on Rte. 111, area lead-silver and copper mines—minor by-product gold.

Jackson Co.

COWDRY, NW 18 mi. on Rte. 125 to Pearl, area base-metal mines—by-product gold.

RAND, SE 9 mi., at Teller on Jack Cr., alt., 9,000–10,000 ft. (an active early-day boom camp), many copper-silver mines—by-product gold.

Jefferson Co.

EVERGREEN: (1) ½ mi. above town, on Cub Cr., small mines—lode gold; (2) S 1½ mi. on Rte. 73: (a) the Augusta Mine—by-product gold; (b) SW ¾ mi., on NW side of Cub Cr. along Rte. 334, and (c) ¼ mi. W of Rte. 74, mine—gold, with fluorite and silver; (3) the Malachite Mine, notable copper producer—by-product gold.

GOLDEN: (1) W, along Clear Cr. to E, many good placers; (2) area sand and gravel pits (source of most gold in recent years)—placer gold.

Lake Co.

Colorado's most important mining districts are in this high-altitude county, with gold production through 1959 totaling 2,983,000 ozs.

AREA: (1) in T. 10 S, R. 80 W: (a) along Box Cr., large dredger operations—placer gold; (b) along lower Box Cr. (a most productive dredging operation)—placer gold; (c) along Lake Cr., many productive placers; (2) the Arkansas R. Valley, many placers (tot. prod. through 1959, about 41,000 gold ozs.); (3) all tributaries of the Arkansas R., gravel deposits—placer gold.

CLIMAX: (1) SE 1 mi., in Arkansas R. Valley (10–12 mi. NE of LEADVILLE), the Alicante Mine—lode gold, in pyrite; (2) 4 mi. farther S, the Birdseye Mine on E side of valley, base metals—by-product gold.

LEADVILLE (district, tot. prod. through 1959 of 2,970,000 gold ozs.):
(1) area great mines—by-product gold; (2) Iowa and California gulches,
rich early placers; (3) old placer camp of Oro City (ghost town once
boasting 10,000 residents): (a) numerous residual placer deposits; (b)
area lead-silver mines, opened in 1868—by-product gold; (4) Breece
Hill, area mines—by-product gold; (5) the California, Evans, Iowa, and
Empire mines (alt., 10,150 ft., on W slope of the Mosquito Range)—
lode gold; (6) N, the Kokomo mines—gold, with pyrite; (7) W 4 mi.,
at St. Kevin-Sugar Loaf mines (early producers, with dumps still showing
abundant auriferous pyrite, galena, and sphalerite; (8) SW 22 mi. and
5 mi. W of GRANITE in Chaffee Co., a district that includes all of the
Lake Cr. drainage W of Twin Lakes, many area mines—lode gold.

La Plata Co.

AREA, along the Animas R., all bars, benches, terraces—placer gold.

DURANGO: (1) Animas R. gravel bars, benches, terraces; many placer
operations near towns; (2) N 25 mi. to Needleton (flagstop), then E 6
mi. to the Needle Mts., the Chicago Basin district (Tacoma, Florida
River, Vallecito camps), alt. 11,000–12,000 ft., many area mines—lode
gold.

LA PLATA (district in La Plata Mts., 15 mi. dia., between the San Juan
Mts. on the E and the Colorado Plateau on the W; tot. prod., about
215,000 ozs.): (1) La Plata R. gravels, small placers; (2) area of head of
Junction Cr. on E flank of mts. (separated from town of EAGLE PASS,
alt., 11,700 ft.)—gold tellurides, pyritic gold.

Larimer Co.

BELLVUE (on U.S. 287), SW 3 mi. at Empire (Howes Gulch) and 6
mi. SW of FORT COLLINS, area copper mines—minor by-product gold.

FORT COLLINS, W 45 mi., at Manhattan, then N 3–4 mi. on steep
rd. off Rte. 14, area placer deposits.

Mineral Co.

CREEDE: (1) area near town, the Amethyst (Big) Mine, Last Chance,
and New York mines (primarily lead-silver-zinc), early major producers

of the county's 149,200 ozs.—by-product gold; (2) W 2 mi., at Sunny-side, numerous early gold claims—lode gold; (3) Willow Cr.: (a) area silver mines—by-product gold; (b) N ¾ mi., on Left Fork of West Willow Cr., the Commodore Mine—by-product gold; (c) W of West Willow Cr., the Amethyst Lode (numerous mines)—by-product gold; (d) between the Amethyst Lode and the Commodore, all cr. gravel bars, benches—placer gold; (4) at end of Rte. 149, the King Solomon and Sunnyside mines (alt., 9,000–11,000 ft.)—by-product gold.

Moffat Co.

CRAIG: (1) SW 12 mi., at Round Bottom, on N side of the Yampa R., area placer workings; (2) W 19 mi., at LAY, on U.S. 40, with latest workings 6–10 mi. N along Lay Cr., productive placers; (3) N 35 mi., at Fourmile Cr. and Timberline Cr., close to Rte. 13 with BAGGS, Wyo., 5 mi. N to area of dry, rolling plains along the W base of the Elk Mts. (30-by-40-mi. coverage), dry placers. The fine gold can be winnowed or recovered by blower-type machines and is 885–935 fine, with about 600 colors comprising a cent (at $35 a troy ounce).

Montrose Co.

NATURITA, area of T. 46 N, R. 15 W, along the San Miguel and Uncompahgre rivers, all sand- and gravel bars and benches—placer gold.

PARADOX (on Rte. 90), 6 mi. out, at La Sal Cr.: (1) numerous area copper-silver mines—by-product gold; (2) all area watercourse beds, benches, and terraces—placer gold.

Ouray Co.

Ouray County produced 1,911,000 ozs. of gold between 1873 and 1959, of which 1,058,774 ozs. came from the famed Camp Bird Mine as a by-product of copper, lead, and silver refining.

OURAY: (1) just N, on E side of valley a few mi. off U.S. 550 by steep grades, to Uncompahgre (Upper Uncompahgre, Ouray), an area of about 15 sq. mi. with most mines in canyon walls of the Uncompahgre R., tot. prod., about 200,000 ozs. before 1900: (a) the American Nettie Mine (phenomenally rich producer)—lode gold; (b) the Bachelor Mine

—by-product gold; (c) many other area mines—by-product gold; (2) S 6 mi., the Treasury Tunnel near U.S. 550—gold, in sulfides; (3) WSW 8–12 mi. on Rte. 361, at Sneffels (Imogene Basin), the Sneffels-Red Mountain district (tot. prod., 1,723,000 ozs. of by-product gold): (a) many area lead-silver-zinc mines—major gold by-product; (b) at head of Canyon Cr. in Imogene Basin to W of Hayden Mt., at Sneffels (camp), many mines, especially the Camp Bird Mine—by-product gold; (c) at head of Red Mt. Cr., to E of Hayden Mt., the Red Mountain camp, many area mines—by-product gold; (4) S 12 mi., at Red. Mt. Pass (alt., 11,018 ft.), the Longfellow Mine—by-product gold.

Park Co.

Gold is the principal metal mined in Park County, with a total production of 1,364,430 ozs. produced between 1859 and 1959, most of it from lode mines along the east slope of the Mosquito Range in the northwest part of the county.

ALMA (district, E of LEADVILLE, tot. prod. of 1,320,000 ozs. of lode gold and 28,000 ozs. of placer gold): (1) NE: (a) ¼–½ mi., along E side of the South Platte R., placers—gold (coarse, nuggets to several ozs., tot. prod., 28,000 ozs.): (b) all area tributary crs. and gulches, in gravel deposits—placer gold; (2) Mt. Lincoln and Mt. Bross, rich area lode mines, especially the London Mine (richest producer, operating through 1942)—lode gold, silver; (3) near head of the South Platte R. along the Continental Divide, North Star Mt., all area gulches and drainage courses (probable source of gold in the Alma placers)—lode and placer gold; (4) NW 2–6 mi., small mines—gold, with lead, silver, and zinc; (5) N 5–10 mi., the Consolidated Montgomery Mine (alt., 11,500–13,500 ft.)—by-product gold; (6) in Buckskin Gulch, the Sweethome Mine—by-product gold; (7) all general area of Mosquito Cr. and its tributaries (alt., 10,500–12,500 ft.), numerous mines—by-product gold.

BAILEY: (1) area of Beaver Cr., alt., 10,000–10,500 ft., in outwash gravels from the South Platte glacier, placers; (2) W 13–14 mi., area mines—by-product gold.

COMO, NW, along upper reaches of Tarryall Cr. and its tributaries, the Tarryall district (tot. prod., minimum 67,000 ozs. of placer gold and

250 ozs. of lode gold): (1) Tarryall Cr., on E slope of Silverheels Mt., extending for several mi. SE of town, extensive placers; (2) area small mines—lode gold; (3) in Montgomery and Deadwood gulches, near headwaters, lode mines probable sources of Tarryall Cr. placer gold—lode gold.

FAIRPLAY (district, tot. prod., 202,000 ozs. of placer gold): (1) along the South Platte R., the Snowstorm and Fairplay placers; (2) along Sacramento and Beaver crs., placers; (3) area SE of town, in glacial outwash gravels—placer gold; (4) W by S 12 mi., at head of Fourmile Cr., the Horseshoe (alt., 11,500–12,500 ft.), area lead-silver mines—by-product gold; (5) SW 12 mi. and N of Fourmile Cr. and S of Mosquito Cr., the Sacramento Mine—by-product gold; (6) SW, at Weston Pass on crest of the Mosquito Range (alt., 11,900 ft.), on Lake Co. line, area lead-silver mines—by-product gold.

GRANT, N, at Geneva Cr. on Collier Mt., head of West Geneva Cr. (and continuous with the Montezuma district to NW in Summit Co., q.v., alt., 10,250–12,000 ft.), area base-metal mines—by-product gold.

HARTSEL, E on U.S. 24 to summit of Wilkerson Pass (alt., 9,525 ft.), then 2 mi. E: (a) the St. Joe Tunnel, and (b) ½ mi. farther E, an old mine (copper)—by-product gold.

Pitkin Co.

ASPEN: (1) area of Roaring Fork (including town and Richmond Hill, Lenado, principal mines within 1 mi. of ASPEN), the Mollie Gibson and Smuggler mines, and others—by-product gold; (2) N 10 mi., at ghostly Ashcroft on Castle Cr., area mines—by-product gold; (3) SE 15 mi., at Lincoln Gulch (mineralized area at head of gulch and 10 mi. from Rte. 82 via poor rd.), on W side of Ruby Mt., area base-metal mines—minor gold by-product; (4) SE 20 mi., the Independence Pass district, mines on West Aspen Mt. (tot. prod., 25,000 ozs.)—lode and by-product gold.

Rio Grande Co.

In the 1880s this county ranked third among Colorado's gold-producing counties, with a total of 257,600 ozs. produced through 1947.

DEL NORTE, W about 8 mi., along Embargo Cr. on both sides of the co. line, various mines—lode gold.

Monte Vista, SW 30 mi., to Jasper (Decatur), area mines and prospects ½ mi. W on Alamosa Cr.—lode gold.

Summitville (district, in SW cor. of co. high in the San Juan Mts., tot. prod., 1873–1959, about 257,600 ozs.): (1) the Little Annie Group of mines, principal producers—lode gold; (2) head of Whightman Fork (tributary of Alamosa Cr.), alt., 11,000–12,000 ft., area mines—by-product gold; (3) on South Mt., both sides area mines (vein exposures throughout 1½ mi. N–S and 1 mi. E–W)—lode gold.

Routt Co.

Columbine, area old mines—by-product gold.

Hahns Peak (district 4 mi. N of Columbine, tot. prod. between 15,000 and 20,000 gold ozs.): (1) area stream gravels, placers; (2) area small mines—lode gold (minor).

Saguache Co.

Area, extreme SW part of co., along Embargo Cr. and extending into the Del Norte area of Rio Grande Co., q.v.: (1) area watercourse gravels, placers; (2) area base-metal mines—by-product gold.

Moffat: (1) NE 10 mi., on Cotton Cr., at Blake (Mirage, Cotton Creek), at head of cr. on W slopes of the Sangre de Cristo Range, many old mines—by-product gold; (2) E 15 mi., at Crestone (Baca Grant), area 3–6 mi. wide, scattered mines—by-product gold.

Saguache, N 12 mi., in NE part of co., the Bonanza (Kerber Creek) district, tot. prod., 17,000 gold ozs. as by-product of base-metal and silver mines: (1) Bonanza camp: (a) numerous area mines, especially (b) the Rawley Mine—by-product gold; (2) N part of district, many small mines—gold tellurides.

Villa Grove, W, toward Bonanza along Kerber Cr., in the Cochetopa Hills at NW end of the San Luis Valley, alt., 9,500–10,000 ft., area copper-zinc mines—by-product gold.

San Juan Co.

A major producer, San Juan County yielded 1,665,000 ozs. of lode gold through 1959.

Silverton: (1) just NE on U.S. 550, at Animas (district), a belt of rich mines several mi. wide along the S rim of the Silverton caldera, with mines along both sides of the Animas R.—lode gold (est. tot. prod., 874,000 to 1,000,000 ozs.); (2) W 3 mi., area copper mines—by-product gold; (3) in Arrastre Basin, Silver Lake Basin, and Cunningham Gulch, such major mines as the Shenandoha-Dives, Aspen, Silver Lake-Nevada, and Highland Mary—lode gold; (a) upper part of Cunningham Gulch, major base-metal mines—by-product gold; (b) near mouth of gulch, major mines—lode gold; (4) NE 4 mi., the Senorita Mine, copper— by-product gold; (5) N, at Eureka (Cement Cr., Mineral Cr., Animas Forks), the Eureka district, tot. prod. at least 500,000 gold ozs. through 1959: (a) area base-metal mines—by-product gold; (b) beyond Eureka, the Sunnyside and Gold King mines, 8.1 mi. NE of SILVERTON—gold in sulfides; (6) E 19 mi., mostly by trail (50 mi. W of CREEDE IN Mineral Co.), at Bear Cr., mines—gold tellurides.

San Miguel Co.

Between 1875 and 1959 San Miguel County produced 3,837,000 gold ozs., to become the third-ranking county in Colorado in gold production.

AREA: (1) many regional placer deposits scattered over the co., 1878– 1959, produced 9,700 gold ozs.; (2) in SE part of co., on W spur of Mt. Wilson, the Mt. Wilson district, at head of Big Bear Cr. (tot. prod., 24,800 ozs.): (a) the Silver Pick Mine, major producer—lode gold; (b) many area small mines—gold showings.

OPHIR (district, in E part of co., including area S of the San Miguel R., W of Bridal Veil Cr., and the Ophir Valley on the S, including the Ames (ghost camp), Iron Springs, and South Telluride mining areas): (1) S 1 mi., at Ames, area old mines extending 6 mi. E to Iron Springs, base-metal ores—lode gold (in quartz); (2) Ophir Valley, many area mines, especially the Alta Mine—lode gold.

TELLURIDE (district, along E border of co. immediately SW of the Sneffels-Red Mountain district in Ouray Co., q.v., tot. prod. through 1959 of 3,000,000 gold ozs.): (1) the Liberty Bell Mine, producer of 633,021 ozs, 1898–1921—lode gold; (2) the Smuggler-Union Mine (closed in 1928 after 52 years) and the Tomboy Mine (closed in 1927) —lode gold; (3) E, to head of the San Miguel R., the Upper San Miguel district, alt., 11,000–12,000 ft., many area mines—lode and by-product gold; (4) NW 14 mi., the Lower San Miguel district (Placerville, Sawpit, and Newmire)—productive placers.

Summit Co.

Ranking first in Colorado for placer-gold production and tenth for lode gold, Summit County produced an estimated minimum of 739,511 ozs. from placer mines and 271,159 ozs. from lode mines through 1959.

AREA, near the N co. line, along Rte. 9, the Big Four Mine (16 mi. S of KREMMLING in Grand Co.), area lead-zinc mines—by-product gold.

BRECKENRIDGE (or Blue River district, including the upper valley of the Blue R. between the Front Range on the E and the Tenmile Range on the W, tot. prod. of 1,000,000 gold ozs., of which 735,000 ozs. came from placer deposits): (1) all regional stream channels, gulches, benches, etc., placers; (2) E and NE of town, in area of about 5 sq. mi., many lead-silver mines—by-product gold; (3) in Georgia Gulch, on N side of Farncomb Hill: (a) rich area placers discovered in 1859; (b) area lode mines, discovered in 1880, especially the Wellington Mine, chief producer (so rich in native gold that pockets supplied collectors and museums throughout the world with specimens of wire and leaf gold); (4) along the Swan and Blue rivers, many deep placers worked by dredges after 1900.

FRISCO (4 mi. SW of DILLON, the Tenmile district, including old camps of Kokomo and Robinson in the Tenmile Valley on the W side of the Tenmile Range, tot. prod., 1861–1959, of 52,000 gold ozs., mostly as by-product of base-metal ores): (1) area: (a) old mine dumps—gold, with galena and pyrite; (b) various local small cr. placers; (2) Kokomo: (a) in McNulty Gulch, rich placers discovered in 1861; (b) many mines along valley of Tenmile Cr. for 2–3 mi. NE and 5 mi. SW to Robinson; (3) on E side of valley, in the Tenmile Range, on W slopes, many base-metal mines—lode gold.

Teller Co.

Lying west of Colorado Springs in the southern part of the Front Range, Teller County produced 19,100,867 ozs. of gold between 1859 and 1959.

CRIPPLE CR. (district, 45 mi. SW of COLORADO SPRINGS, near Pikes Peak, leading gold-producing district in Colorado and second in the United States, next to the Homestake Mining Co. in LEAD, South Dakota, q.v.): (1) very many huge mines, e.g.: (a) the Washburn, Independence, Portland, Granite, Strong, Vindicator, Golden Cycle, Victor, Isabella, and Cresson—lode gold; (b) regional pits and prospects, very many—gold tellurides; (2) on Mt. Pizgah, several early-day prospects, low-grade—gold showings.

GEORGIA

The second great gold rush in America, involving thousands of miners as well as whiskey-town derelicts, took place in Georgia between 1828 and 1850, after Benjamin Parks kicked over a rock about three miles south of Dahlonega, in Lumpkin County, and discovered a chunk of gold "as yellow as the yolk of an egg." Almost overnight, some 4,000 prospectors were searching the mountains of Georgia, building shanty-towns and opening up the hinterlands to settlement. The influx of gold hunters was so great in 1830 that Governor Gilman wrote: "I am in doubt as to what ought to be done with gold-diggers. They with their various attendants, foragers, and suppliers make up between six and ten thousand persons." In 1831 a newspaper correspondent wrote: "I can hardly conceive a more amoral community than exists around these mines; drunkenness, gambling, fighting, lewdness, and every other vice exist here to an awful extent. . . ."

A prospector wrote home in 1833, declaring that he had "never before been amongst such a complete set of lawless beings. I do really believe that for a man to be thought honest here would be a disadvantage to him." These early accounts sound prophetic of the far western states a generation later, with a difference: There were no outlaws or desperadoes, and only a single murder was recorded during the entire Georgia rush.

Gold prospecting and mining has continued in Georgia to the present. However, there are no public lands within the state, and any gold hunter must obtain permission from a landowner to prospect on his land, or be subject to laws of trespass. That raw gold is still to be found in some abundance is attested to by the fact that in 1958 a historical society sent a wagon train carrying 46 ounces of new gold from Dahlonega to the state capital of Atlanta. The gold, hammered out into gold leaf to cover the capitol dome, was in remembrance of the 870,665 ozs. taken from the Georgia earth in the 130 years since Benjamin Parks' original discovery.

For further information, write: Department of Natural Resources, Earth and Water Division, 19 Hunter St. S.W., Room 400, Atlanta, Georgia 30334.

Cherokee Co.

BALL GROUND: (1) area mines and prospects in exposures of the residual mantle of decomposed bedrock—free gold; (2) SE 7 mi., on the Etowah R., the Creighton-Franklin Mine, tot. prod., 1840–1909, estimated at 35,400–48,500 ozs.—lode gold.

HOLLY SPRINGS: (1) W 1 mi., the 301 Mine—lode gold; (2) W 4 mi., adjacent to the Little R., the Cherokee Mine, once large scale—lode gold; (3) 2 mi. N of the Cherokee, the Sixes Mine, major producer—lode gold.

Forsyth Co.

CUMMING: (1) area stream and bench gravels—placer gold; (2) E 2 mi., a small area of placer grounds once hydraulicked.

SUGAR HILL, the Simmons Mine, productive—lode gold.

Hall Co.

GAINSVILLE, regional creeks emptying into the Chattahoochee R., many prospects and placers, productive.

Haralson Co.

DRAKETOWN, access to a mineralized belt that occurs in a diagonal belt entirely across the co., from the NE cor. to the SW boundary with Alabama, very many prospects and mines—lode gold (with placer gold in all regional watercourse gravels).

Lincoln Co.

AREA, N tip of co., NE of DANBURG in Wilkes Co., many old mines —lode gold.

AMITY, NW, in the SW part of co., many old mines and placer prospects along regional creeks—lode and placer gold.

LINCOLNTON (a gold belt extends from the South Carolina boundary (Clark Hill Reservoir) SW across the lower portion of co., along the boundary between adjoining Wilkes and McDuffie cos., into the NE tip of Warren Co.): (1) all regional creek and bench gravel deposits—placer gold; (2) many regional old mines—lode gold.

Lumpkin Co.

First among the gold-producing counties of Georgia, Lumpkin County produced between 400,000 and 500,000 ozs. of placer and lode gold, 1828–1959, with production continuing to the present. The earliest auriferous deposits were stream placers. Later, the free-milling gold found in the saprolite (decomposed, weathered bedrock mantle) was obtained by hydraulic methods.

AREA, all co. stream gravels, benches, terraces, and hillsides—placer gold (exceeding $40,000,000).

AURARIA (oldest and once most prosperous settlement in co. and one of the wildest from 1829 to 1835): (1) E, area gravel deposits along Yahoola Cr., rich placers; (2) N ⅞ mi., the Whim Hill saprolite deposits, many gopher holes down shoots of rich ore—free-milling gold; (3) W: (a) near the Etowah R., the Battle Branch Mine—free gold, with galena; (b) all area cr. gravels (pay fee to landowners)—placer gold; (4) S 2 mi., along Baggs Branch, stream gravels—placer gold; (5) S 3 mi., at Turkey Hill (original gold discovery site), many area mines and prospects—free gold; (6) SE, in bottom of the Barlow Cut: (a) the Barlow Mine (largest of the old saprolite workings)—free-milling gold, gold sulfides; (b) SE, the Bunker Hill Mine—saprolite gold; (7) W, on McClusky Cr., close to the Etowah R., the Topabri Mine (a major hydraulic operation during the 1930s)—placer gold.

DAHLONEGA: (1) E: (a) on Yahoola Cr., the Lockhart Mine (initially a hydraulic saprolitic operation with extensive gophering for rich pockets; later an underground mine)—placer and lode gold; (b) adjacent to the Chestatee R., the old Boly Field Property (deep shaft and wide surrounding saprolite area)—free-milling gold; (2) S, on the Chestatee R., the Briar Patch Placers; (3) SE, at jct. of Long Branch with the Chestatee R., the Long Branch Mine—placer gold; (4) on the N end of Findley Ridge, the Findley Mine (one of richest early district producers, $200,000 before the Civil War)—lode gold; (5) NE several mi., the McDonald Mine in saprolite, high-grade—free-milling gold, with galena.

McDuffie Co.

THOMSON, NW 12 mi., adjacent to the Little R.: (1) many area workings (placers, gopherings, saprolite operations), mostly by the regional farmers on their own lands; (2) the Fluker Property, many area mines, which produced well in the past, e.g., the Columbia, Park, Hamilton, and Seminole (McGruder) mines—free gold.

Paulding Co.

CARTERSVILLE-DALLAS (about halfway between), the Burnt Ridge district: (1) many area placer and saprolite workings—placer and free-milling gold; (2) the old Twillery Mine, shallow trenches—saprolite gold.

YORKVILLE, E 2½ mi., the Yorkville Mine—free-milling gold.

Union Co.

BLAIRSVILLE: (1) S 4–5 mi., area prospects and gravels—placer gold; (2) NW 5 mi., in Teece Valley, 1 mi. below hwy. crossing, gravel deposits—placer gold.

White Co.

The first gold found in White County was picked up by black slaves in 1828, followed by so many placer discoveries that in 1831 the state legislature held a lottery for 35,000 plots of 40 acres each, with drawings made by some 130,000 prospectors; total production of gold between

1829 and 1959 is estimated between 35,000 and 52,000 ozs. Gold deposits lie in a mineral belt considered to be an extension of the Dahlonega Gold Belt.

CLEVELAND: (1) numerous area mines and prospects (including those around nearby HELEN)—free gold; (2) N 4 mi.: (a) the Dunbar Mine, major producer—lode gold; (b) on the Little Tesnatee Cr., the Cox Bottoms Mine (a dragline operation)—placer gold; (3) NW 5 mi., on Land Lot. 47, Fourth District: (a) the Poland and Beach Mine ("Sprague Vein")—saprolite gold; (b) W, the adjoining lots, numerous workings—saprolite gold; (c) the adjoining Blake Property—saprolite gold.

NACOOCHE, at jct. of the Nacooche and Chattahoochee rivers, the Nacoochee district, in E-central part of co.: (1) along Bean Cr.: (a) all cr. and tributary gravel deposits—placer gold; (b) the Bean Cr. Mine, a major placer operation; (2) S 3 mi. and just W of the CLEVELAND-HELEN hwy.: (a) many area placers—source of sizable nuggets, abundant; (b) along Dukes Cr., the Hudson Mine, operated by hydraulic equipment—placer gold; (3) the Horshaw Mine, good early producer—placer gold.

Wilkes Co.

WASHINGTON, E on U.S. 378 to co. line, turn onto first right fork rd., then take a left fork to crest of a hill, the Magruder Mica Mine—by-product gold, with copper.

IDAHO

The earliest known discovery of gold in Idaho was on a sandbar in the Pend Oreille River in 1852. Then in 1860 Captain E. D. Pierce found gold near what became the town of Pierce in Clearwater County, with other rich placer deposits soon being mined in adjacent areas. Placers were the major source of gold prior to 1900, since in the preceding 36 years an estimated 1,000,000 ozs. of gold had been produced. As the placer deposits were commercially worked out, lode occurrences in the huge Idaho batholith were put into production, so that by 1965 a total of around 9,300,000 ozs. of gold, including the original placer output, placed Idaho ninth among America's gold-producing states.

For further information, write: Director, State of Idaho Bureau of Mines and Geology, Moscow, Idaho 83843.

Ada Co.

BOISE, E. 10 mi., near the Elmore Co. line, the Black Hornet district (tot. prod., 1880–1959, at least 21,431 gold ozs.), area mines—lode gold.

Adams Co.

CUPRUM, very many regional old mines scattered throughout co. (mainly copper)—by-product gold.

Bingham Co.

BLACKFOOT, along the Snake R. for a considerable distance, in sand and gravel river deposits (tot. prod., 1885 to mid-1930s, of 24,242 ozs.) —placer gold.

Blaine Co.

Although rich silver mines were found in Boise Basin in 1864, carrying minor gold values, the most active production of gold did not begin until after 1942. Total gold production through 1959 was 212,638 ozs.

BELLEVUE: (1) SW cor. of co., the Wood district: (a) along the Wood R. and its tributaries, in T. 1 S, R. 17 and 18 E, placers; (b) area copper-lead mines—minor by-product gold; (2) the Camas, Croesus, and Tip Top mines, all base-metal—by-product gold.

HAILEY, SW 5–15 mi., the Camas (Hailey, Mineral Hill) district, discovered in 1865, tot. prod. through 1959 of 102,000 gold ozs.: (1) area base-metal mines—by-product gold; (2) SW, in T. 1 N, R. 16 and 17 E, the Camas 2 Mine (major producer, closed in 1898 after $1,250,000) —by-product gold; (3) old camp of Mineral Hill, area mines produced 14,180 gold ozs., 1902–59—by-product gold.

KETCHUM: (1) in T. 4 N, R. 17, 18, and 19 E (between lat. 43°35' and 43°50' N, long. 114°10' and 114°30' W), the Warm Springs district, tot. prod., 1864–1959, of 76,639 gold ozs.: (a) area lead-silver mines, numerous—by-product gold; (b) the Triumph Mine (reopened in 1927 to become a major producer and closed in 1957)—by-product gold; (2) the Sawtooth district, area silver mines—lode gold.

Boise Co.

Rich placers were discovered in the 300-sq.-mi. Boise Basin 25 miles northeast of Boise, with many lode mines found shortly afterward at the heads of the regional streams. The placer gravels are still productive, and many amateur gold hunters seasonally prospect them. Between 1863 and 1959 a total of 2,891,530 gold ozs. were produced.

AREA: (1) the Banner silver mines—by-product gold; (2) in Deadwood Gulch—placer gold, garnets; (3) the Willow Cr. district, at (a) Checkmate, (b) Gold Hill, and (c) many other co.-area base-metal mines—by-product gold.

CENTERVILLE (the Boise Basin district, including IDAHO CITY, q.v., Moore Cr., Gambrinus—all in the middle of the basin—with alluvial gravel deposits covering much of the region, tot. prod. through 1959 of 2,300,000 gold ozs., mostly from placers): (1) along Grimes Cr. on way toward PLACERVILLE, and (2) along the Boise R. near Twin Springs, richly productive placers; (3) all regional watercourse gravels, benches, terraces—placer gold; (4) Gambrinus: (a) many area lode mines—by-product gold; (b) the Illinois and Gambrinus mines, big producers—by-product gold.

GRIMES PASS, PIONEERVILLE, area between lat. 44°00' N, long. 115° 50' W, the Pioneerville Summit Flat, Grimes Pass, tot. prod., 1895–1959, of about 25,000 gold ozs.: (1) the Golden Age and Mammoth mines—lode gold, with some lead and silver; (2) the Comeback Mine, primary producer of electrum—50 per cent gold, 50 per cent silver; (3) the Grimes Pass area, numerous mines—lode gold.

IDAHO CITY: (1) area alluvial gravel deposits (stream and terrace), many productive placers; (2) vicinity of town: (a)along Mine cr. and (b) near Horseshoe Bend of the Payette R., rich placers; (c) on upper level and older gravels along ridges between regional watercourses—placer gold.

QUARTZBURG (district, including Gold Hill, Granite, and PLACERVILLE, in T. 7 N, R. 4 E, tot. prod. through 1932 of about 400,000 gold ozs.): (1) the Gold Mine, 1863–1938—lode gold; (2) the Mountain Chief and Belshazzar mines, big producers—lode gold.

Bonneville Co.

BONE, SE, to Mt. Pisgah in the Caribou Range, in T. 4 S, R. 44 E, the Mt. Pisgah district, tot. prod. of 16,600 ozs. of mostly placer gold: (1) area placer mines; and (2) area lode mines (free gold in pyrite).

Camas Co.

CARRIETOWN: (1) along Little Smoky Cr., in T. 2 and 3 N, R. 14 and 15 E, rich placers; (2) numerous mines in T. 4 and 5 N, R. 13 and 14 E—lode gold. The town, not on most maps, lies in the E part of the co. in the Big and Little Smoky-Rosetta district embracing about 150 sq. mi., with many mines, most of which were abandoned prior to 1900; tot. prod. of district was about 10,000 gold ozs.

Cassia, Jerome, and Minidoka Cos.

AREA, very many gravel bars and bench deposits along the Snake R. are credited with a tot. prod. of 22,000 ozs. (Cassia Co.), 1,736 ozs. (Jerome Co.), and 133 ozs. (Minidoka Co.)—placer gold.

BURLEY (Cassia Co.), area placer workings along the Snake R.: (1) in T. 9 and 10 S, R. 24 and 25 E, and (2) lead-silver mines in T. 15 and 16 S, R. 21 E, on old dumps—gold showings.

Clearwater Co.

Placer deposits in Clearwater County lie mainly in stream channels and on terraces as much as 500 feet above existing streams. A total of 30,137 gold ozs. has been recorded through 1959.

AREA: (1) along Kelly and Morse crs. in T. 39 N, R. 10 and 11 E, numerous placers; (2) N fork of the Clearwater R. and its tributaries in T. 37, 38, and 39 N, R. 1–4 E, rich placers.

PIERCE, in T. 36 and 37 N, R. 4 and 5 E, the Pierce district; (1) many area placer operations; (2) along Orofino Cr., between OROFINO and PIERCE (T. 36 and 37 N, R. 1–4 E), rich placers; (3) the Wild Rose Mine, most productive around 1905—lode and by-product gold.

Custer Co.

Between 1881 and 1959 a total production of 329,586 ozs. of gold came largely from the western part of the county.

AREA: (1) Bayhorse, in T. 12 and 13 N, R. 18 E, area copper-zinc mines—by-product gold; (2) NW part of co., between lat. 44°20′ and 44°30′ N, long. 114°40′ and 114°45′ W, the Yankee fork (of the Salmon R.) district, tot. prod., 1875–1959, of 266,600 gold ozs.: (a) the General Custer and Bonanza mines—lode gold; (b) the Lucky Boy Mine, reopened in 1937—lode gold; (c) along the Yankee fork from mouth of Jordan Cr. almost to the mouth of the Yankee fork, most productive placers in district; (d) along Jordan Cr., stream, bench, and terrace gravel deposits—placer gold; (3) the Nicholia district, rich early-day lead-silver mines—minor by-product gold; (4) the Loon Cr. drainage area (lat. 44° 32′ and 44°38′ N, long. 114°45′ and 114°52′ W), the Loon Cr. district, tot. prod., 1869–1959, of at least 40,000 gold ozs.: (a) many area mines, especially the Lost Packer Mine—by-product and lode gold; (b) along Loon Cr., near ghost town of Casto, many placers.

CLAYTON: (1) area of Squaw and Thompson crs., in T. 11 and 12 N, R. 16 E, numerous mines—lode gold; (2) in T. 11 and 12 N, R. 17 E, several lead mines—by-product gold.

MACKAY, area copper mines—gold in pyrite.

STANLEY: (1) in the Stanley Basin, all regional stream gravels—placer gold; (2) in T. 10 and 11 N, R. 12 and 13 E, all stream, bench, terrace gravels—placer gold; (3) along the Salmon R., especially between Robinson Bar and CLAYTON, q.v., many rich placers; (4) along Stanley Cr., the Willis placers, productive.

Elmore Co.

Between 1862 and 1959 a total of 441,696 gold ozs. were mined in Elmore County, largely from districts clustered above the granitic rocks of the Idaho batholith in the northeastern part of the county, where both placer deposits and lode veins contributed gold.

AREA: (1) along the Boise R.: (a) the middle fork, in T. 6 N, R. 10 E; (b) the S fork, in T. 2 N, R. 10 E; and (2) at Twin Springs, in T.

3 and 4 N, R. 5–7 E—all placer gold.

MAYFIELD, NW, near the Arrowrock Dam about 15 mi. SE of BOISE in Ada Co., in T. 2 and 3 N, R. 5 E, the Neal district (tot. prod., around $2,000,000, mostly prior to 1911): (1) area small placers, intermittently worked to the present on all regional crs.; (2) numerous area mines (major sources)—lode gold.

MOUNTAIN HOME (access point to NE part of co., where the principal mining districts lie 25–50 mi. NE): (1) ATLANTA (district, including Hardscrabble, Middle Boise, and Yuba, in T. 5 N, R. 11 and 12 E, 65–70 mi. E of BOISE, tot. prod., about 385,000 gold ozs.): (a) area cr. beds and terraces, intermittently worked to the present—placer gold; (b) numerous mines, especially on Atlanta Hill, important—lode gold; (2) FEATHERVILLE (district, on S fork of the Boise R., in T. 3 N, R. 10 E, tot. prod., 32,777 gold ozs.), all area streams and benches—placer gold; (3) PINE (old townsite, where the author's father and mother were married in 1910, now under the waters of the Anderson Reservoir; in T. 1 and 2 N, R. 9 and 10 E, the Pine Grove district): (a) many area small placers on regional crs. and terraces; (b) numerous mines, especially the Franklin (producer of $750,000 before being closed in 1917)—lode gold; (4) ROCKY BAR (district, in T. 4 N, R. 10 E): (a) Bear Cr. and tributaries, bed and terrace gravels—placer gold (about $2,000,000 to 1882); (b) area mines, especially the Elmore and Pittsburg, major producers of about $3,500,000 to 1939—lode gold.

Gem Co.

The 20,000 gold ozs. attributed to Gem County came almost entirely from the Westview (Pear-Horseshoe Bend) district, which sprawls across the Boise Co. line about 18 mi. NNW of BOISE.

PEARL, extending NE to co. line and including parts of T. 6 N, R. 1 E, many old mines, especially the Red Warrior—lode gold.

Idaho Co.

Idaho County is the state's largest gold-producing county, with 2,176,550 gold ozs. coming from placers and lode mines between 1862 and 1942.

AREA: (1) along Lolo, Mussell, Eldorado crs., many once-rich placer grounds; (2) Salmon R. bars, benches, and terrace gravels, rich placers.

BURGSDORF-WARREN: (1) entire reach of the Salmon R., especially in T. 22 N, R. 4–6 E, productive placers; (2) S part of co., in T. 20 and 24 N, R. 4 and 8 E, the Warren-Marshal (Resort) district, tot. prod. through 1959 of 906,500 gold ozs.: (a) Warren Meadows (gold discovered in 1862), big placers worked by bucket dredges; (b) all area bench gravels and high-meadow deposits—placer gold; (c) area mines (tot. prod. 1,765 ozs., 1929–35)—lode gold.

DIXIE (20 mi. S of ELK CITY) district, in T. 25 and 26 N, R. 8 E (tot. prod., between 40,000 and 50,000 gold ozs.): (1) area mines, big producers—lode gold; (2) N side of the Salmon R., along Sheep and Crooked crs., extensive placers.

ELK CITY (district, in parts of T. 29 and 30 N, R. 8 E, tot. prod. between 550,000 and 800,000 gold ozs.): (1) along the American R. and S fork of the Clearwater R.: (a) many placer workings; (b) above the placers, many mines—lode gold; (2) the Buster Mine (discovered in 1870 and largest mine in district after 1902)—lode gold; (3) regional high-level Tertiary terrace gravels, rich placers.

GRANGEVILLE, out about 42 mi., in T. 25 N, R. 3 and 4 E, the French Creek-Florence district (one of most productive in Idaho, tot. prod. through 1959 estimated at 1,000,000 gold ozs.): (1) all regional gulches and streambeds once swarmed with placer miners and, later, by Chinese —placer gold; (2) above the placers, several mines—minor lode-gold output.

KOOSKIA: (1) S, along the S fork of the Clearwater R. (including areas around STITES and HARPSTER), rich placers; (2) E, along the Middle Fork of the Clearwater R.: (a) area rich placers; (b) along Maggie Cr., numerous productive placers.

LUCILLE-RIGGINS: (1) along the Salmon R. and its tributaries, many regional rich placers; (2) between RIGGINS and FREEDOM, in T. 24–28 N, R. 1 E, the Simpson Camp, Howard-Riggins district (along the Salmon R., tot. prod. of 9,578 gold ozs., 1860s–1959), numerous placers intermittently worked to the present: (a) streambed and bench gravels, and (b) terrace gravels several hundred feet above existing streams— placer gold.

OROGRANDE (12 mi. SW of ELK CITY; district, still producing gold but most productive between 1902 and 1959, tot. prod., 32,000 gold ozs.): (1) local crs., gulches, and benches, small placers; (2) the Orogrande-Frisco and the Gnome mines, largest producers—lode gold; (3) SW 5–8 mi., the Buffalo Hump district, tot. prod., 1898–1941, of 27,000 gold ozs.: (a) many area mines around several ghost camps—lode gold; (b) the Big Buffalo Property, major mine—lode gold; (4) immediately N of Buffalo Hump Mt. and W of the Elk City district, the Tenmile district (tot. prod., about 147,000 ozs., mostly placer gold): (a) Newsome Basin, in gravels of Newsome Cr. and its tributaries, rich placers discovered in 1861; (b) many regional mines developed after 1888 (tot. prod., about 18,400 ozs. through 1932)—lode gold.

WHITE BIRD, regional gravel deposits along the Salmon and Wind rivers to 6 mi. N of town, rich placers.

Kootenai Co.

COEUR D'ALENE (district), many old mines going back to 1860: (1) along regional streams, placer workings with panning gold still obtainable; (2) area lode veins—minor by-product gold.

Latah Co.

AREA, NE part of co., in T. 42 N, R. 1 and 2 W, the Hoodoo district (about 28 sq. mi. in the Hoodoo Mts., tot. prod., about 17,165 ozs., 1904–55): (1) along Palouse R. and Poorman Cr., early placers worked by Chinese after original miners left; (2) S fork of the Palouse R., extensive placers dredged in 1930s and ending in 1942.

Lemhi Co.

Prospectors discovered gold in the N-central part of Lemhi County in 1866 and founded Leesburg (long since a ghost town), which served as a base for other discoveries at the abandoned camps of Gibbonsville, Moose Creek, Bohannon Bar, and Yellow Jacket. The total gold production through 1959 is uncertain, but it is somewhere between 570,725 and 720,000 ozs.

AREA, on W slope of the Lemhi Range, the Bluebird district: (a) area mines (lead, copper)—by-product gold; (b) the Ima Mine, in sec. 23, (14N–23E)—by-product gold.

CARMEN, in NE part of co., along flanks of the Beaverhead Mts.: (1) contiguous small camps of Carmen Cr., Eldorado, Pratt Cr., and Sandy Cr. (tot. prod., about 24,500 gold ozs.): (a) many area stream and bench placers, major producers, and (b) a few lode mines—lode and by-product gold; (2) Bohannon Bar, productive placers worked by Chinese in 1870s; (3) the Oro Cache Mine on Carmen Cr.—lode gold; (4) on Pratt Cr., the Goldstone Mine—lode gold.

COBALT (35 mi. SW of SALMON, in central part of co.), the Blackbird district (first worked for gold in 1893, then primarily for cobalt and copper): (1) area cobalt mines in the Blackbird block (5 mi. long, 2 mi. wide)—gold traces only; (2) the Calera Mine (copper-cobalt, but major gold by-product producer); (3) in T. 20 and 21 N, R. 21–23 E, many regional copper mines—minor gold by-product; (4) SW 17–27 mi. via dirt rds., in W-central part of co., about lat. 44°58′ N, long. 114°31′ W, the Yellow Jacket district (gold the chief metal, tot. prod. about 25,000 ozs., 1868–1959): (a) regional mines in the Yellow Jacket Formation— lode gold; (b) the Yellow Jacket Mine, major gold producer.

GIBSONVILLE (district at N apex of co., tot. prod., about $2,000,000 in gold): (1) area extensive placers; (2) area lode mines, most productive being the A. D. and M. Mine (destroyed by fire in 1907); (3) in T. 25 and 26 N, R. 21 E, numerous area mines and prospects—gold showings; (4) along the Salmon R. and all tributary crs., sand and gravel deposits (including benches and old channels)—placer gold.

GILMORE, in parts of T. 13 and 14 N, R. 26 and 27 E, the Texas district (tot. prod. 21,745 ozs., 1903–59): (1) the Pittsburg-Idaho Mine (lead-silver)—by-product gold; (2) the Martha Vein—only lode gold.

LEADORE, in T. 13 N, R. 27 E, area lead mines—minor by-product gold.

SALMON: (1) E 6 mi., in T. 21 and 22 N, R. 23 E, the Kirtley Cr. district (once most productive placers in Idaho): (a) along Kirtley Cr., area placers produced 24,300 gold ozs.; (b) at head of cr., lode quartz mines—gold; (2) W 10 mi., near lat. 45°14′, long. 114°10′ W, the Mackinaw district (tot. prod., about 271,200 gold ozs. through 1959):

(a) Wards Gulch, along Napias Cr., large-scale placers discovered in 1866; (b) Moose and Beaver crs., and in tributaries of these and of Napias Cr., productive placers; (3) NW 28 mi. by dirt rd., the Leesburg district (the ghost town boasted 7,000 residents in 1870), area mines, once very productive—lode gold.

SHOUP (in NW part of co.), in T. 23 and 24 N, R. 17–19 E, the Mineral Hill and Indian Cr. district (primarily silver and base-metal mines, but gold prod., 1932–59, was 21,937 ozs.): (1) S, along Panther Cr. and its tributaries, productive placers; (2) Mineral Hill: (a) numerous area mines developed in 1880s—lode gold; (b) Kentuck and Grunter mines, big producers—lode gold; (3) at Indian Cr., the Kittie Burton and Ulysses mines, 1901–10—lode gold.

Nez Perce Co.

LEWISTON, all regional streams, benches, and terraces, especially along the Clearwater R. upstream from town for about 100 mi.—placer gold.

Owyhee Co.

Although placers along the Snake River as it crosses the northeastern part of Owyhee County produced some gold, most of the total county production of 1,103,545 gold ozs. between 1863 and 1959 came from the Silver City district in the northwestern part of the county.

BRUNEAU, far to the SW via dirt rds., the abandoned camps of Ruby City, Fairview, Booneville, and Wagontown, many area mines—lode gold.

GRAND VIEW, all bars, benches, and tributary gravels along the Snake R., productive placers.

SILVER CITY (district, original county seat of Owyhee Co. until it was moved to Murphy about 1935), in T. 3–5 S, R. 1–4 W, the abandoned camps of De Lamar, Flint, Florida Mountain-War Eagle, and Silver City, tot. prod. of lode gold, over 1,000,000 ozs.): (1) area great mines, e.g., the Black Jack, Trade Dollar, War Eagle, and Florida—lode gold, silver; (2) NW a short distance, the De Lamar Mine (tunnels penetrate clear through Florida Mt. into the Silver City mines)—lode gold; (3) the

Poorman and Orofino mines, discovered in 1865, in quartz-cemented breccias—lode gold; (4) area of Jordan Cr. and its tributaries, numerous placer deposits (first gold discoveries in 1863, traced to origins in the lode veins of War Eagle Mt.).

Power Co.

AMERICAN FALLS, along the Snake R., numerous placer deposits worked from the 1880s into the 1950s. The estimated total production of 18,485 ozs. probably represents only a small proportion of the actual recovered gold. Much gold may still be obtained by amateur gold hunters by panning and sluicing.

Shoshone Co.

Between 1881 and 1959 Shoshone County produced 434,201 gold ozs., almost entirely from the 500-sq.-mi. Coeur d'Alene region containing eight districts.

CALDER, W, in the St. Joe (Benewah Co.) district: (1) many area mines in the Shoshone part of the district—by-product gold; (2) along the St. Joe R., in T. 45 N, R. 3, 4, and 9 E, many mines—lode gold.

KINGSTON, area of the Pine Creek district in T. 48 N, R. 1 and 2 E, numerous lead-silver mines—gold, in pyrite.

MULLAN-WALLACE-KELLOGG (including the districts of Beaver, Eagle, Evolution, Hunter [Mullan], Lelande [Burke], Placer Center [Wallace], Summit, and Yreka): (1) at KELLOGG, the great Bunker Hill and Sullivan mines (now combined as one of the world's largest lead-silver producers)—by-product gold; (2) the Mammoth, Tiger, Morning, Poorman, and Granite mines—substantial lode and by-product gold; (3) many other area mines, e.g., the Sunshine Mine at WALLACE, most hidden back in timbered mts.—lode and by-product gold; (4) in T. 47 and 48 N, R. 4 and 5 E, area stream and bench gravels—placer gold; (5) in T. 47 N, R. 3 and 4 E, the "Silver Belt" (an area extending 6½ mi. E–W by 2½ mi. N–S), very many lead-silver mines—substantial gold by-product.

MURRAY (district, original county seat, now nearly a ghost town): (1) area right in town, lately especially behind Kris Kristofferson's tavern, in surface soil—colors, nuggets to 6 ozs. (one found in 1974); (2) along Prichard and Eagle crs., very rich placer grounds of the 1880s, revived in 1930s and intermittently worked to the present; (3) S fork of the Coeur d'Alene R., many rich lead-silver lode mines (peak production in 1911)—important by-product gold; (4) all regional watercourse and, especially, bench, gravel deposits, productive—placer gold.

PRICHARD, area cr. beds and bench gravels—placer gold.

Valley Co.

With most of its total gold production credited to pre-existing counties from which Valley County was created in 1917, the records do show 324,460 gold ozs. produced between 1917 and 1958, largely from the northeastern part of the county, in a triangular area with EDWARDSBURG at the apex, and the Yellow Pine and Thunder Mountain districts at the southwest and southeast corners, respectively. Each side of the triangle is about 15 miles long.

BIG CR. (extreme N-central part of co.): (1) area stream and bench gravels in T. 20 and 21 N, R. 9 and 10 E, productive placers; (2) NW, to Ramey Ridge (district, extending into S part of Idaho Co.), many area copper mines—by-product gold; (3) E, into NE part of co., in T. 19 N, R. 11 E, the Thunder Mt. district, ghost camps of Belleco and Roosevelt (business center during height of boom after 1896), the Dewey and Sunnyside mines—lode gold (17,500 ozs. through 1959).

EDWARDSBURG (ghost town S of BIG CREEK, in T. 20 N, R. 8 E), area lode mines—lode and by-product gold.

STIBNITE, area silver-antimony mines—minor gold by-product.

YELLOW PINE (district, at lat. 44°50' N, between long. 115°00' and 115°30' W, tot. prod. through 1959 of 309,734 gold ozs.): (1) in T. 18 N, R. 8 E, productive placers; (2) in T. 19 N, R. 8 and 9 E, area mines (antimony)—important gold by-product; (3) the Meadow Creek Mine (gold-antimony lode) and Yellow Pine Mine (major district producer of gold, 1937–52). These two mines produced 101,437 gold ozs. through 1945. During World War II, the Yellow Pine Mine was in continuous operation as the largest producer of strategic tungsten in America. (4) in T. 18 and 19 N, R. 11 E, near Thunder Mt., along Monumental Cr.,

numerous hard-rock mines—lode gold; (5) SW about 40 mi., to Warm
Lake in T. 15 N, R. 7 E, regional streambeds and bench gravels—placer
gold.

INDIANA

Gold occurrences in Indiana are the result of Pleistocene glacial
movements, which deposited gold in varying amounts in glacial moraines
from unknown sources in Canada. No natural veins bearing gold or
natural placers occur in Indiana. Nevertheless, gold panning of many
glacial morain deposits and the gravels of streams annually produce small
amounts of gold.

For more information, write: Department of Geology, Indiana University, Bloomington, Indiana 47401.

Brown Co.

AREA: (1) all glacial drift deposits throughout co.—placer gold, with
an occasional diamond; (2) along Greenhorn Cr., auriferous sands.

Morgan Co.

AREA: (1) all regional stream and cr. gravels and all glacial moraines
—panning gold; (2) along Highland Cr. (7 mi. NNW of MARTINSVILLE
in Johnson Co.), area gravel deposits and stream sandbars—panning
gold, with diamonds; (3) along Gold Cr., in gravel bars—panning gold,
with diamonds, zircons, sapphires, and corundum.

WILBUR: (1) all co. glacial morains, cf. area; (2) along Greenhorn Cr.:
(a) cr. gravels and (b) all surrounding tributary watercourse gravel and
sandbars—panning gold; (3) along Highland Cr., numerous glacial drift-
bars—panning gold.

MAINE

Although gold mining has not been important in Maine's extensive
mineral industry, the yellow metal has been reported in streambed and
bench gravels in many parts of the state. Since no broad systematic study

has ever been made of auriferous areas, all gold hunters are advised to investigate, at least briefly, any stream or creek in which the rate of flow and streambed configuration appear attractive for the concentration of gold particles. The most promising streams are those adjacent to the valleys of the Chaudière and Rivière-du-Loup in the eastern townships of Quebec, where considerable gold has been found in the past.

There are no specific laws relating to prospecting on private lands, but all interested gold hunters should at least obtain permission from landowners as a courtesy to enter their property to prospect for gold. Topographic maps covering the auriferous areas described below may be obtained at regional sporting goods stores or bookstores, as well as from the U.S. Geological Survey at 1200 South Ends St., Arlington, Virginia 22220 upon prepayment of 75 cents.

For additional helpful information, write: State Geologist, Bureau of Geology, Department of Conservation, Augusta, Maine 04330.

Cumberland Co.

AREA, the Piscataqua Mine—lode gold, with copper, nickel.

Franklin Co.

BYRON: (1) the E branch of the Swift R., in area near Tumbledown Mt. via Rte. 17, in low-water gravel bars—abundant placer gold; (2) all area watercourse gravels—placer gold. Local area residents have been panning or sluicing for gold for generations. Gravel caught in potholes and between upended ledges of rock are especially productive.

CHAIN OF PONDS, along Gold Brook, numerous placer workings.

DALLAS, RANGELEY TWPS., along Nile Brook, numerous placers.

KIBBY TWP.: (1) along Gold Brook, numerous placers; (2) along Kibby Stream, area placers.

MADRID-NEW SHARON TWPS., along the Sandy R. between towns, numerous placer localities.

Hancock Co.

BLUE HILL: (1) the Atlantic Mine (copper-silver)—gold in pyrite; (2) the Blue Hill Mine—gold in copper minerals and pyrite; (3) the Douglas

Mine—by-product gold in copper minerals.

BROOKLIN, the Brooklin Mine—gold, with silver.

CASTINE, the Castine (Castine Head) Mine, in copper-lead-silver ores —by-product gold.

Knox Co.

ROCKPORT, the Porterfield Mine—by-product gold (from copper, nickel, silver ores).

SOUTH THOMASTON, the Owl's Head Mine—gold, with silver.

Oxford Co.

BOWMAN (twp.), along Gold Brook, numerous placers.

BYRON (twp.), along the East Branch of the Swift R., productive placers.

HIRAM, area of Cutler Mt., the Frenchman's Gold Mine (mainly feldspar and mica)—gold traces in quartz.

RUMFORD (twp.), along Black Mt. Brook, area placers.

WOODSTOCK, an area mine—by-product gold.

Somerset Co.

AREA, along Gold Brook: (a) in Chase Stream Tract Twp., placers; (b) in T5–R6 and Appleton twps., numerous placers; (2) along the South Branch of the Penobscot R., especially in Sandy Bay, Bald Mt., and Prentiss twps.—placer gold.

ST. ALBANS, the St. Albans Mine, primarily lead—gold traces.

Waldo Co.

KNOX, the Stone Mine—gold, with silver.

PROSPECT, the Fort Knox Mine—by-product gold, from copper-lead-silver ores.

Washington Co.

BAILEYVILLE (twp.), along the St. Croix R., gravel and sandbars— placer gold.

CHERRYFIELD, the Cherryfield Mine (mainly lead-silver-zinc)—by-product gold.

MARYLAND

Following the Civil War, Maryland was extensively prospected for gold, and gold-quartz showings were found throughout much of the Piedmont Plateau in the central part of the state. Although gold mining never proved profitable, one can still pan gold from the regional Piedmont stream and creek gravels. Since all land is privately owned, the gold hunter must obtain permission to prospect on any land not his own.

For more information, write: Department of Geology, Mines, and Water Resources, The Johns Hopkins University, Baltimore, Maryland 21233.

Baltimore Co.

CATONSVILLE, area old prospects in quartz veins—gold showings.

Frederick Co.

LIBERTYTOWN, area old mines and dumps—gold traces.

Howard Co.

SIMPSONVILLE, the Maryland Mine (once richest gold property in Maryland)—gold in quartz, with silver and pyrite.

WOODBINE, several area mines (make inquiry)—by-product gold.

Montgomery Co.

GREAT FALLS, the Great Falls Mine—gold, with tetradymite.

MICHIGAN

Michigan's first state geologist, Dr. Douglass Houghton, is thought to have found the first gold in an unrecorded streambed in the Upper Peninsula in the 1840s but died before he could name the location. Since then, some 29,000 gold ozs. have come from the state's auriferous localities, but the list of gold prospects is as long as the list of failures to mine gold profitably.

For more information, write: State Geologist, Geological Survey Division, Department of Natural Resources, Stevens T. Mason Building, Lansing, Michigan 48926.

Allegan Co.

ALLEGAN, all area stream gravels, once panned—placer gold.

Antrim Co.

AREA, all along the Antrim R., gravel deposits, placers—sizable gold nuggets occasionally.

Charlevoix Co.

AREA, along the Boyne R., low-water gravels—placer gold.

Emmet Co.

AREA, along the Little Traverse, tributary cr. gravels—pannable gold.

Grand Traverse Co.

WALTON, area cr. low-water gravel bars—placer showings.

Ionia Co.

AREA, along the Mapre R., numerous placer gold deposits.
LYONS, below, along the Grand R., area placer deposits.

Kalkaska Co.

KALKASKA, gravel bars along the Rapid R.—placer gold.

Kent Co.

AREA, along Ada Cr., gravel bars—placer gold.
LOWELL: (1) area glacial moraines—pannable gold; (2) along the Grand R., localized mineral-rich gravel deposits—placer gold.

Leelanau Co.

SOLON: (1) area stream gravels—placer gold; (2) N, along shore and stream tributaries of Leelanau L., gravels—placer gold.

Manistee Co.

AREA: (1) all along the Little Sable R., many placer deposits; (2) low-water gravel bars of the Manistee R.—placer gold.

Marquette Co.

ISHPEMING: (1) all regional stream gravels and moraine debris—placer gold; (2) area prospects in T. 48 N, R. 27 W—gold showings; (3) WNW 3–5 mi., on N side of the Marquette Range: (a) in NW¼ sec. 29, the Ropes Mine (only major gold mine in Michigan, tot. prod., about $625,000)—free-milling gold, with 4 per cent silver; (b) 2½ mi. W, in sec. 35, (48N–28W), the Michigan Gold Mine (producer of some of the finest museum specimens of native gold, ore so rich that "high graders" made a practice of stealing it during the peak production year of 1890); (c) just W of the Michigan, the Gold Lake Mine—fine native gold specimens; (d) in NW¼ sec. 35 (just E of the Michigan), the Superior Gold Mining Co. Mine, little production but excellent specimens—native gold; (e) in SW¼ sec. 25, the Peninsula Mining Co. Mine—gold in granite; (f) the Grummett, Swains, Mocklers, Grayling, and Giant mines (all prominent during the 1890s)—lode and native gold; (2) 2 mi. N and E of the Ropes Mine, in iron-ore formations—pannable gold; (4) N 8 mi., the Dead R. district in the Dead R. Valley (extending N to L. Superior and for several mi. W): (a) in the S range, many productive mines—by-product gold; (b) in the N range (spur), a few area prospects —gold showings; (c) in sec. 35, (49N–27W), the Fire Centre Mining Co. mines—free-milling gold.

Montcalm Co.

GREENVILLE, regional stream gravels, once panned—placer gold.
HOWARD CITY, cf. GREENVILLE.

Newaygo Co.

NEWAYGO-OXBOW, all gravel deposits of the Muskegon R. running diagonally across co.—placer gold.

Oakland Co.

BIRMINGHAM, area cr. and stream gravel beds—placer gold.

Oceana Co.

MONTAGUE to HESPERIA, all along the White R., local placer deposits.
WHITEHALL, area low-water stream gravels—placer gold.

Ontonagon Co.

AREA, along the Flat R., many placer locations.
VICTORIA, the Victoria Copper Mine—gold showings, occasional nuggets of native gold.

Ottawa Co.

GRAND HAVEN, area stream gravels once panned—placer gold.

St. Joseph Co.

BURR OAK, area stream gravels—placer gold (but gold showings may be pyrites, i.e., "fool's gold").
MARCELLUS, area stream gravels—placer gold.

Wexford Co.

AREA: (1) West Summit, local streambeds and glacial outwash gravel deposits—pannable gold; (2) NW part of co., along the Manistee R., local low-water gravel bars—placer gold.

MINNESOTA

All reported gold discoveries in Minnesota have been minor occurrences. Stream gravel deposits derived from glacial drift areas do show small amounts of placer gold.

For information, write: Director, Minnesota Geological Survey, University of Minnesota, 1633 Eustis St., St. Paul, Minnesota 55108.

Fillmore Co.

Spring Valley, area glacial drift deposits—pannable gold.

Scott Co.

Jordan, area glacial drift gravels—pannable gold.

MISSOURI

Gold occurs in very small amounts in Missouri's northern counties, dropped by the Pleistocene glaciers, but because of its extreme fineness, no commercial recovery is feasible today. Hunting gold in Missouri is considered merely as an interesting part-time avocation.

For information, write: Office of the State Geologist, Mineral Resources Section, Department of Natural Resources, P.O. Box 176, Jefferson City, Missouri 65101.

Adair Co.

Kirksville, numerous area glacial gravel deposits—fine gold.

Macon Co.

Elmer: (1) NE 1 mi., in NW¼ sec. 36, (60N–16W), in Murray Gulch gravels—"scale" gold (extremely fine) and reportedly corn-grain-size nuggets, with a little silver; (2) NE 1½ mi., in NW¼ sec. 31, (60N–15W), along Sand Cr. in sand and gravel deposits—fine gold (valued at $0.13 per yard at $35 per oz. price).

Gifford (6 mi. NW of Murray's Gulch), area glacial drift deposits —2.34 gold grains plus silver per ton.

Macon, Adair, and Schuyler Cos.

Region, all along the Chariton R. and its tributaries, in glacial drift deposits—pannable gold.

MONTANA

The first gold found in Montana was in gravels of Gold Creek in 1852 in what is now Powell County, but not until 1863, when enormously rich gold deposits were found along Alder Gulch, near Virginia City in Madison County, did the big rush begin. Subsequently, Montana produced 17,752,000 gold ozs. to become seventh in rank among the gold-producing states. Four districts yielded more than 1,000,000 gold ozs. each, while 27 of the 54 districts that produced more than 10,000 ounces recorded amounts between 100,000 and 1,000,000 ozs. of gold.

By 1975 only a few small operations were producing only gold in Montana. About five eighths of the state's annual gold production comes as a by-product from the Anaconda Company copper mines at Butte, and most of the rest is a by-product of other base-metal refining. Amateur gold hunters should be aware that experienced prospectors have been exploring all of Montana's likely areas very thoroughly for more than a century, and they missed few potentials. Nevertheless, gold panning in any of the once commercially productive districts may bring its reward in colors and an occasional nugget.

For more information, write: Director, Information Service, Montana Bureau of Mines and Geology, Butte, Montana 59701.

Beaverhead Co.

Beaverhead County probably produced much more than its recorded 370,000 gold ozs. between 1862 and 1959, since early figures were not compiled.

ARGENTA (district, 12 mi. N and W of DILLON, tot. prod., 65,350 gold ozs., 1904–57): (1) in Rattlesnake, French, and Watson crs., gravel and bench deposits—placer gold; (2) area mines, discovered in 1865—lode gold.

BALD MOUNTAIN (20 mi. W of DILLON and 7 mi. W of BANNACK), area placers, first worked in 1869 and intermittently since.

BANNACK (district and ghost town 22 mi. SW of DILLON, first territorial capital; tot. prod., 240,000 gold ozs., of which 108,400 ozs. came from lode mines: (1) along Grasshopper Cr.: (a) best-known early-day placer workings, large-scale (1862); (b) all tributary crs. and benches—placer gold; (2) area lode mines along both sides of Grasshopper Cr.—lode gold.

BEAVERHEAD (Big Hole, Dark Horse, Mulchy Gulch; 45 mi. E of DILLON and 30 mi. S of WISDOM), area placers worked intermittently since 1862.

BRENNER, SE 6 mi., at Horse Prairie, in Colorado Gulch, recent placer workings.

DILLON, NW: (1) in Frying Pan Basin, area cr. and bench gravels (worked in 1930s)—placer gold; (2) along Camp Cr., area gravels—placer gold.

MELROSE, W 12 mi., in NE part of co., the Bryant (Hecla) district, tot. prod., 1873–1959, of 17,440 gold ozs.: (1) area base-metal mines—by-product gold; (2) all area watercourse beds and benches, minor placers.

MONIDA, E, numerous N-flowing crs.—placer gold (found in recent years).

WISDOM, W 14 mi., along Pioneer Cr., bed and bench gravels—placer gold.

Broadwater Co.

Some of Montana's richest placers were found in Broadwater County. Placer production before 1940 is not known, but was probably very large. From 1901 through 1959 a total of 1,000,000 gold ozs. were produced, of which only 34,500 ozs. came from placer deposits.

RADERSBURG (10 mi. W of TOSTON, on E side of the Elkhorn Mts. in S part of co., the Radersburg [Cedar Plains] district, largest lode-gold producer in co., tot. prod. of 325,000 gold ozs., 1886–1959): (1) area lode mines, especially the Keating Mine—lode gold; (2) along Crow Cr. and its tributaries, placers; (3) below town, along Eagle and Sam crs.—placer gold (potentially abundant to amateur panners and sluicers); (4) in Johnny Gulch, area placers.

TOWNSEND: (1) area lode mines, worked until World War II—lode gold; (2) S 3 mi., along Deep Cr., placer workings; (3) N 15 mi., the

Confederate Gulch district (Backer, Canyon, Diamond City), along the gulch tributary of the Missouri R. in N part of co.; tot. prod. between 550,000 and 600,000 gold ozs., of which only about 10,000 ozs. came from placers: (a) area gulches and benches—placer gold; (b) all area ancient stream channels, bars, and benches actively placered in 1930s —placer gold; (4) NW of Confederate Gulch, the White Creek district (including the drainage basins of White Cr., Avalanch Cr., and upper Magpie Gulch, tot. prod. between 68,000 and 92,000 gold ozs.): (a) upper part of White Cr. and its tributary Johnny Gulch, many placers; (b) below mouth of Johnny Cr., for a mile or more, numerous drift mines —placer gold; (c) in Magpie Gulch, rich placers produced some $330,000; (5) N 25 mi., at Hellgate Cr. and tributaries, placers.

WINSTON, SW, in the Beaver Cr. drainage basin in N part of co., the Winston (Beaver Creek) district, base-metal mining with tot. gold prod., 1866–1953, about 106,000 ozs. from lode mines and 12,000 ozs. from placers: (1) Beaver Cr. and all its tributaries—placer gold; (2) main gulches of area between WINSTON and RADERSBURG, especially those above and below Confederate Gulch, promising placers for modern gold hunters.

Carbon Co.

BELFRY, S, along the Clark fork (of the Yellowstone R.), near the Wyoming line, gravel bars—placer gold (finely divided).

Cascade Co.

A total of 67,000 gold ozs., primarily from the Neihart area between 1881 and 1959, have been recorded.

CASCADE (Great Northern RR sta.), 5 mi. above on the Missouri R., gravel bars yield "a day's pay with a rocker."

NEIHART, E, in SE cor. of co., in the central Little Belt Mts.: (1) small area placer deposits; (2) the Montana (Neihart) district, primarily lead-silver, copper, zinc—by-product gold (most of county's total production); (3) area S of the mts.: (a) regional watercourse beds—placer gold; (b) area mines—lode gold.

Deer Lodge Co.

Disregarding early unrecorded amounts of gold, a total of 470,000 placer and lode-gold ozs. came from Deer Lodge County through 1959.

ANACONDA: (1) S 5 mi., at Heber (RR stop), along Mill Cr., quite productive placers; (2) S about 12 mi., in S part of co., the French Cr. district, tot. prod. estimated 50,000 to 250,000 ozs. of gold, 1864–1940: (a) along French Cr., and (b) all tributaries, in watercourse gravels and bench deposits—placer gold; (3) NW 10–15 mi. (about 10 mi. SE of PHILLIPSBURG in Granite Co.), the Georgetown (Southern Cross) district in upper drainage basin of Warm Springs Cr., tot. prod. at least 460,000 gold ozs., 1866–1959: (a) the Cable Mine, principal producer —lode gold; (b) nearby, the Cable placers, a "bonanza deposit" worked for many years; (c) the Southern Cross Mine, major producer—lode gold.

CHAMPION, E 5 mi., area stream gravels—placer gold, sapphires.

DEER LODGE, S 8 mi., at Oro Fino (RR sta.): (1) in Oro Fino and Caribou gulches and their tributaries, watercourse and bench gravels— placer gold; (2) S, along Dry Cottonwood Cr., area placer deposits worked to the present.

GEORGETOWN: (1) the Georgetown placers, early-day producers of 1,935 gold ozs.; (2) area NW of Georgetown Lake, promising placers.

Fergus Co.

Gold accounts for almost 99 per cent of the metals produced in Fergus County, with total production recorded as 653,000 ozs. between 1888 and 1959.

LEWISTON: (1) NE 10 mi., near S end of the Judith Mts., the Warm Springs (Gilt Edge, Maiden) district, tot. prod. about 200,000 gold ozs., 1879–1954; (a) area placer gravels—minor placer gold; (b) area lode mines—lode gold; (c) in the Judith Mts., the Spotted Horse Mine—gold tellurides; (2) NNW 15–18 mi., in W-central part of co., in the North Moccasin Mts., the North Moccasin (Kendall) district, tot. prod., 1880s–1959, of 425,000–450,000 gold ozs.: (a) in Bedrock, Iron, and McClure gulches, productive placers; (b) area lode mines, especially the

North Moccasin Mine (closed in 1922), most productive in district—
lode gold; (c) the old Barnes-King Mine, good producer through 1942
—lode gold.

Flathead Co.

KALISPELL, SW, in Chief Cliff Draw (Elmo), area gravel deposits—
gold showings.

Gallatin Co.

BOZEMAN, SW 32 mi., West Gallatin (Spring Hill), in the W fork
of the Gallatin R., gravel bars and benches, recently worked by sluicing
—placer gold.

Granite Co.

Of a total production of 710,000 gold ozs. produced in Granite
County, 334,000 ozs. are credited to placer mines. Most of the lode gold
has come as a by-product of silver mining in several districts.

AREA: (1) in the Philipsburg Mts. (Flint Creek Range), all regional
streams and gulches—placer gold; (2) in the Sapphire and Anaconda
ranges, the "forks" and tributaries of Rock Cr.: (a) Copper, Wyman,
Welcome, and Ranch crs:, etc., promising placers; (b) tributaries of the
Clark fork, between Bearmouth and Rock Cr., promising placers.

GOLD CR., SW 12 mi., the Gold Cr. district, as the SW extension
of the Pioneer district in Powell Co., area gravel deposits—pannable
gold (coarse), quite productive by sluicing.

PHILIPSBURG: (1) gravel deposits around town—placer gold; (2) just
E, along Flint Cr. in 3-sq.-mi. area, the Flint Cr. district (including the
Red Lion camp), tot. prod., about 260,000 gold ozs., 1864–1959: (a) area
lode mines—gold, with manganese and silver; (b) the Algonquin Mine
—gold by-product of manganese; (c) the Granite-Bimetallic and Hope
mines, major producers—gold, silver; (3) N of Bearmouth (RR sta.), the
Garnet (First Chance, Bear Gulch) district and mines—pyritic gold,
with silver: (a) gulches of Bear, Packer, Fealan, Tenmile, Klondike,
Secret, Deep, Cayuse, Gambler, Kearns, Chicken Run, Cave, First
Chance, and Williams, rich sluicing and drift mining—placer gold

($3,500,000); (b) near mouth of Bear Gulch and "Top-o'deep" area, promising placer deposits; (4) NE 7–9 mi., in E-central part of co., the Boulder Cr. district, tot. prod., about 58,450 gold ozs., 1885–1959: (a) along Boulder Cr., area lode mines, especially the Royal Mine as chief producer—lode gold, with silver; (b) gravels along Boulder Cr. and all tributaries, including benches—placer gold; (5) W 10 mi., Rock Cr. district (including Alder, Basin, and Eureka gulches), placer deposits worked intermittently since 1870; (6) N 10–12 mi., along Henderson Cr., the Henderson placers, tot. prod., 1866–1959, about 81,000 gold ozs.: (a) Henderson Gulch gravels, worked through the 1930s—$1,000,000 in placer gold; (b) area lode mines, major gold producers; (7) NE 17 mi., the Princeton (South Boulder, Maxville) district, along Boulder, South Boulder, Princeton, and Gold crs., drift mining and sluicing operations under 20-ft. overburden—coarse placer gold.

QUIGLEY, W 12 mi. (25 mi. SW of Bonita RR stop), at Welcome Cr., all bar, bench, and other gravels—placer gold, with magnetite.

Jefferson Co.

Between 1864 and 1959 Jefferson County produced 575,000 ozs. of lode gold and 125,000 ozs. from placers.

AREA: (1) 5 mi. S of EAST HELENA (Lewis and Clark Co.), along McClellan and Mitchell crs., productive placers; (2) in N part of co., 10 mi. S of HELENA (Lewis and Clark Co.), the Clancy district, tot. prod., 2,000 lode-gold ozs. and 101,000 ozs. of placer gold. (a) area placers along Prickly Pear Cr., Clancy Cr., and in Lump Gulch (Lump Gulch City a ghost camp)—placer gold; (b) area small mines, heavily mineralized dumps—gold showings; (c) 8 mi. W of Clancy, at Buffalo, area cr. and bench gravels—placer gold.

BASIN (Cataract, Comet), in central part of co., near headwaters of the Boulder R., the Basin and Boulder district (including Basin Cr., Cataract Cr., Lowland Cr., and the upper Boulder R.), tot. prod., 1870–1959, about 188,200 gold ozs., mostly lode gold: (1) area lode mines—lode gold and silver, with base metals; (2) along Basin, Boulder, Cataract, Lowland, Jack, and Rocker crs., active placers worked by hydraulicking, sluicing, and panning (especially in the 1930s); (3) along upper Boulder R., silver mines—by-product gold.

BOULDER, along Boulder Cr. and in Galena and Boomerang gulches, productive placer grounds.

ELKHORN: (1) area mines: (a) especially the Elkhorn Queen, Skyline, and Dolcoath—by-product and telluride gold. This district in the Elkhorn Mts. NE of BOULDER, q.v., produced 70,015 gold ozs., 1870–1953; (b) of 16 clustered mines, the richest gold producers were the Elkhorn, Golden Curry, and Swissmont; (2) immediately N of the Elkhorn district and 20 mi. SE of HELENA, the Tizel (Wilson Creek) district, tot. prod., around 10,500 gold ozs.: (a) area cr. and bench gravels, especially along Wilson and Crow crs., placers; (b) the Callahan and Center Reef base-metal mines—by-product gold.

GOLCONDA (S of JEFFERSON CITY), along Prickly Pear and Wilson crs., placers.

PIPESTONE: (1) N, to I-90, then W 7 mi. (to milepost 19 E of BUTTE), turn N for 5 mi.: (a) the Homestake district, area mines—by-product gold; (b) along Homestead Cr., placers; (2) along Pipestone Cr., numerous placers.

WHITEHALL (S-central part of co., on S end of Bull Mt.), the Whitehall (Cardwell) district, tot. prod., 1890–1959, at least 100,000 gold ozs.: (1) the Golden Sunlight Mine, primary producer, closed in 1957; (2) other area mines in sedimentary rocks—all lode or by-product gold, with lead, silver, and zinc; (3) area surrounding the Homestake and in the highlands E of Elk Park, favorable gravel deposits for placer-gold prospecting.

WICKES (20 mi. S of HELENA), the Wickes (Colorado) district, tot. prod., 1864–1959, about 220,000 gold ozs. (district active to the present): (1) the Gregory lode, e.g., the Gregory, Alta, Comet, and Minah mines, primary producers of lead and silver—by-product gold; (2) area watercourse and bench gravels, especially along Warm Springs and Clancy crs., and in Lump Gulch—placer gold.

Judith Basin Co.

HOBSON, SW 25 mi., at Yogo, along Yogo Cr., productive placers—waterworn nuggets.

STANFORD, SW 14 mi., along Running Wolf Cr., minor placers.

UTICA, SW 15 mi., in Yogo Gulch, in igneous dikes intruded through limestone outcrops on hillsides above gulch—gold, with rubies and sapphires.

Lewis and Clark Co.

Between the first discoveries of placer gold in Iowa Gulch in 1863 and the continuing gold production of today, Lewis and Clark County has produced between 4,000,000 and 5,000,000 gold ozs., divided about equally between placer and lode-mine operations. The chief lode concentrations lie in and near the Boulder batholith, which runs about 60 miles north and south, with an average width of 18 miles, in which the Marysville district has been the most productive of lode gold.

AREA, Landers Fork and its tributaries, Little Prickly Pear Cr. and tributaries, middle and S forks of the Dearborn R., Smith Cr. (tributary of the S fork of the Sun R.), Blacktail Cr., and W fork of Falls Cr. (both tributary to the Dearborn R.), all area gravels, bed and bench, excellent placer-gold prospects.

AUSTIN, along Greenhorn Gulch, numerous rich placers.

EAST HELENA: (1) along Prickly Pear Cr., rich placers; (2) cf. AREA in Jefferson Co.

GARDINER, area mines—lode gold, with pyrites.

HELENA: (1) area, the Last Chance Gulch (Helena-Last Chance district, richest, most famous, most productive placers in Montana, tot. prod., 1864–1959, of 940,000 gold ozs.): (a) intensively worked placers in gulch, and (b) all surrounding gulches and benches, placers; (c) area mines, tot. prod., at least 345,000 ozs.—lode gold; (2) very many other area producers: (a) mines along the N contact of the Boulder batholith —lode gold; (b) all regional gravel deposits, benches, terraces, of all watercourses—placer gold; (c) American Bar (noted for its sapphires)— placer gold; (d) Emerald, Metropolitan, and Spokane bars—placer gold, with sapphires; (2) the ghost camps of Park, Spring Hill, Unionville, in Last Chance, Oro Fino, Grizzly, and Nelson gulches and all their tributaries, rich placers; (3) NW 4 mi., just N of Sevenmile Cr., the Sevenmile-Scratchgravel district (including the Scratchgravel Hills), tot.

prod., about 108,000 gold ozs.: (a) in N part of hills, at Iowa Gulch and along Sevenmile Cr. and its tributaries and along Greenhorn Cr., very productive placers; (b) area mines, especially the Franklin and Scratchgravel—lode gold; (3) SE 6 mi., at French Bar, placers—gold, with sapphires; (4) SW 10 mi., in Bear and Tenmile gulches: (a) all area gravelbars and bench deposits, and (b) all tributaries—placer gold; (5) NE 12 mi., the Eldorado Bar, placers—gold, with sapphires; (6) SW 14 mi., in S tip of co., the Rimini-Tenmile (Vaughn) district, tot. prod., 1864–1957, of 194,000 gold ozs.: (a) area crs. and gulches, tributaries, and benches—placer gold; (b) the Lee Mountain Lode, several mines along the Eureka vein—lode gold; (c) at Red Mt., all area streams and glacial gravel deposits—placer gold; (7) E 16 mi., in Magpie Gulch, area gravels of Magpie and Cave gulches and their tributaries, good placers worked by drifting and hydraulicking, prod., several million dollars; (8) NE 16 mi., in SE cor. of co. on W side of the Belt Mts., the Missouri-York district (including the Missouri R. tributaries of Trout Cr., York, Clark, Oregon, Cave, and Magpie gulches; tot. prod. of 335,000 gold ozs., two thirds from placers): (a) at mouth of York Cr., upstream ½ mi., original placer discovery, 1864; (b) along Trout, Rattlesnake, York, Dry, Kingsbury, Kelly, and Oregon gulches, bar and benches, rich placers; (c) ghost camp of York: many area lode-gold mines and, in area of Dry Gulch and nearby tributaries of Trout Cr., such mines as the Old Amber, Golden Messenger (in Dry Gulch), etc.—lode gold; (d) along all the Missouri R. terraces, rich placer gravels; (9) NW 18 mi., near headwaters of Silver Cr.: (a) the Marysville-Silver Creek district (one of most productive precious-metal mining districts in state., tot. prod., 1864–1959, about 1,310,000 gold ozs., with about 164,000 ozs. coming from placers); (b) at Bald Butte, area lode mines and bench placers, especially the Drumlummon Mine (most productive and steadily mined operation in the district, closed in 1956, about half the district's total prod. of 1,145,800 gold ozs. being from the lode mine); (10) NW 28–35 mi., in drainage basin of Virginia Cr., the Stemple (Gould)-Virginia Cr. district, tot. prod., 1878–1959, of 216,000 ozs. of lode gold and 29,200 ozs. of placer gold; (a) along Virginia Cr., from old camp of Stemple (Fool Hen, Gould, Poorman) 8 mi. to its mouth, rich placers; (b) around

Stemple, area mines, especially the Homestake and the Jay Gould (major district producer)—lode gold.

LINCOLN (in W part of co.): (1) area: (a) along Lincoln, Seven Up Pete, McClellan, Sauerkraut, and other regional gulches and all tributaries, rich placers (worked by sluicing and dredging produced millions of dollars in gold); (b) along Arrastre, Stonewall, and Poorman crs., good continuing placer potentials; (2) the Lincoln district (including Lincoln Gulch and several tributaries of the Blackfoot R., tot. prod., 1865–1959, about 342,000 ozs. of placer gold (200 ozs. of lode gold mined in the 1930s); (3) S 8 mi., the McClellan district, tot. prod., 1864–1959, around 340,000 ozs. of placer gold: (a) along McClellan Gulch, all stream and bench gravels, rich placers; (b) on slopes at head of gulch, outcrops of low-grade gold quartz (no mines), probable source of all placer gold below.

Lincoln Co.

EUREKA (RR sta.), along the Tobacco R., area bars, bench gravels, placers.

JENNINGS: (1) W, both N and S tributaries of the Kootenai R., extending W from town, along streambeds, potentially productive placers; (2) SE 15 mi., along Wolf Cr., many placers.

LIBBY: (1) NW 4 mi., the Rainy Cr. district: (a) quartz mines, mainly copper—by-product gold; (b) area stream and gulch gravels—placer gold; (2) S 12 mi., at Snowshoe: (a) along Libby, Cherry, Little Cherry, and Howard crs., (b) their tributaries, (c) West Fisher R. and its tributaries—placer gold (abundantly recovered by rocker, sluice, drifting old channels, and hydraulicking); (d) area bench gravels and glacial deposits, not commercially productive—placer gold; (3) S 20 mi., in drainage basin of Libby Cr., the Libby (Snowshoe) district, largest gold-producing area in co., tot. prod., 1867–1959, of 16,300 ozs. of lode gold and 3,225 ozs. of placer gold; (a) area lead-silver mines—minor by-product gold; (b) along Libby Cr., major lead-silver-zinc mines—by-product gold; (c) in cr. bed and in tributaries, placers; (d) at Cabinet, several small area mines—minor lode or by-product gold.

SYLVANITE (ghost town in NW part of co., 16 mi. N of TROY), the

Sylvanite district, tot. prod., about 10,800 ozs. of lode gold through 1940; (1) area mines in quartz veins in sandstone—lode gold; (2) along the Yaak R. and tributaries: (a) many streambed placers, and (b) all bench gravel deposits—placer gold.

YAAK, area small mines and prospects—gold showings.

Madison Co.

By far the richest placer deposits in Montana occurred in Madison County, third in rank among the state's gold-producing counties. Between the first discoveries in 1863 through 1959, at least 2,605,000 ozs. of placer gold and 1,141,000 ozs. of lode gold came from the auriferous districts.

AREA: (1) NE part of co., in drainage basin of Willow Cr., including Mineral Hill and S Boulder (Mammoth), the Pony district, tot. prod., about 346,000 ozs. of lode gold, 1870–1959; (2) along Pole Cr. and Ruby R. and its tributaries, placers (rubies and sapphires in Pole Cr. also); (3) the Tobacco Root Mts., area stream and bench gravels—placer gold.

ALDER (8 mi. W of VIRGINIA CITY, q.v.): (1) in Alder Gulch (20 mi. long), extremely rich early placers, discovered in 1863, with literally thousands of auriferous vein outcroppings above the placers; (2) in California Gulch, many rich placers; (3) in Junction, Nevada, Fairweather, Highland, Pine Grove, Summit, and Browns gulches, and all tributary watercourses, very rich, long-enduring placers; (4) S, along Williams Cr., productive placers.

MELROSE, NE 13 mi., at Red Mt., area creek beds, placers worked by sluicing and hydraulicking.

NORRIS (district, in NE part of co., including Norwegian, Lower Hot Springs, and Washington, tot. prod., 1864–1959, of 265,000 ozs. of lode and placer gold): (1) area lode mines (sites of 8 mills before 1870), especially the Revenue Mine, most productive in district—lode gold; (2) in Norwegian Gulch and South Meadow Cr., very rich placers, worked fairly large-scale, 1936–42; (3) SE 5 mi., at Lower Hot Springs, Hot Springs Cr., and tributaries—rich placers; (4) SW 6 mi., at Upper Hot Springs, along Hot Springs Cr., numerous productive placers; (5) SW 10 mi., Washington (Meadow Creek, McAllister camps): (a) along

Washington Cr., steady yields of placer gold; (b) along Meadow Cr., placers (not extensively worked).

PONY: (1) area mines, especially the Clipper Mine and the adjoining Boss Tweed—lode gold; (2) along Pony and North Willow crs., area beds, bars, benches, etc., rich placers worked by sluicing and hydraulicking.

SAPPINGTON, S, at Sand Cr., along the Jefferson R. and Antelope Cr., productive placers.

SHERIDAN (district, in W part of co., including Ramshorn, tot. prod., 1864–1959, about 33,500 gold ozs., of which 2,100 ozs. came from placers): (1) area mines (chiefly gold, but also copper, lead, and silver) —lode gold; (2) Mill Cr., Brandon, Quartz Hill, Indian Cr., area placers worked by sluicing and hydraulicking (especially along Mill and Indian crs.); (3) N 3 mi., along Wisconsin and Wet Georgia crs., productive placers; (4) SE 6 mi., the Ramshorn (Union) district: (a) along Bivens, Harris crs., and their tributaries, placers—coarse gold, irregularly distributed; (b) in Ramshorn and California gulches, especially productive placer potential.

TWIN BRIDGES, NW 10 mi., in NW part of co., W of the Jefferson R., the Silver Star-Rochester (Rabbit) district, tot. prod., around 185,-000 gold ozs., 1860s–1959: (1) at Rochester camp: (a) gravels and benches along Rochester Cr., promising placers; (b) the Watseca Mine, major district producer—lode gold; (2) Silver Star camp, the Green Campbell, Iron Rod, and Broadway mines, all major producers—lode gold; (3) NE 5 mi., along W slope of the Tobacco Root Mts., the Tidal Wave (Twin Bridges) district, tot. prod., 1864–1959, about 33,400 gold ozs.: (a) several silver mines—lode gold; (b) in Dry Georgia, Goodrich, and Bear gulches, productive early placers; (4) at N end of the Tobacco Root Mts., in Bose Basin, the Renova district, tot. prod., about 162,000 gold ozs., 1896–1942: (a) numerous regional mines in oxidized veins cutting the Belt Series—free gold; (b) the Mayflower Mine, major producer (closed in 1905 and reopened from 1935–42)—lode gold.

VIRGINIA CITY, at S end of the Tobacco Root Mts., including Summit and ALDER, q.v., the Virginia City-Alder Gulch district, tot. prod., 1863–1959, at least 2,617,000 gold ozs., of which only about 142,000 ozs. came from lode mines: (1) area mines, especially the Oro Cache and Kearsarge (in the Summit camp)—lode gold; (2) SE 16 mi., along Cherry Cr., locally productive placers; (3) S 35 mi., in S end of the Gravelly Range, at Madison, along the W fork of the Madison R., productive placers.

Meagher Co.

AREA, in the Big Belt Mts., on tributaries of the Smith R., promising placers.

WHITE SULPHUR SPRINGS: (1) NE 14 mi., at Musselshell (Copperopolis), area cr. gravels and benches—placer gold; (2) NW 25 mi., at Watson (Elk Cr.), along Benton and Elk crs., productive placers; (3) NW 30 mi., along Beaver Cr., productive placers.

Mineral Co.

Production records prior to 1914 were credited to Missoula County, but Mineral County is estimated to have produced at least 120,000 gold ozs., mainly from placer deposits.

HUSON, W, along tributaries of the St. Regis and Clark Fork rivers, productive placers.

SUPERIOR: (1) S, on E slopes of the Bitterroot Mts., on Cedar and Trout crs. (6 mi. apart), the Cedar Cr.-Trout Cr. district, tot. prod., 1869–1950, of 120,000 gold ozs. (notably fine, 960 to 982), all stream and bar gravel deposits—placer gold; (2) on Quartz Cr.: (a) major placer deposits, worked by sluicing and hydraulicking of stream and benches; (b) along Oregon, Snowshoe, Windfall, Deep, Tucker, Lost, and Prospect crs. (and all other area streambeds and benches), all excellent potentially productive placers today.

Missoula Co.

GREENOUGH (about 30 mi. E of MISSOULA, in SE cor. of co. on N side of the Garnet Range), the Elk Creek-Coloma district, tot. prod., 1865–1959, between 70,000 and 117,000 gold ozs. (including 52,000 to 100,-000 ozs. from placers): (1) along: (a) Elk Cr., many rich placers intermittently worked until 1940; (b) Union Cr., promising placers; (2) the Coloma area (including Elk, Garnet, Potomac, and Washoe), not developed until 1897, numerous area mines, especially the Comet and Mammoth mines—lode gold.

LOLO, W 10 mi., along Mormon Cr., placers.

STARK (6 mi. NW of Ninemile RR sta.), in NW cor. of co., along Ninemile Cr. tributary of the Clark Fork R., the Ninemile Cr. district, tot. prod., 1874–1959, between 100,000 and 125,000 gold ozs.: (1) along Ninemile Cr., many rich placers: (a) in creek bed and benches, and (b) in glacial moraines (alleged in 1910 to contain great amounts of gold); (2) all regional streams, including McCormick, Kennedy, Upper Kennedy, Pine, Dry, Marion, Dutch, and Butcher crs., placers; (3) S of Ninemile, along Patty Cr., productive placers.

Park Co.

Placer gold was found in Park County just north of Yellowstone National Park in 1862, when mining began. More than half the 295,000 gold ozs. recorded was produced between 1933 and 1953.

BIG TIMBER, SW 45 mi., in headwaters of Boulder R., at Cowles, discovered in 1866: (1) along r., numerous placers; (2) all tributaries of the Boulder R., productive placers.

COOKE CITY (New World, Blackmore) district, on SW flank of the Beartooth-Snowy Mt. anticlinorium, in SE part of co., tot. prod., about 66,000 gold ozs. from mostly base-metal ores as a by-product: (1) area stream gravels, minor placer showings; (2) area mines in contact metamorphic veins—by-product gold.

EMIGRANT (Chico), S 4 mi., in Emigrant Gulch, area placers discovered in 1864 and quite active in 1930s.

GARDINER: (1) NE 5 mi., in S part of co. at JARDINE, the Jardine (Sheepeater, Bear Gulch) district, tot. prod., between 190,000 and 200,-000 gold ozs., 1865–1959: (a) area crs., minor placers; (b) hidden ancient placer channels beneath area basalts (should be investigated); (2) E 7 mi., at Crevasse, stream gravels and benches along: (a) Crevasse Gulch and its tributaries and (b) E of the gulch, along Slough Cr.—placer gold.

Phillips Co.

Phillips County is the easternmost gold-producing area in Montana. Between 1893, when lode gold deposits were found in the Little Belt

Mountains, through 1951, a total of about 380,000 gold ozs. were mined.

LANDUSKY (50 mi. SW of MALTA, in SW part of co.), the Little Rockies (Whitcomb, Zortman) district, many area gold-silver lode mines, especially the Ruby Gulch Mine, major producer—lode gold.

ZORTMAN (on E side of mts. from LANDUSKY, in Alder Gulch): (1) area stream channels and benches: (a) several placers worked extensively in 1880s; (b) the August Mine, discovered by Peter Landusky in 1893; (2) such lode mines as the Goldbug, Pole Gulch, Independent, Mint, Alabama, Fergus, Ella C, and Hawkeye, all closed by World War II— lode gold.

Powell Co.

Between the initial discovery of placer gold in 1852, through 1959, Powell County produced 517,000 gold ozs. from its rich placers and about 50,000 gold ozs. from lode veins.

AREA: (1) all tributaries of the Blackfoot R., E of Monture Cr., reportedly promising placers in stream and bench gravels; (2) at Race-track (RR sta.), along Racetrack Cr., placers.

AVON, N 8 mi., along Carpenter (Ophir) Cr., the Ophir (Avon) district, including Nigger Hill, tot. prod. variously estimated between 188,200 and 242,000 ozs. of placer gold, 1865–1959: (1) in Carpenter, Warm Springs, Snowshoe, and Deadwood gulches, rich placers; (2) NE 3 mi., along Ophir Gulch, placers dredged in 1930s along an 8-mi. stretch; (3) N 16 mi., in Washington Gulch (Finn): (a) in Washington and Jefferson gulches, good placers worked by sluicing, drifting, and hydraulicking; (b) in Nevada and American gulches, extensive placers; (4) a few area small mines—lode gold (tot. prod., 8,250 ozs.).

DEER LODGE, ESE 8 mi., on W slope of the Continental Divide in S part of co., the Zosell (Emery) district, tot. prod., about 39,450 ozs. of lode gold and 3,625 ozs. of placer gold: (1) area gravel deposits: (a) especially the Emery Mine, and (b) some 20 other lode mines, 1888–1951—lode gold.

DRUMMOND, NE 17 mi., the Big Blackfoot (Helmville) district, worked by drifting, sluicing, and hydraulicking: (1) tributaries of the

Blackfoot R. and Nevada Gulch, rich placers; (2) area of Ogden Mt., Douglass and Yourname Crs., productive placers.

ELLISTON, S 6 mi., the Elliston (Ontario, Nigger Hill) district: (1) area streams and bench deposits—placer gold; (2) area small mines and prospects—lode gold.

GARRISON: (1) adjoining community of GOLDCREEK, area stream and bench gravels, placers; (2) W, on Gold and Pikes crs. in SW part of co., the Pioneer district (first gold discovered in Montana in 1852; tot. prod. through 1959 estimated at 246,200 ozs.): (a) in Gold Cr., Pioneer and Pikes Peak gulches, numerous gravel bars worked by dredging, sluicing, and hydraulicking; (b) high-terrace gravels along Pikes Peak Cr., placers; (c) Pioneer Bar on Pioneer Cr., and (d) gravels in French Gulch, worked by large-scale hydraulicking; (e) all other area gravel deposits in benches, terraces, glacial moraines, etc.—placer gold.

Ravalli Co.

DARBY: (1) tributaries of the Bitterroot R., especially those entering from the E, in beds, benches, and terrace gravels, locally small placers; (2) S 30 mi., at Hughes Cr. (Alta, Overwich) district, along Hughes, Cow, and Overwich crs., with most productive placers along Hughes Cr. for 15 mi. from its mouth to the top of the Continental Divide.

FLORENCE, S 7 mi.: (1) in Threemile Cr. and tributaries, productive placers; (2) in all gravel beds extending from the river to the hillsides —placer gold.

STEVENSVILLE, area cr. gravel deposits and terraces around Pleasant View—placer gold.

Sanders Co.

TROUT CREEK (RR sta.), along Prospect, Vermillion, and Trout crs., stream and terrace gravels, placer possibilities.

Silver Bow Co.

As the leading mining county in Montana, Silver Bow County produced approximately 2,800,000 ozs. of gold, along with tremendous amounts of copper, lead, silver, and zinc, between 1864 and 1959.

BUTTE (Summit Valley) district, one of largest copper producers in the world, with tot. prod. of 2,725,000 ozs. of by-product gold: (1) city mines (Alice, Allie Brown, Lexington, Leonard, and Rainbow reach depths of 4,000 ft. on "the richest hill on earth," some mines open for public visits, with more than 3,000 miles of underground workings—by-product gold; (2) Missoula Gulch (in city limits), site of first discovery of gold-bearing gravels; (3) Silver Bow Cr. and Oro Fino Gulch and tributaries, productive placers of 1870s; (4) N 4 mi., at Lost Child in Yankee Doodle Gulch, recent placers; (5) in Summit Valley, area placers and mines (minor by-product gold); (6) S 14–15 mi., in the Highland Mts. 2–3 mi. E of the Continental Divide, the Highland district, tot. prod., more than 50,000 gold ozs.: (a) Fish Cr. and tributaries, good placers; (b) area lode mines (reactivated in 1931 and operating through 1944)—lode gold; (7) SW, at Independence: (a) S, along Silver Bow Cr. and tributaries, stream and bench gravels—placer gold; (b) W and S of Silver Bow Jct., area watercourse gravels—minor placer gold.

DIVIDE (RR sta.): (1) along Divide Cr. and tributaries—placer gold; (2) E 10 mi., at Moose Cr. and tributaries, placers produced several hundred thousand dollars after 1866.

MELROSE, 5 mi. S of Gregory RR sta.: (1) in German Gulch (Siberia), rich placers worked 1864–1930s by sluicing and hydraulicking; (2) along Soap and Camp crs. and tributaries, placers.

WALKERVILLE, along Dry Cottonwood Cr. for about 4 mi. (beginning 12 mi. NW of BUTTE), area stream gravels—placer gold, with sapphires.

Sweet Grass Co.

AREA, headwaters of the Boulder R., productive placers.

Toole Co.

SUNBURST, E 23 mi., at Gold Butte, area stream and bench gravel deposits—placer gold.

NEVADA

Metal mining has dominated the whole of Nevada, which ranks fifth in America for the production of gold, with 27,475,395 ozs. produced between 1859 and 1965, primarily as a by-product of silver and base-metal refining; the amount of silver produced greatly exceeds the total gold production in value.

In compiling auriferous districts within the state, except for a few purely gold districts, like Goldfield, and placer mining areas, virtually all mines and districts produced gold only as a by-product of other metallic ores. Nevertheless, in exploring old mine dumps, whether the mines produced mainly copper, lead, silver, zinc, or other base metal, the gold hunter may very well find specimen minerals showing the glittering presence of gold in native form.

For more information, write: Director, Nevada Bureau of Mines and Geology, Macakay School of Mines, University of Nevada, Reno, Nevada 89507.

Carson City Co. (Ormsby Co. until 1969)

CARSON CITY: (1) W, in foothills of the Sierra Nevada Range, the Voltair district, the Washoe and Eagle Valley mines—by-product gold; (2) E 9 mi., along the Carson R., area mines—by-product gold.

DELAWARE, in E part of co., in Brunswick Canyon, the Sullivan Mine by product gold.

Churchill Co.

One of Nevada's nine original counties when created in 1861, Churchill County has produced a total of 164,605 ozs. of by-product gold through 1958 and 12 ozs. of placer gold between 1904 and 1937.

EASTGATE (on Rte. 2, 5 mi. E of U.S. 50 and about 60 mi. E of FALLON): (1) area lead-silver mines—by-product gold; (2) SSW 18 mi. on Rte. 23, the Gold Basin district, area mines—lode gold.

FALLON, SE 23–30 mi., the Sand Springs district (tot. prod., 1905–1951, of 20,875 gold ozs.), the Dan Tucker Mine—by-product gold.

FRENCHMAN (on U.S. 50 about 42 mi. SE of FALLON, then S 20–30 mi. on Rte. 31 in adjoining Mineral Co. and SE on Rte. 23 in Mineral and Nye cos., a very considerable mining region): (1) E 6 mi. to N-trending dirt rd. (to DIXIE VALLEY): (a) NE 3 mi. to 5-way crossrds., take NE-trending rd. 11 mi. to old camp of Wonder (Hercules), on W slope of the Clan Alpine Range, the Wonder district (tot. prod. of

73,890 ozs. of by-product gold, mostly from the Nevada Wonder Mine); (b) old camps of Alpine (Clan Alpine) in area of Dry L., and Bernice —area lode mines—by-product gold; (c) NE 3 mi. to crossrds., turn N 12–20 mi., old camps of Boyer, Cottonwood Canyon, Bolivia (all around Table Mt.)—by-product gold; (2) S about 20 mi. on Rte. 31, to Shad Run, many mines—by-product gold; (3) E 11 mi. and S on Rte 23, on W slope of Fairview Peak, the Fairview district, tot. prod., about 53,100 ozs. of by-product gold: (a) many area mines, (b) the Nevada Hills Mine (major base-metal producer), Eagleville, and Dromedary mines—by-product gold; (c) just S of Fairview, area gravels—placer gold; (4) 7 mi. S of Fairview, old camp of South Fairview, many area mines and 1 mi. E, at Bell Mt., other mines—by-product gold.

WHITE PLAINS (Huxley Sta.), SW 3 mi., the Desert (White Plains) Mine, rich ore body—lode gold.

Clark Co.

Although gold-silver deposits were discovered in Clark County in 1857, the total recorded production of 291,770 gold ozs. (including 200 ozs. of placer gold) came between 1908 and 1959.

ALUNITE (Railroad Pass, Vincent; about 19 mi. SE of LAS VEGAS via U.S. 93), numerous area mines—lode gold.

BUNKERVILLE (5 mi. SW of MESQUITE on rd. S of U.S. 91): (1) the Copper King district, the Bunkerville, Great Eastern, and Key West mines—by-product gold; (2) S about 30 mi. on dirt rds., at Gold Butte, area mines—lode gold.

GOODSPRINGS (8 mi. NW of JEAN on I-5, in SW part of co., an area of several hundred sq. mi. in the South Spring Mt. Range, the Good-springs (Yellow Pine) district, tot. prod., 1856–1959, of 58,815 ozs. of lode gold: (1) area mines, especially the Keystone, Bass, and Clementine (mainly lead-silver-zinc); (2) the Potosi Mine, first in operation but a poor producer—lode gold.

JEAN, SE 15 mi., to Sunset (Lyons), area small mines.

MOAPA, SW 3 mi., in bottoms of the Muddy R. (large area of several sq. mi.), in sand deposits—fine placer gold.

NELSON (about 20 mi. NNE of SEARCHLIGHT or 22 mi. S of BOULDER

CITY via U.S. 95 and Rte. 60), an area 6 mi. wide and 12 mi. long in
N part of the Opal Mts., the Eldorado district, tot. prod. in 2 yrs.
following discovery in 1957 of 101,729 gold ozs. from lode mines and
168 ozs. from placer deposits; (1) area dry cr. beds and benches—placer
gold; (2) the Eldorado and Colorado mines (largest producers) and the
Eldorado-Rand, Techatticup, Crown Queen, Wall Street, Mocking
Bird, Rambler, Rover, and Flagstaff mines—lode gold.

SEARCHLIGHT (district, tot. prod., 1902–59, of 246,997 ozs. of lode
gold and 26 ozs. of placer gold): (1) area mines, especially the Duplex
and Quartzite—lode gold; (2) area watercourse and bench gravels—
placer showings; (3) W 10 mi. on Rte. 68, to Crescent (about 6 mi. E
of NIPTON, Calif.), then 3 mi. ESE and just S of Crescent Peak, area
mines—gold showings.

ST. THOMAS (RR sta.), SE 24 mi., in S end of the Virgin Range and
7 mi. E of the Virgin R., area of many dry-wash placers—fine gold in
large amts. of black sand.

Douglas Co.

AREA: (1) extreme E side of co., the Bucksin district (extending across
the Pine Nut Mts. and best reached N from WELLINGTON in Lyon Co.):
(a) many area mines, and (b) the Buckskin Mine—lode gold; (c) 2½ mi.
NE of the Buckskin, in watercourse and terrace gravels, the Ambassador
placers in a dry ravine ½-mi.-long tributary of Spring Canyon, extensively hydraulicked—placer gold; (d) 1 mi. SE of the Ambassador, on
alluvial fan sloping toward Artesia, the Guild-Bovard placers, well
worked—shot gold; (2) extreme SE cor. of co.: (a) old camp of Silver
Glance, and (b) in S end of the Pine Nut Mts., in area of Topaz Lake,
old camp of Mountain House (Holbrook, Pine Nut), area mines—
by-product and lode gold.

GARDNERVILLE: (1) the Eagle district: (a) area mines in diorite, and
(b) mines in lake sediments—by-product gold; (2) SE 4 mi., in Red
Canyon, the Silver Lake camp—by-product gold; (3) ESE about 18 mi.,
on Mt. Siegel (elev., 9,450 ft.), area watercourse, bench, and slopewash
gravels—placer gold, with platinum.

GENOA (Mormon Station): (1) W, on E slope of the Sierra Nevada

Range: (a) area mines—lode gold; (b) in Tertiary gravel deposits—placer gold (minor); (2) regional stream and slope gravels, in Triassic sediments intruded by Cretaceous granite—placer gold.

MINDEN, E 20 mi., on N side of Mt. Siegel, in the Pine Nut Range at an elev. of 7,100 ft., the Mt. Siegel placers (covering 2,440 acres): (1) all watercourse gravels in a large depression in mts.—placer gold (fine, coarse, nuggets) concentrated in ravines or on a hard clay false bedrock; (2) along Pinto, Dudley, and Black Horse gulches, principal placer areas worked by drifts, tunnels, and open cuts—placer gold.

Elko Co.

Lode mines have produced 554,737 gold ozs. and placers have yielded 6,450 ozs. of gold between 1903 and 1959, but estimating for earlier unrecorded gold yields, the county's total gold production is around 614,000 ozs.

CARLIN, the Carlin Mine, discovered in 1965 by geophysical prospecting and largest gold mine discovered since 1910. No regional surfaces show evidence of gold anywhere. Recoverable gold is finely disseminated, and the mine works on a gross value based on volume, inasmuch as the obtainable gold runs about 0.289 ozs. per ton of rock.

CHARLESTON (95 mi. NNE of ELKO via dirt rds., reached from NORTH FORK ON Rte. 51), area watercourse, bench, and terrace gravels—placer gold.

CONTACT (on U.S. 93 just S of Idaho line), area mines, e.g., the Kit Carson, Porter, Salmon River, mainly copper—by-product gold.

CURRIE (SE part of co. on U.S. 95); (1) SE 8 mi., old camp of Kinsley —by-product gold; (2) NE 25 mi., old camps of Delker and Dolly Varden, the Mizpah and Dolly Varden mines—by-product gold.

DEEP CR. (68 mi. NNW of ELKO on Rte. 11): (1) N 10 mi., on W slope of the Centennial Range, the Edgemont district (tot. prod., about $1,000,000 before 1907 in gold), old camp of Aura: (a) the Bull Run, Columbia, and Lucky Girl mines—lode gold; (b)area gravel deposits, small placers; (2) old camp of Cornucopia, area mines—lode gold; (3) area of Lime Mt. (80 mi. N of ELKO), numerous mines and prospects —gold showings.

DEETH, N 55 mi., on S side of the Jarbridge Mts. (especially on S side of Copper Mt.), the Charleston (Copper Mt., Cornwall) district: (1) 4 mi. N of camp on 76 Cr. near base of Copper Mt., rich placers discovered in 1876, extensively worked; (2) between 76 Cr. and the Bruneau R., in Pennsylvania and Union gulches, Dry Ravine, and Badger Cr., many rich placers worked in early days; (3) all along the Bruneau R. for many miles, all stream, bench, and terrace gravel deposits—placer gold.

ELKO: (1) NW 5 mi., old camp of Good Hope: (a) area mines—gold showings; (b) W 10 mi., in the Burner Hills, old camp of Burner, area mines—minor by-product gold; (2) SSW 27 mi. (12 mi. SE of PALISADE) by dirt rds., area mines—by-product gold; (3) NW 28 mi. on poor rds., old camp of Merrimac (Lone Mt.)—by-product gold.

JARBRIDGE (ghost town in N part of co. and district, tot. prod., 1904–59, about 217,800 ozs. of lode gold), area mines—native gold, electrum, and silver.

MIDAS (W side of co. on Rte. 18 and 35 mi. W of TUSCARORA), the Gold Circle district (gold-silver), tot. prod., 1907–58, of 109,765 ozs. of lode gold and 45 ozs. of placer gold: (1) SW, the Gold Circle and Summit mines, major producers—lode gold; (2) the Elko Prince Mine —placer gold.

MOUNTAIN CITY (Cope) district, in N-central part of co. 75 mi. N of ELKO, in NE part of the Centennial Range on N fork of the Owyhee R. and 1½ mi. E of the Duck Valley Indian Reservation 15 mi. S of the Idaho line: (1) entire length of the Owyhee R., gravel, bench, and terrace deposits (little worked anywhere)—placer gold; (2) N, on Van Duzer Cr., the Van Duzer district (tot. prod., about $100,000); (a) the Cope and Van Duzer mines—lode gold; (b) along cr., rich placers worked by hydraulicking after 1893—gold (dust to nugget size); (3) N of Sugar Loaf Peak: (a) along Grasshopper Gulch, placers; (b) in Hansen Gulch tributary of Grasshopper Gulch, placers; (4) SE, to Island Mt.: (a) the Gold Cr. Mine—lode gold; (b) all area watercourses, benches, and terraces, very rich placers (most prominent in Nevada, with large production in 1870s and 1880s); (c) 8 mi. N of Gold Cr., in the Alder district, area mines—lode gold; (5) SW 20 mi., the Aura (Bull Run, Centennial, Columbia) district, in the Bull Run Basin, extensively worked placers (discovered in 1869).

ROWLAND (about 30 mi. NE of MOUNTAIN CITY and 12 mi. NW of JARBRIDGE), the Gold Basin district: (1) area copper mines—by-product gold; (2) all area watercourse, bench, and terrace gravels, worked intermittently for many years—placer gold; (3) along N fork of the Bruneau R., small gravel deposits—placer gold; (4) S 10 mi., the Alder (Tennesse Gulch) district, ½ mi. N of the Baker Ranch on Gold Run Cr., small placers.

TUSCARORA (district, in W-central part of co. on Rte. 18 about 42 mi. E of MIDAS, q.v., and 45–50 mi. NW of ELKO, near headwaters of S fork of the Owyhee R.; tot. prod., 1867–1959, estimated at least 100,000 ozs. of gold (lode and placer); (1) area mines, especially the Dexter (low-grade but major gold producer) and Grand Prize—lode gold; (2) regional watercourse, bench, and terrace gravels, rich placers (about $700,000 in early days), worked first by Americans and followed by Chinese, tot. prod. to $7,000,000—placer gold (fine, coarse, large nuggets), with horn silver and native silver; (3) W: (a) 1½ mi., the Harris claims, rich placers (worked by Chinese); (b) along Review Gulch, bottom gravels, placers worked by sluicing; (4) NW 8 mi., to divide at head of Dry Cr., area mines—lode gold.

Esmeralda Co.

Prospected originally for silver, Esmeralda County startled the economic world with its sudden enormous production of gold, totaling 4,194,800 lode gold ozs. between 1902 and 1959, plus 2,071 ozs. of placer gold. The heyday years were 1904–12, when GOLDFIELD boasted some 40,000 inhabitants; gold mining was essentially over by 1918 on any large scale.

AREA: (1) 5 mi. S of TONOPAH in Nye Co., the Divide (Gold Mt.) district, tot. prod., 1901–59, of 26,483 ozs. of by-product gold: (a) area mines—by-product gold; (b) 2 mi. S, at Gold Mt., area mines such as the Tonopah Divide, Divide Extension, and Tonopah Hasbrouch, all rich producers—lode gold; (2) SW 12 mi., the Dolly Mine—lode gold.

BLAIR JCT.: (1) N 8 mi., the Castle Rock mines—lode gold, with mercury; (2) ENE via rough rds. to Lone Mt. and old camp of West

Divide, in T. 1 S, R. 41 E, the Lone Mt. district, tot. prod., 1903–49, of 31,961 ozs. of gold, by-product of silver from area mines; (3) SE 13 mi. from U.S. 95 to S flank of Lone Mt., Weepah (a short-lived Model T Ford boom camp of the early 1930s), area prospects—gold showings; (4) SW 25 mi., at Windypah (Fesler), area mines—by-product gold.

DYER (23 mi. S of COALDALE via Rte. 3A), area lead-silver mines— by-product gold.

GOLDFIELD (site of the great gold rush of 1903, tot. prod. through 1942 of 4,194,000 ozs. of lode gold, mostly before 1918): (1) city mines with enormous dumps and tailing piles (a belt of these great shaft and open-cut mines embraces Diamond Peak on the S, including the N end of town and extending to E, such mines as the Mohawk, Florence, etc.), interesting to visit but not worth prospecting; (2) W and SW 7 mi. by various jeep rds. crossing the "Malapai," a large region surrounding Montezuma Mt., many old mines and prospects—gold showings; (3) N 12–15 mi. (about halfway to TONOPAH and to W of U.S. 95), the Klondyke and South Klondyke district (old camps): (a) area copper mines—by-product gold; (b) area watercourse gravels, placers; (4) SSW 20 mi. (on dirt rd. that turns W from U.S. 95 just S of the Goldfield Summit 6 mi. S of town): (a) old camp of Hornsilver (Lime Point), district with tot. gold prod., 1868–1956, about 25,000 ozs. from lode mines; (b) at Railroad Springs, area copper-silver mines—by-product gold.

LIDA (19 mi. W of U.S. 95 from STONEWALL via Rte. 3): (1) area mines—by-product gold; (2) all regional watercourse beds, benches, and slopewash gravel deposits—placer gold; (3) W 10 mi., in the Sylvania Mts., the Pigeon Springs (Palmetto) district of several sq. mi. just E of the California boundary, very many small-scale placers (mostly dry-wash) worked after 1866; (4) S 10 mi., in Tule Canyon at S end of the Silver Peak Range and tributary to Death Valley: (a) many area placer deposits; (b) at head of Tule Canyon, in Nugget Gulch, extensive Chinese placers —gold nuggets to fair size; (c) all watercourse, bench, and slopewash gravels in a 10-sq.-mi. area—placer gold; (5) W 11.7 mi., to ghost town of Palmetto (13.2 mi. E of jct. of Calif. Rte. 168 with Nev. Rte. 3 at OASIS, Calif.): (a) area lead-silver-gold mines, tot. prod., $6,500,000—

by-product gold; (b) N 5 mi., the Palmetto Mine, major producer—lode gold; (c) all regional stream gravels—placer gold.

NIVLOC (8 mi. SW of SILVER PEAK, q.v., the Nivloc (Red Mt.) Mine, lead-silver primarily—by-product gold.

SILVER PEAK (20 mi. S of BLAIR JCT., in central part of co. in T. 5 S, R. 39 E), the Silver Peak (Mineral Ridge, Red Mt.) district, tot. prod., 1864–1959, about 568,000 ozs. of lode gold: (1) numerous area mines, active into 1950s, especially the Drinkwater and Crowning Glory, major producers—lode gold; (2) on Red Mt., area small mines (mostly silver) —gold showings.

TOKOP (Gold Mt., Oriental Wash), 15 mi. W of the ghost camp of BONNIE CLAIR in Nye Co. on Rte. 72 into N end of Death Valley Nat'l. Mon.: (1) the Oriental Mine, rich veins—lode gold; (2) along Oriental Wash, between Slate Ridge on N and Gold Mt. on S, draining into Death Valley, bed gravels, rich placers discovered in 1866.

Eureka Co.

It is estimated that between the first discoveries in 1863 and 1883, at least 1,000,000 ozs. of gold were produced but not recorded in Eureka County. From 1902 through 1959 a total of 203,597 gold ozs. were mined, of which 9,618 ozs. came from placer operations.

AREA: in far N part of co., reached from CARLIN in Elko Co.: (1) NW 9–15 mi., the Maggie Cr. (Schroeder) district: (a) area mines, especially the Copper King—by-product gold; (b) 19 mi. NW to area along co. line, in the Tuscarora Mts., watercourse, bench, and terrace gravels— placer gold; (2) 20 mi. NW of CARLIN, the Lynn district (entirely placer until 1962; tot. prod., about 9,000 gold ozs.): (a) a broad area, in water-course, bench, and terrace gravels—placer gold; (b) along Lynn, Simon, Rodeo, and Sheep crs., placers (prod., about 6,800 ozs. of gold through 1935); (c) lower end of Lynn Cr., the Clemans Claim, productive placers; (d) on W slope of mts., in Sheep Canyon, the Arrowhead placers, consistent producers; (e) at head of Lynn Cr., the Bulldog placers (original discovery site), consistent producers; (f) side hills adjacent to Lynn Cr. Canyon (especially on S side), all slopewash gravels— placer gold (by panning and dry-wash methods).

CORTEZ (36 mi. S of BEOWAWE via Rte. 21 and dirt rd., on the Lander Co. line, district, tot. prod., 1863–1958, about 48,720 gold ozs., by-product of silver ores): (1) area: (a) small mines, and (b) the Garrison Mine, major producer—by-product gold; (2) NE 5 mi., in S end of the Cortez Mts., the Buckhorn (Mill Canyon) district, tot. prod., 1910–59, of 39,632 gold ozs. (by-product of silver), the Buckhorn Mine.

EUREKA (district and at one time the greatest lead-silver camp on earth, tot. gold prod., 1869–1959, estimated at 1,230,000 ozs. as a by-product): (1) area mines, e.g., the Pinto, Prospect, Secret Canyon, Silverado, and Spring Valley—by-product gold; (2) on outskirts of town, Ruby Hill (mine and district)—by-product gold.

PALISADE (9 mi. SW of CARLIN in Elko Co.), W 6 mi., old camp of Safford (Barth, Palisade), area mines—by-product gold.

Humboldt Co.

Although gold mining began in Humboldt County in the 1860s, the first really rich mine was not found until 1907. A total of 811,712 ozs. of gold were produced through 1959, from lode veins, and another 36,720 ozs. from placer operations.

DENIO (extreme N part of co. on Rte. 140), W about 25 mi., the Warm Springs district: (1) such mines as the Vicksburg, Ashdown, and Pueblo (tot. district prod., 1863–1959, about 24,000 ozs.)—lode gold; (2) all area watercourse, bench, and terrace gravels—placer gold; (3) W, to far NW cor. of co., at S end of the Pine Forest Mts., the Maryville (Columbia, Leonard Cr.) district about 100 mi. N of WINNEMUCCA): (a) along Leonard Cr., area of 1,500 acres above the Montero Ranch extending for 2 mi., the Leonard Creek placers, worked somewhat in the 1930s; (b) on Teepee and Snow Cr. tributaries of Leonard Cr., placers worked by panning, sluicing, and hydraulicking, as well as dry-wash methods during arid seasons; (c) several area mines (probable sources of the placer gold)—lode gold.

GOLCONDA: (1) ESE 3 mi., area small mines—by-product gold; (2) E 5 mi., the Preble (Potosi) Mine—lode gold, with silver; (3) S 9 mi., on E slope of mts., the Ontario-Nevada Mines, Inc., placers (site of wash mill); (4) S 10–12 mi., on E slope of the Sonoma Range, in the Gold

Run (Adelaide) district, tot. prod., 1866–1959, of 23,747 gold ozs.: (a) the Adelaide Mine—lode gold; (b) the Gold Run, Bonanza, and other area operations, productive placers; (c) all regional watercourse, bench, and terrace gravels—placer gold; (5) NE 26 mi. (16 mi. on Rte. 18 to jct., turn N 20 mi.), the Potosi district, in T. 38 and 39 N, R. 42 E, in the North Osgood Mts., tot. prod., 1934–62, about 485,700 gold ozs. (primarily lode gold, but during World War II mines were operated for abundant tungsten): (a) the Getchell Mine (large-scale, open-cut, leading Nevada gold mine 1939–41)—lode gold (then a major tungsten producer); (b) the Riley Mine—by-product gold from tungsten ores.

McDermitt (about 75 mi. N of Winnemucca on U.S. 95 and just S of the Oregon line): (1) general area, so inquire locally: (a) W, old camp of Disaster, veins and placers—gold, with platinum; (b) W, old Jackson Cr. district, area mines and prospects—gold showings; (2) SW 11 mi., the Cordero Mine—by-product gold; (3) SE 18 mi., on W slope of the Santa Rosa Range, the National district (tot. prod., 1860s–1959, of 177,000 gold ozs.): (a) numerous area mines—lode gold, abundant silver; (b) the National Mine, leading producer—coarse electrum.

Orovada (on U.S. 95 about 43 mi. N of Winnemucca), N 11 mi., the Rebel Cr. (New Goldfields, Willow Cr.) district: (1) area: (a) along Pole Cr., productive placers; (b) the Pole Cr. area mines, site of 50-stamp mill—free gold; (c) from Pole Cr. on N to Rebel Cr. on S (about 17 mi.), all watercourses, benches, and terrace gravels—placer gold; (d) along Willow Cr. and Canyon Cr., extensive placers worked by Americans and Chinese in early days; (2) NW 45 mi., near the Oregon line, the Kings R. district (worked primarily by Chinese), in gravels of Horse and Chinese crs. (last sluiced in 1935)—placer gold.

Paradise Valley (44 mi. NNE of Winnemucca via U.S. 95 and Rte. 88): (1) area: (a) the Mt. Rose and Spring City mines—lode gold; (b) regional watercourse, bench, and terrace gravels—placer gold; (2) NW 8–11 mi., on E slope of the Santa Rosa Range, the Paradise Valley district (tot. prod., 1868–1959, about 70,000 gold ozs.): (a) area big mines—by-product gold; (b) regional watercourse, bench, and terrace gravels—placer gold.

Sulphur (60 mi. W of Winnemucca on Rte. 49): (1) the Black Rock

Mine, lead-silver—by-product gold; (2) numerous other regional mines accessible by rough dirt rds.

WINNEMUCCA: (1) NW 4 mi., to Winnemucca Mt., the Winnemucca (Barrett Springs) district (tot. prod., 1863–1959, about 35,000 gold ozs.): (a) the Winnemucca (Barrett Springs) Mine, rich early-day producer—lode gold; (b) the Pride of the Mountain, rich producer—lode gold, silver; (c) near Barrett Springs, area mines created a short-lived boom, the Pansy Lee Mine (reactivated 1937–43)—lode gold; (2) NW 10 mi., the Ten Mile Mine—lode gold, silver; (3) W on Rte. 49: (a) 17 mi. to Pronto, area mines, and (b) another 17 mi. to JUNGO, area lead-silver mines—by-product gold; (4) NE 18 mi. (18 mi. N of GOLCONDA, q.v.), the Dutch Flat district, tot. prod., 1893–1959, about 10,000 gold ozs., the Dutch Flat placers, an area 1½ mi. long and 300 to 2,000 ft. wide—placer gold, with considerable scheelite and cinnabar; (5) N 23 mi., the Sherman Mine—lode gold; (6) W 25 mi., the New Central Mine—gold, with lead and silver; (7) S 26 mi. (into Pershing Co.), old camp of Grandpap, area mines—lode gold, silver; (8) N 28 mi., the Shone Mine—lode gold, silver; (9) NW 45 mi., in the Slumbering Hills, the Awakening district (early production unknown, but tot. prod., 1935–59, of 25,648 gold ozs.): (a) old camp of Amos, area mines (lode gold) and placer workings; (b) the Jumbo Mine, most important of several area mines—lode gold.

Lander Co.

Although noted primarily for its great silver production, Lander County also produced 607,000 ozs. of gold between 1862 and 1959, of which 48,899 ozs. came from placers between 1902 and 1936.

AUSTIN: (1) NE on Rte. 21, old camp of Spencer, area mines—by-product gold, with antimony and silver; (2) W 6–10 mi., along the Reese R., the Reese R. district (overwhelmingly silver, but with tot. prod. of 2,816 gold ozs., 1862–1936)—by-product gold; (2) NW 9 mi., the Skookum Mine—by-product gold; (3) S 10 mi., the mines along Big Cr.—gold showings (4) W 16 mi. on U.S. 50, then NW 11 mi. on dirt rd., on E slope of the New Pass Range near the Churchill Co. line, the New Pass district, tot. prod., 1865–1959, about 16,000 gold ozs., numerous area mines—lode gold; (5) N 16 mi., the Iowa Canyon district: (a) in canyon near the Joseph Phillips Ranch, placers (little worked); (b) 8 mi. NW of old Camp Raleigh, the Mud Springs district (including Mud Springs Gulch, its tributary Rosebud Gulch, and S of Mud Springs Gulch in Tub Springs Gulch, rich placers—native gold (becomes finer farther down), often encrusted with quartz; (6) NNW 20 mi. (7 mi. W of the Silver Cr. RR siding, the Ravensgood (Shoshone) Mine—lode gold; (7) S 24 mi. (via 2 mi. W on U.S. 50, 8 mi. SW on Rte. 2, and 5 mi. on dirt Rte. 21, turn S on rough rd.), the Kingston district, the Bunker Hill, Sante Fe, Summit, and Victorine mines—lode gold, silver; (8) SW 36 mi. via Rte. 2 to the Campbell Cr. Ranch, the Gold Basin district (cf. under EASTGATE in Churchill Co.), old camp of Carrol on co. line, area silver mines—by-product gold.

BATTLE MOUNTAIN (district, including the Battle Mt. Range, an area 15 mi. long by 12 mi. wide, tot. prod., 1866–1959, of 149,372 gold ozs., of which 47,633 ozs. came from placers, 1902–36): (1) area: (a) big copper mines near town—by-product gold; (b) all regional watercourse, terrace, and slopewash gravels—placer gold; (c) SW, along the Reese R., the Amador, Austin, and Yankee Blade mines—by-product and lode gold; (2) W, in the Galena Range near the Humboldt Co. line and W of the Copper Basin district, the Buffalo Valley area mines (17 mi. S of VALMY in Humboldt Co.)—lode gold: (a) the Bannock, Copper Basin, Copper Canyon, Cottonwood Cr., Rocky Canyon, and Galena mines— lode and by-product gold; (b) all regional watercourse, bench, and slope- wash gravels—placer gold; (c) near Bannock at mouth of Copper Can- yon and in Black Canyon, the Wilson, Dahl, and Christensen placers (most productive of numerous area placer operations using power equip- ment—coarse gold, large nuggets), and the Grand Hills Mining Co. claims, extensively worked placers; (3) SE 17 mi., in SE¼ (30N–45E), the Lewis district, tot. prod., 1867–1959, of about 51,124 gold ozs. (half from placers): (a) the Pittsburg and Morning Star mines—lode gold; (b) the Betty O'Neal Mine, large-scale silver producer—lode gold; (c) the Dean and the Mud Springs mines—lode gold, with copper-lead-silver; (4) NE 18 mi., on NW slope of Shoshone Peak, in secs. 3–6, (29N– 46E), the Hilltop district, tot. prod., 1907–59, of 17,834 lode gold ozs.

and 119 ozs. of placer gold: (a) area watercourse and slopewash gravels, placers; (b) the Hilltop and Pittsburg Red Top mines (major producers) —lode gold; (5) SE 20–50 mi., on E slope of the Shoshone Range (23 mi. SW of BEOWAWE in Eureka Co.), in secs. 8, 9, 16, and 17, (28N–47E), the Bullion district, tot. prod., 1870s–1959, of 146,154 ozs. of lode gold and 10,373 ozs. of placer gold: (a) the Campbell, Lander, Cortez, and Mt. Tenabo mines—lode gold; (b) all watercourse and slopewash gravels around the Mt. Tenabo mine, placers; (c) the Gold Acres open-pit mine (largest gold operation in Nevada until 1958)—placer gold; (d) near Camp Raleigh, in Mill and Triplett gulches, most productive placers of the Tenabo district, worked by rockers (water hauled for 2½ mi.) and dry-wash methods—placer gold (fine, angular, nuggets); (6) SSW 30 mi., the McCoy and Horse Canyon mines—lode gold; (7) S, and about halfway to AUSTIN, at old narrow-gauge RR stop of Bobtown, the Steiner Canyon (Bobtown) district, discovered in early 1870s, area deep (to 40 ft. down) placer operations.

Lincoln Co.

Between 1869 and 1959, Lincoln County produced 556,800 gold ozs., almost all of it from the Delamar mines and from the Pioche district.

AREA, Wedge of co., in the Timpahute Mts., old camp of Tem Piute, area mines—lode gold, with copper, silver, and zinc.

CALIENTE: (1) NNW 8 mi., in the Chief Range, the Chief (Caliente) Mine—lode gold, with base metals; (2) W 16 mi. on U.S. 93, turn onto SW-trending dirt rd. 6 mi., then branch left 6 mi., to old camp of Delamar, area mines—lode gold, silver; (3) SW 29 mi., on W slope of the Meadow Valley Range, the Delamar district (tot. prod., 1891–1957, of 217,240 ozs.): (a) the Delamar Mine, largest producer in Nevada before closing in 1909, except for a few mines in GOLDFIELD in Esmeralda Co.—lode gold; (b) the Magnolia Mine and the Jumbo Claim, big producers after 1932—lode gold; (4) WNW 30 mi., the Ferguson (Delmar) Mine—lode gold.

PANACA, E 8 mi. on Rte. 25: (1) SW 10 mi. on dirt rd., old camp of Crestline; and (2) 13 mi. farther, the old Acoma district, area mines—lode gold.

PIOCHE (district, 19 mi. W of the Utah line, primarily lead-silver-copper-zinc, bonanza years 1869–75 and high production 1911–58; tot. gold prod., 1906–59, of 104,583 ozs.): (1) area mines—by-product gold; (2) NW 3 mi. on U.S. 93, turn W on dirt rds.: (a) 4 mi. to old camp of Mendha, and (b) 11 mi. to Comet (Mill), area mines—by-product gold; (3) WNW 7 mi., the Highland Mine—by-product gold; (4) NW 12 mi. on U.S. 95, turn W on dirt rd. 10 mi., the Bristol (Jack Rabbit) Silver Mine—by-product gold; (5) E 15 mi., old camp of Ursine: (a) area mines—lode gold, with lead and silver; (b) S 2 mi. on dirt rd., the Eagle Valley district, the Fay and State Line mines—lode gold; (6) N 28 mi. on U.S. 93, then NE 21 mi. on dirt rd., the Atlanta district, area mines, especially the Silver Park and Silver Springs—lode gold, silver; (7) N 52 mi. on U.S. 93, turn W 11 mi. over Patterson Pass (elev., 7,400 ft.), the Patterson district at S end of the Shell Creek Range, the Cave Valley and Geyser mines—lode gold, with lead; (8) W 75 mi., the Worthington (Freiberg) Mine, best reached by dirt rds. from Rte. 25 near the Nye Co. line—lode gold, with silver.

Lyon Co.

The earliest gold discovery anywhere in Nevada was made in Lyon County in 1849 in Gold Canyon in the Silver City district, a discovery that ten years later sparked the great Comstock Lode find in adjoining Storey County. Between 1903 and 1959 Lyon County produced 254,-722 ozs. of gold.

DAYTON: (1) nearby, on E slope of the Virginia Range, in Gold Canyon, the Gold Canyon (Chinatown, Silver City, Devils Gate, Dayton) district: (a) sandbars along the Carson R. at mouth of Gold Canyon (original discovery site), productive placers; (b) N from town, in broadening canyon alluvial fan, all gravels—placer gold; (2) SW of Gold Canyon, on terrace sloping toward the Carson R., the Rae placers, productive; (3) SE 10 mi., the Como (Palmyra, Indian Springs) district, tot. prod., 1916–40, between 10,000 and 15,000 gold ozs.: (a) old camp of Como, area mines, some worked through 1936—lode gold; (b) the Star of the West Mine, minor producer through 1939—lode gold.

FERNLEY, S 14 mi., the Talapoosa Mine—lode gold, with copper and silver.

SILVER CITY (district, 5 mi. S of VIRGINIA CITY in Storey Co., in W tip of co. in T. 16 N, R. 21 E, tot. prod., 1849–1959, about 190,000 gold ozs.): (1) along Gold Canyon, bed and bench gravels, huge placers worked by dragline and floating washer plant, prod., 1920–23, of 14,625 ozs. of placer gold, and 1941–43, $1,115,752 in bullion; (2) the Silver City vein and mines (occupying a fault parallel to Gold Canyon)—lode gold; (3) the S extension of the Comestock Lode, area mines such as the Chinatown, Dayton, and Devils Gate—lode gold.

SILVER SPRINGS (16 mi. S of FERNLEY), NW a few mi., old camp of Ramsey, area mines—free gold.

WELLINGTON, N, along the Pine Nut Mts. (straddling the Douglas Co. line; not a mining district): (a) area small mines—by-product gold; (b) all area watercourse, bench, terrace, and slopewash gravels—placer gold.

YERINGTON (Ludwig, Mason) district, primarily copper: (1) W 2 mi.: (a) the Mason and Ludwig mines—lode gold; (b) area gravel deposits—placer showings; (2) NW 8 mi., on W side of Mason Valley between Mason Pass and Gallagher Pass, in Big Canyon: (a) the Adams-Rice placers (400 acres in 2-mi.-long canyon 300–600 ft. wide), in ancient stream channel and surface gravels—placer gold; (b) in alluvial fan at canyon mouth, many prospects (little worked)—placer gold; (c) 2 mi. SW, on Lincoln Flat in the Singatze Range, the Penrose Placer (in ancient channel of Tertiary gravels)—fine gold in black sand; (3) S 11 mi. on Rte. 3, turn SSE 14 mi. on dirt rd. W of the Walker R. to crossrds. of W-trending rough rd.: (a) in T. 9 and 10 N, R. 25 and 16 E, the Wilson (Pine Grove, Rockland, Cambridge) district, tot. prod., 1866–1959, of 408,000 gold ozs., many area mines, especially the Wilson and Wheeler mines; (b) W 5 mi. on rough rd., the Pine Grove Mine, and (c) S 5 mi., the Rockland Mine—lode gold, with silver and platinum; (4) S 30 mi., old camp of Washington, area copper-silver mines—by-product gold.

Mineral Co.

Gold has been the most valuable mineral product from Mineral County, among many minerals mined, with a total production between 1910 and 1959 recorded as 266,122 ozs., although prior to 1910 a very considerable amount of gold was produced but not listed.

AREA: (1) extreme NW wedge of co., just S of the Churchill Co. line and 16 mi. S of U.S. 50 via Rte. 23, at Broken Hills (with the Quartz Mt. district to S just over the Nye Co. line), area old mines—gold, with lead and silver; (2) old camp of Acme (Fitting); (a) area mines—lode gold; (b) regional watercourse beds and slopewash gravels—placer gold.

BABBITT, NE 5 mi., to THORNE: (1) area mines—gold, with lead and silver: (2) SE 5 mi., in Ryan Canyon, area mines—by-product gold (of lead-silver ores).

BASALT: (1) SW 10 mi., to N end of the White Mts. and about 4 mi. E of the old camp of Queens, the Mt. Montgomery and Oneota (Buena Vista) district, tot. prod., 1862–1959, above 10,000 gold ozs.: (a) the Tip Top and Golden Gate (Oneota) mines—lode gold; (b) in the Mt. Montgomery area, the Indian Queen and Poorman mines—lode gold; (c) other area mines, such as the Basalt, minor producers—lode gold; (2) N 17 mi. on Rte. 10 and E 8 mi. on dirt rd. (22 mi. S of MINA), in the Candelaria Mts., the CANDELARIA district (including Bellville and Columbus, tot. prod., 1863–1958, of 13,024 gold ozs.): (a) the Candelaria, Northern Belle, Mount Diablo, Holmes, and Argentum mines—by-product gold; (b) the Columbus area mines—by-product gold.

HAWTHORNE: (1) area: (a) local small mines—lode gold; (b) the La Panta Mine—lode gold; (2) at S end of Walker L., the Hawthorne district (gold production, 1880–1936, not recorded, but tot. prod., 1936–59, of 5,067 gold ozs., with prod., 1904–35, of 4,700 ozs. of lode gold and 155 ozs. of placer gold), area mines—lode gold; (3) at Laphan Meadows near Mt. Grant, the Murray placers, worked by dragline after 1932—coarse gold and nuggets; (4) several mi. S of the Murray diggings, in Baldwin Canyon, not well prospected—placer gold (good potential); (5) ESE 10 mi., the Lucky Boy Mine, major early producer, and the Pamlico Mine, good producer—lode gold; (6) NW 10–15 mi.: (a) W of Walker L. on E slope of the Wassuk (Walker R.) Range, the Walker L. district (Buckley, Cat Cr. mines)—by-product gold, with copper and silver; (b) on W slope of the Wassuk Range, the West Walker district, area mines—lode gold, silver; (6) SE 16 mi., the Whiskey Flat Mine—gold, copper, silver; (7) SE 18 mi., the Sulphide Mine—lode gold, with tungsten; (8) SW 30 mi. on rough rds., old and famous ghost town of Aurora (Cambridge, Esmeralda), more often reached 8 mi. E from Bodie, California, area mines—by-product gold. Aurora was the sister camp to Virginia City, 1859–99; Aurora's 10,000 inhabitants were so lawless that Mark Twain feared for his life after one night's visit and returned posthaste to the quieter city.

Luning: (1) area mines—lode gold, silver; (2) E, the Santa Fe district, the Luning and Kincaid mines—by-product gold, with copper-lead-silver. During World War II, this district was noted for its production of magnesium.

Mina: (1) SE 4 mi. to Sodaville and E 18 mi. to Pilot Mt.: (a) area mines, and (b) area watercourse and slopewash gravels—placer gold; (2) NE 7 mi., in the Excelsior Mts., the Gold Range (Silver Star, Camp Douglas) district, tot. prod., 1893–1948, about 97,000 gold ozs., area mines—free gold, with pyrite and comb quartz; (3) NW 10 mi., the Garfield district (tot. prod., 1882–1959, estimated between 10,000 and 50,000 gold ozs.): (a) the Garfield Mine—lode gold; (b) the Mable Mine (20 rd. mi. NW of town), a later discovery—lode gold, with lead, silver; (4) NE 23 mi. by rough rds., in the Cedar Mts., the Bell (Cedar Mountain) district, tot. prod., 1879–1959, about 34,000 gold ozs.: (a) the Simon Mine (original lead-silver discovery)—by-product gold produced much later; (b) the Copper Contact Mine and the Bell and Olympic (Omco) mines, major producers—32,000 ozs. of 500-fine gold; (5) NE 30 mi. (over the Nye Co. line, cf. under Gabbs for the Golddyke and Warrior mines), the Athena Mine—lode gold, silver.

Pine Grove (Rockland, Wilson) district, 20 mi. S of Yerington in Lyon Co., on E flank of the Smith Valley Range, on slopes of Sugar Loaf Mt., near mouth of Pine Grove Canyon, low-grade placers.

Rawhide (district, 24 mi. SSW of Frenchman in Churchill Co. via Rte. 31 and 5 mi. on dirt rd. S and W from the Nevada Scheelite Mine, at S end of the Sand Springs Range and 29 mi. E of Schurtz, or 50 mi. SE of Fallon; tot. prod., 1906–59, of 50,707 gold ozs. from lode mines and 2,065 ozs. of placer gold): (1) area: (a) all regional gravel deposits (watercourses, benches, terraces, slopewash), rather well worked for placer gold, most productive area on SE slope of Hooligan Hill in area 1 mi. long by ½ mi. wide; (b) numerous area small mines—lode gold only; (c) the Regent Mine—gold by-product of copper, lead, silver, and tungsten; (2) SE, the King Mine—lode gold, lead, silver; (3) SE 11 mi. by rough rd., to Hot Spring on E side of Alkali Flat (best reached by dirt rd. 34 mi. N of LUNING or 23 mi. W from GABBS in Nye Co.), area mines—lode gold, silver; (4) SE 14 mi., the Hot Spring (Sunnyside) Mine—lode gold, silver.

SCHURTZ: (1) W 8 mi., the Granite Mt. district, the Mt. View and Reservation mines (copper, lead, silver)—by-product gold; (2) N 9 mi., the Benway Mine—lode gold, silver; (3) NE 12 mi., old camp of Holy Cross, the Fallon and Terrell mines—lode gold, with silver and manganese; (4) E 28 mi., the Bovard district, the Copper Mountain and Rand mines (cf. also under RAWHIDE as access point)—gold by-product of lead, manganese, and silver ores.

SODAVILLE (cf. also under MINA), E 5 mi., on W side of the Pilot Mt. Range: (1) near mouth of Telephone Canyon, productive placers discovered in 1931; (2) 6 mi. up canyon, at its head, the Belleville Mine—lode gold.

Nye Co.

The most important metal mined in Nye County's 17,000 square miles of heavily mineralized desert playas and mountain systems is gold and silver. A total of 2,975,034 ozs. of lode and placer gold were produced between 1903 and 1959, of which 298,593 ozs. are attributed to placer operations.

AREA, NW part of co.: (1) 42–54 mi. ESE of FALLON in Churchill Co., the old Westgate district—lode gold, lead, silver; (2) 45 mi. S of AUSTIN in Lander Co., the old Millet (North Twin R.) district, area mines (pockets in limestone)—lode gold; (3) on W slope of the Shoshone Mts., at lat. 39°07' N, long. 117° 33' W, the Jackson (North Union) district, inactive since 1911 but estimated gold production, 1878–1911, between $500,000 and $1,000,000, area mines—lode gold.

BEATTY: (1) the Bullfrog district, including BEATTY, Bullfrog, Bonanza, and Rhyolite (tot. prod., 1904–59, of 120,401 gold ozs.): (a) above and below town, along the Amargosa R., minor placers; (b) nearby, old camp of Pioneer, area mines—lode gold; (c) N 4 mi., ghost town of Rhyolite, area mines—lode gold, silver; (d) W 8 mi., old site of Bullfrog, area mines—lode gold; (2) E 6 mi., old camp of Fluorine (Bare Mt. and Telluride mines)—lode gold, with mercury and silver; (3) S 8 mi. on U.S. 95, ghost camp of Carrara: (a) area mines—lode gold; (b) on W slope of Bare Mt., watercourse and slopewash gravels (deep to bedrock), placer potentials; (4) NE, old camp of Johnnie (cf. under JOHNNIE); (5) WNW 22 mi., old camp of Grapevine, area mines—lode gold; (6) E 30 mi., the old Wahmonie Mine—lode gold, with silver.

BONNIE CLAIRE (ghost camp and mill on Rte. 72 into N end of Death Valley, 6 mi. W of Scotty's Jct. on U.S. 95), area mines—gold, with copper and silver.

CLARK STATION (33 mi. E of TONOPAH on U.S. 6): (1) the Clifford Mine—lode gold, silver; (2) the old Blakes camp (32 mi. ENE of TONO-PAH), the Golden Arrow Mine—lode gold, silver, (3) N 48 mi. on dirt rd. to Crockers Ranch (old stop of Morey), then W into the Hot Cr. Range, area mines (lead-silver)—by-product gold.

CURRANT (NE part of co., at jct. of U.S. 6 with Rte. 20): (1) E, in the Grant Range, area mines—gold, copper, lead; (2) S 30 mi., at Troy Peak (elev., 11,268 ft.): (a) old camp of Troy at base, the Irwin Canyon-Nyala Mine—lode gold, with lead, silver, (b) W 8–12 mi., the Nyala Mine (39 mi. SSW of CURRANT)—lode gold.

GABBS: (1) N 3 mi., turn NE on dirt rd. from Rte. 23 to old camp of Downieville, area mines—by-product gold; (2) SSE: (a) 10 mi., old camp of Golddyke, and (b) 8 mi. farther, the Warrior Mine, all area mines—lode gold, with lead, silver; (3) N 12 mi. on Rte. 23, turn right on dirt rd. to old camp of Quartz Mt.: (a) area mines—by-product gold; (b) nearby, old camp of Westgate (possibly over Mineral Co. line), area mines—gold; (c) NW 3–4 mi., the Broken Hills district (straddling co. line), area mines—lode gold, lead, silver; (d) in T. 13 N, R 36 E, the Lodi (Quartz Mountain) district, the Granite, Marble, and other mines in the Lodi Hills, tot. prod., 1874–1940, of $809,905 in lode gold, and 1932–59, of 1,079 ozs. of by-product gold; (4) SSE 14 mi., the Fairplay and Atwood mines (near Golddyke)—by-product gold; (5) E 16 mi. to dirt crossrds.: (a) E 4 mi., old camp of Grantsville, area copper-silver mines—by-product gold; (b) N 4 mi., turn E on old rd. to ghost camp of Berlin, the Union Mine, cf. under IONE.

IONE (Union, Berlin) district, on W slope of the Shoshone Range at lat. 38°55' N, long. 117°35' W (co. seat in 1864), tot. prod. through 1909, at least 10,000 ozs. of lode gold and another 743 ozs. through 1959: (1) area gravel deposits 1 mi. SW of the camp, dry-wash placers (several feet of surface debris) worked in 1930s; (2) area mines—lode gold; (3) NW 11 mi., old camp of Bruner (lat. 39°05' N, long. 117°46' W, the Bruner (Phonolite) district, tot. prod., 1906–59, of 17,213 gold ozs.: (a) the Paymaster Mine, minor producer—lode gold; (b) the Penelas Mine, major producer opened in 1936—lode gold, silver.

JOHNNIE (district, in extreme S triangle of co., lat. 36°26' N, long. 116° 04' W, 25 mi. NE of Death Valley and 14 mi. SSE of Amargosa in the NW end of the Spring Mts.; tot. prod., 1890–1959, about 40,000 gold ozs.): (1) area old mines, especially the Johnnie Mine—lode gold; (2) area watercourse, bench, and slope wash gravels—placer gold, extending from 2–12 mi. N of the mine and 1 mi. wide.

MANHATTAN (35 mi. N of TONOPAH, at S end of the Toquima Range) district, tot. prod., 1905–59, of 280,022 ozs. of lode gold and 206,340 ozs. of placer gold): (1) in Manhattan Gulch: (a) in deep gravels near bedrock, and (b) in surface gravels above streambed, extensively worked placers—coarse gold, nuggets; (2) S, in Giffen Gulch, productive rich placers; (3) N, in Slaughterhouse Gulch, productive placers; (4) area lode mines, important after 1908—lode gold.

POTTS (44 mi. SE of AUSTIN in Lander Co. and just S of the Nye Co. line): (1) area mines surrounding old camp of Jackson (Gold Park), overlapping into Lander Co.—lode gold; (2) SE on rough rd. into the Monitor Range, old camp of Danville, area mines—lode gold, silver.

ROUND MT. (district, 55 mi. N of TONOPAH and 8 mi. N of MANHATTAN, on W flank of the Toquima Range, tot. prod., 1906–59, of 537,000 gold ozs.): (1) area: (a) very rich mines, worked till 1935—lode gold; (b) even richer nearby placers, large-scale operations to the present by the Round Mt. Gold Dredging Corp., mainly on the S and W slopes of Round Mt.—angular, coarse gold; (2) N 6 mi. in N part of the Toquima Range, in T. 11 N, R. 44 E, the Gold Hill district and mine, tot. prod.,

1931–32, of 24,725 ozs. of lode gold from a single quartz vein in rhyolite.

SCOTTY'S JCT. (35 mi. S of GOLDFIELD in Esmeralda Co. on U.S. 95, entrance to N end of Death Valley), ESE about 10 mi., old camp of Tolicha (20 mi. E of BONNIE CLAIRE, q.v.), the Monte Cristo Mine—lode gold, silver.

STONEWALL (16 mi. S of GOLDFIELD on U.S. 95 and 1 mi. E of jct. with Rte. 3): (1) area mines—lode gold, with silver, some mercury; (2) E 20 mi., old camp of Wellington (O'Brien's), area presently prohibited entry as on the Nellis Air Force Range—lode gold.

TONOPAH: (1) great area mines in and around town, primarily silver —by-product gold; (2) nearby, in S part of the San Antonio Mts., the Tonopah district (tot. prod., 1900–42, of 1,880,000 ozs. of by-product gold; (3) E a few mi., in T. 2 N, R. 43 E, the Ellendale district, tot. prod., 1909–48, above 10,000 ozs., the Ellendale Mine—lode gold; (4) on E slope of the Montgomery Mts.: (a) the old Congress Mine—lode gold; (b) in gulches below the mine, dry-wash placers (worked with rockers during snowmelt runoff); (c) on W slope of the mts., area gravels, placers; (5) SSW 14 mi., in S part of the Klondyke Hills, the Klondyke district: (a) area mines—lode gold; (b) area watercourse and slopewash gravels, placers worked by Chinese in 1870s; (6) E 20 mi., the Hannapah district, the Silverzone and Volcano mines—lode gold, with silver and mercury; (7) NW 20–25 mi., in the San Antonio Mts., the San Antonio district, the San Antonio and Royston mines—by-product gold; (8) S 27 mi. to COLDFIELD in Esmeralda Co. (as access point to the Ralston Desert, now prohibited entry as a military reservation extending 70–80 mi. E of U.S. 95 and 70–80 mi. S of TONOPAH. The Nye Co. line lies 2 mi. E of GOLDFIELD.): (a) E 24 mi., in NW end of the Cactus Range, old camp of Cactus Springs, area mines—lode gold, silver; (b) SE 27 mi., the Gold Crater Mine—lode gold; (c) ESE 30 mi., the Antelope Mine —lode gold, silver; (d) SE 46 mi., the Silverbow Mine—lode gold, silver; (e) E 54 mi., the Kawich (Gold Reed) district, area mines—lode gold, with mercury; (9) N 42 mi. via Rtes. 8A and 82, turn SW onto the MANHATTAN-BELMONT back-entry dirt rd., old camp of Spanish Belt (silver only), nearby, the equally old camp of Arrowhead, area mines—lode gold, silver; (10) NW 42 mi. via Rte. 89, the Cloverdale district:

(a) the Republic and Golden mines—by-product gold; (b) area water-course and slopewash gravels—placer gold; (11) NNE 46 mi. via Rtes. 8A and 82, ghost town of BELMONT: (a) area mines, e.g., the Philadelphia and Silver Bend—by-product gold; (b) N 18 mi., then W, to area about 6 mi. NE of ROUND MT., q.v., at lat. 38°43' N, long. 117°00' W, the Jefferson Canyon district (tot. prod., 1866–71, about $1,000,000), the Concordia and Green Isle mines—lode gold, silver; (12) N approx. 57 mi. on Rte. 8A (about 44 mi. S of AUSTIN in Lander Co.), the Jackson (Gold Park) district, cf. under POTTS; (13) NE 76 mi. (25 mi. N of BELMONT, on E side of the Toquima Range), the Northumberland district, tot. prod., 1866–1959, of 35,353 gold ozs., area open-pit mines in 60–70-ft.-thick shale, low-grade—lode gold.

WARM SPRINGS (49 mi. E of TONOPAH on U.S. 6): (1) the Bellehelen Mine—lode gold, silver; (2) E about 6 mi., the Eden (Gold Belt) mines —lode gold, silver; (3) NE 8 mi. on U.S. 6, turn NW 4 mi. on dirt rd. to Tybo, lat. 38°23' N, long. 116°23' W, the Tybo district, in the Hot Cr. Range (tot. prod., 1872–88 and 1929–58, of 27,183 gold ozs.): (a) the Tybo Mine, principal producer—lode gold; (b) the Hot Creek, Keystone, Empire, and other mines—lode gold; (4) SE 40 mi. on Rte. 25, at S end of Railroad Valley, the Willow Cr. district (copper-silver mainly), area mines—by-product gold.

Ormsby Co. (cf. Carson City Co.)

Pershing Co.

The youngest of Nevada's seventeen counties, Pershing County produced 162,109 ozs. of lode gold and 16,233 ozs. of placer gold between its creation in 1919 and 1958.

AREA, far NE part of co.: (1) 22 mi. S of GOLCONDA in Humboldt Co., the Black Diablo Mine—by-product gold; (2) 36 mi. S of WINNEMUCCA by dirt rd., the Goldbanks Mine—by-product gold; (3) 55 mi. S of WINNEMUCCA by dirt rd., on E side of Granite Mt., the Kennedy Mine (lead-silver)—by-product gold.

IMLAY (4 mi. W of MILL CITY, q.v., on I-80): (1) on N end and W flank of the Humboldt Range, the Humboldt (Imlay, Eldorado) district, tot. prod., 1860–1959 (mostly after 1932), of 35,483 gold ozs.: (a) the

Imlay Mine, major producer of silver—by-product gold; (b) the Star Peak Mine, major producer till 1935—lode gold; (c) the Prince Royal, Humboldt, and Eldorado mines—by-product gold; (2) W 23 mi. on dirt rd. to crossrds., then N 6 mi. on rough rd., the Haystack Mine (7 mi. S of Jungo in Humboldt Co.)—lode gold; (3) W 29 mi. on dirt rds.: (a) the Rosebud Mine, and (b) another 4 mi., the Rabbithole Mine, with (c) other area mines, cf. under Scossa, and under Sulphur in Humboldt Co.—lode and by-product gold.

Lovelock: (1) SE a few mi., on W flank of the Humboldt Range: (a) the Sacramento Mine—lode gold, silver; (b) area watercourse and terrace gravels—placer gold; (2) NE 10 mi., in the Humboldt Range, the Loring district, the Lovelock and Willard mines—by-product gold; (3) W 10 mi., the Velvet Mine—lode gold; (4) NW 12 mi., then 8 mi. on improved rd., the Eagle Pitcher Mine, mainly tungsten—by-product gold, with copper and silver; (5) SW 20 mi., then NW 20 mi. (from Huxley in Churchill Co.), in the Juniper Range, area mines—by-product gold (from silver-tungsten ores); (6) E 22 mi., at Antelope Springs, the Relief Mine—by-product gold; (7) NW 24–30 mi. on dirt rds., on W slope of the Seven Troughs Range, the Seven Troughs district (tot. prod., 1907–59, of 160,182 gold ozs., with by-product silver): (a) the Seven Troughs Mine, minor producer—lode gold; (b) the Vernon and Mazuma Hills mines, major producers—lode gold; (8) N 36 mi. on Rte. 48 and dirt rd. that turns N to Placeritas, a large mining district (cf. under Scossa)—lode and by-product gold, some placer; (9) NW 45 mi. (10–15 mi. E of Gerlach in Washoe Co.), the Farrell (Stone House) Mine—lode gold.

Mill City: (1) E 10 mi., at N end of the East Range, in NE part of T. 33 N, R. 36 E, the Sierra (Dun Glen, Chafey) district, tot. prod., 1863–1959, about 241,000 gold ozs.: (a) the Sunshine and Oro Fino mines—lode gold; (b) the Tallulah, Auld Lang Syne, Munroe, Mayflower, and Auburn mines—lode gold; (c) in Auburn, Barber, Spaulding, Wright, and Rock Hill canyons, productive placers; (2) S 10 mi. on Rte. 50, to the Star Creek Ranch, then S 6 mi. and W 4 mi., on E slope of the Humboldt Range: (a) the Unionville district, the Buena Vista Mine—by-product gold; (b) S 30–35 mi., in Indian Canyon, on E flank of the Humboldt Range, the Indian Mine—lode gold, silver; (c) watercourse, bench, and slopewash gravels about the Indian Mine, productive placers; (3) NE 9 mi., the Dun Glen Mine (in the Sierra district)—lode gold; (4) E 11 mi., the Straub Mine—lode gold; (5) SE 15 mi., the Rockhill Mine—by-product gold; (6) W 20 mi., in the Antelope Range, the Antelope (Cedar) Mine—by-product gold.

NIGHTINGALE (extreme SW corner of co., approx. 40 mi. WSW of LOVELOCK), on E flank of the Nightingale Range, area mines—by-product gold.

OREANA (14 mi. NE of LOVELOCK on I-80): (1) E 5 mi., in NW flank of the Humboldt Range, in Sacramento Canyon (and district), placer prospects, little worked; (2) W 5 mi., the Trinity district, the Arabia and Oreana mines—by-product gold.

ROCHESTER (22 mi. NE of LOVELOCK), in the Humboldt Range: (1) in Rochester Canyon, rich placers (discovered in 1911); (2) in Gold Springs Gulch, near mouth of Rochester Canyon, placers discovered in 1931 worked by dry-wash equipment; (3) N 1 mi., in Limerick Canyon (6 mi. long), many productive placers from surface to bedrock 2–38 ft. down—gold (fine, coarse, angular); (4) upper end of Limerick Basin, ancient channel gravels drift-mined over 2,000 ft.—placer gold; (5) S 1 mi., in Weaver Canyon (2 mi. long): (a) along the canyon, several placer prospects; (b) at head of canyon, the Colligan Mine, and others nearby —lode gold; (6) S 2 mi., in the central Humboldt Range, the Lower Rochester district, opened in 1860s, the Nenzel Mine on Nenzel Hill, high production, 1912–28—lode gold; (7) E 9–10 mi., on E side of the Humboldt Range, the Spring Valley district (about 28 mi. NE of LOVE-LOCK): (a) along Limerick and American Canyons, extensive placers (first skimmed by Americans, then worked by Chinese, and considered to be richest placers in Nevada, tot. prod., about $11,000,000, with abundant placer gold remaining); (b) in Walker Canyon, productive placers; (c) 1 mi. S of American Canyon, in S American Canyon (and district), extensive placers worked by Chinese; (d) halfway up S American Canyon, a bowl-shaped depression 1,500 ft. wide, rich placers worked by panning and rockers; (e) the Bonanza King Mine, rich producer—lode gold; (f) the American and Fitting mines—by-product gold; (8) all regional watercourse and slopewash gravels—placer gold.

RYE PATCH DAM, E 4 mi., on W flank of the Humboldt Range, the Rye Patch (Echo) district, tot. prod., 1864–1959, of 9,453 gold ozs.: (1) the Rye Patch Mine, mainly silver—minor lode gold; (2) the Gold Standard Mine, major producer, 1935–59—lode gold.

SCOSSA: (1) the Rosebud district (NW of SCOSSA and 12 mi. S of SULPHUR in Humboldt Co.), in Rosebud Canyon, numerous rich placers worked by Chinese in 1870s; (2) S 8 mi. (47 mi. N of LOVELOCK), in low hills in T. 32 N, R. 29 E, the PLACERITAS district, discovered in 1870s (at least 5 area gulches have productive placers extensively worked by dry-wash methods); (3) NE 5 mi. and W of Rabbithole Spring, the Rabbithole district (adjoining the Rosebud district): (a) in Coarse Gold Canyon, rich placers—gold nuggets, often found on surface after every cloudburst; (b) all tributary gulches and ravines of Rosebud Canyon, placers—coarse gold and nuggets; (c) W of Coarse Gold Canyon, at Barrel Springs in Barrel Springs Canyon, the Jancke placers (3,000 acres), operated for many years—small to large gold nuggets; (4) N 12 mi., the Sawtooth district (near Humboldt Co. line 25 mi. E of SULPHUR in Humboldt Co.), in area about 6 sq. mi., many rich placers rather extensively worked in early 1930s—gold (coarse, rough, small nuggets).

UNIONVILLE (30 mi. NE of LOVELOCK), on E slope of the Humboldt Range, the Unionville (Buena Vista) district, organized in 1861 and scene of the big silver rush described by Mark Twain in *Roughing It:* (1) area mines—lode gold, silver; (2) above town, rich placer paystreak 2–3 ft. thick and 6–10 ft. down, discovered in 1931; (3) in Buena Vista Canyon and adjoining canyon, rich placers; (4) S 8 mi., in Indian Canyon, early-day placers.

Storey Co.

There are only two hundred square miles in this smallest of Nevada's counties, but a total of 8,560,000 ozs. of gold were produced, largely from the great Comstock silver mines.

VIRGINIA CITY: (1) the Comstock Lode district, discovered in 1859 and one of the richest mining districts of the world, many great silver mines worked to depths of 3,000 ft., e.g., the Comstock, Gold Hill, Silver Star, Ophir, Gould, Curry, Crown Point-Belcher, Big Bonanza,

Consolidated Virginia, Yellow Jacket, and Mexican—by-product gold; (2) E, in adjoining mt. range, the Flowery district, area mines—by-product gold; (3) in Gold Canyon, rich initial placer deposits in 1852 lead to discovery in 1859 of highly productive veins on nearby Gold Hill beneath the placer overburden, area mines—lode gold.

Washoe Co.

Between 1902 and 1959 Washoe County produced a total of 46,107 ozs. of gold.

AREA: (1) W side of Mt. Davidson (reached from VIRGINIA CITY in Storey Co.), the West Comstock (Jumbo) Mine—lode gold, silver; (2) 15 mi. W of Renard, the Sheephead Mine—lode gold; (3) E of Sano, the Cottonwood (Round Hole) Mine—gold, lead, silver.

FRANKTOWN, SW 5 mi., in Little Valley (a 4-mi. trough on the E slope of the Sierra Nevada Range at 6,500 ft. elev.), many profitable placers worked in early days.

GERLACH: (1) SW, in the Smoke Cr. Desert, area of Deep Hole, placers; (2) N 38 mi. on Rte. 34, ghost town of Leadville: (a) area mines —by-product gold; (b) N 1 mi., the Donnelly (Gerlach) Mine—lode gold, silver.

RENO: (1) NE 4 mi., the Wedekind Mine—by-product gold; (2) NW 6 mi., on N slope of the Peavine Mts., the Peavine district, in narrow ravine 1,500 ft. long and 2–3 ft. deep, rich placers worked 1876–1900s (first gold shipped to Sacramento, Calif., from Nevada came from here about 1863); (3) NW 10 mi. in the Peavine district: (a) the Reno and Crystal Peak mines—by-product gold; (b) area watercourse and bench gravels, placers; (4) N 34 mi. on Rte. 33 and dirt rd., on W side of Pyramid L., the Pyramid Mine—by-product gold.

WADSWORTH, NNW 9 mi. (to within about 12 mi. S of Pyramid L.), the Olinghouse (White Horse) district at lat. 39°40' N, long. 119°25' W (tot. prod., 1902–59, about 36,000 gold ozs.): (1) the Olinghouse (White Horse) Mine, major producer—lode gold; (2) in several ravines tributary to Olinghouse Canyon, numerous rich placers worked prior to 1900 with pans and rockers and again in mid-1930s—fine to coarse gold; (3) all regional watercourse, bench, and slopewash gravels, placers (evidences of considerable work).

WASHOE CITY (17 mi. S of RENO on U.S. 395), N 1 mi., the Galena Mine—by-product gold.

White Pine Co.

Between 1903 and 1959, White Pine County produced 2,049,895 ozs. of lode and placer gold.

AREA: (1) far W part of co., at Bald Mt. (about 75 mi. S of ELKO in Elko Co. and N of Pancake Summit on U.S. 50): (a) many area mines —by-product gold; (b) regional watercourse and slopewash gravels, placers; (2) far NW part of co., on W slope of the Ruby Range at 7,400 ft. elev., the Bald Mt. (Joy) district, in Water Canyon and tributary ravines: (a) productive placers extending up canyon to 1 mi. W of old camp of Joy; (b) richest placers at lower end of canyon—coarse gold, nuggets.

BAKER (4 mi. W of the Utah line and 5 mi. S of U.S. 50): (1) W, in the Snake Range, area mines—lode gold; (2) the Eagle district, the Kern, Pleasant Valley, Regan, and Tungstonia mines—by-product gold; (3) NW 13 mi. to W-trending dirt rd. across mts. to Spring Valley, about halfway across to old camp of Osceola (and district), also reached 35 mi. SE of ELY via U.S. 50 and turnoff in Spring Valley, on W slope of the Snake Range (tot. prod., 1877–1959, of 91,555 ozs. of placer gold and 40,145 ozs. of lode gold): (a) area mines—lode gold; (b) along Dry Gulch, extensive placers worked mainly 1877–1907.

CHERRY CR. (district, 45 mi. N of ELY on U.S. 93, turn W 9 mi. on Rte. 35), in the Egan Range, tot. prod., 1861–1959, of 36,197 gold ozs.: (1) area mines, such as the Egan Canyon, Gold Canyon (by-product gold, with lead and silver), Teacup, Star, Exchequer, and Cherry Creek (major producers)—lode and by-product gold; (2) SE 18–36 mi. (accessible also NW from McGILL, q.v.) and 10 mi. E of old camp of Melvin, many area mines, e.g., the Aurum (Muncy Cr.), Queen Springs, Ruby Hill, Schellbourne, Shell Cr., Siegel, Silver Canyon, Silver Mountain, etc.—by-product gold.

ELY-EAST ELY: (1) area, the Dely (Robinson) district, tot. prod., 1868–1959, of 1,959,659 gold ozs. (largely as by-product of copper ores): (2) S 19 mi., on S side of Ward Peak (elev., 10,936 ft.), the Ward (Taylor) Mine—by-product gold. This is the site of the Ward Charcoal

Ovens Historic Monument; (3) SE 40 mi., cf. also Osceola under BAKER: (a) adjoining Osceola, in Dry and Grub (Wet) gulches, very productive placers; (b) E of the divide, in Weaver Cr. on crest of the divide the Summit Placers, extensively worked—fine to coarse gold; (c) at jct. of the two gulches, very rich placers (a nugget found in 1878 weighed 24 lbs., possibly the largest gold nugget ever found in Nevada)—fine to coarse gold, nuggets; (d) 3 mi. SW of Osceola, in Mary Ann Canyon, the Hogum Placers—fine gold; (e) N of the Hogum, on W slope of the mts., covering 4,000 acres, 19 placer claims quite well worked; (f) 3½ mi. E of Osceola, on E slope of the mts., the Black Horse Gold Mining Co. Mine (lode gold) and adjoining area gravel deposits—placer gold.

HAMILTON (about 35 mi. W of ELY and 10 mi. S from turnoff 9 mi. E of Little Antelope Summit; a noted silver camp of the late 1880s, now a ghost town), area old mines and prospects—by-product gold.

McGILL: (1) NW 3 mi., the Duck Cr. (Success) Mine—by-product gold; (2) S 3 mi. on U.S. 93 to W-trending dirt rd.: (a) NW 21 mi., old camp of Steptoe, the Granite Mine—by-product gold; (b) NW 28 mi., turn SE toward the Magnuson Ranch, the Warm Springs Mine—lode gold. This rd. also leads to CHERRY CR., q.v.

STRAWBERRY (29 mi. NE of EUREKA in Eureka Co. via Newark Pass), on E slope of the Diamond Mts., the Newark Mine—by-product gold.

NEW HAMPSHIRE

Gold occurrences in New Hampshire have been reported on and off for at least two hundred years but never in commercially valuable deposits. The source of the gold is not known, but it probably reflects similar occurrences in the Chaudière River watershed of southern Quebec and is, therefore, relatable to gold occurrences in nearby Maine. Panning for gold is a regular summer hobby activity in the state's northern streams, especially around the headwaters of Indian Stream in the extreme northern part of Coos County.

For information, write: Department of Geology, University of New Hampshire, Durham, New Hampshire 03824, or Dartmouth College, Hanover, New Hampshire 03755.

Coos Co.

MILAN, area lead-copper-zinc mines—by-product gold, silver.

PITTSBURG, N, into extreme NW part of co. (no rds. on most maps), headwaters of Indian Stream (including the W, middle, and E branches), all gravel bars, beds, benches, and terraces—placer gold.

Grafton Co.

BATH, numerous area base-metal mines—by-product gold.

LISBON, area mines (copper, lead, silver)—by-product gold.

LITTLETON: (1) area base-metal mines—by-product gold; (2) a mineralized belt containing many mines extends SW along Rte. 10 for 12–15 mi., to include LYMAN, LISBON, and BATH—by-product gold.

LYMAN: (1) area copper-lead-silver mines—by-product gold; (2) the Dodge Mine, tot. prod., 1865–75, of some $75,000 in native gold before the veins pinched out into barren slate at 100-ft. depth

TINKERVALE, NW, on Gardner's Mt. (elev., 2,330 ft.), area base-metal mines—by-product gold.

NEW MEXICO

The first discovery of gold in New Mexico was in 1828, when the Old Placers were opened in the Ortiz Mountains south of Santa Fe to inaugurate an era of placer mining which, along with later lode-gold mining, was to produce a total of 2,267,000 gold ozs.

The areas considered most promising for gold hunting are placer gravel deposits that have eroded from the known lode-gold mines. One serious drawback to commercially profitable placering is the severe shortage of water for sluicing and hydraulicking. However, waterless areas that show good gold concentrations by panning may, sometimes, be profitably worked with a dry rocker, a wind or blown-air device, or by winnowing. Unfortunately, the finer colors are usually lost.

For more information, write: Director, State Bureau of Mines and Mineral Resources, New Mexico School of Mines, Socorro, New Mexico 87801.

Bernalillo Co.

AREA, far E part of co., in Tijeras Canyon (now within the Sandia military base and inaccessible to outsiders), the Tijeras Canyon district, tot. prod., 1882–1903, estimated at 34,488 gold ozs. from a number of lode mines in area.

Catron Co.

Catron County ranks third in New Mexico, with a gold production of 362,225 ozs., entirely from the Mogollon district.

AREA, in SW cor. of co., 85 mi. NW of SILVER CITY in Grant Co. by good rd. into the Mogollon (pronounced "Muggy-own") Mts., the Mogollon (Cooney) district, discovered in 1875 but not operable until the defeat of Geronimo 10 years later, area mines in quartz, moderately active through 1946—lode gold.

Colfax Co.

The metal-mining districts of Colfax County lie just south of the Colorado boundary in the Cimarron Range, extending along the west edge of the county.

CIMARRONCITO (ghost town near ELIZABETHTOWN), area mines—by-product gold (from copper ores).

ELIZABETHTOWN, on W side of the Cimarron Range, the Elizabethtown-Baldy district (tot. prod., 1866–1959, of 368,380 ozs. of lode and placer gold): (1) W side of mts.: (a) the Moreno Cr. Valley, especially the drainage headwaters area (including Grouse and Humbug gulches), very productive placers; (b) along Willow Cr., very rich placers (to bring water, the "Big Ditch" was dug 41 mi. long, from headwaters of the Red R. 11 mi. W of town, completed in 1868); (2) on E side of mts., area of Baldy Peak (elev., 12,441 ft.): (a) along Baldy (Ute) Cr., rich placers; (b) area lode mines, with est. prod. through 1941 of 221,400 ozs.—lode gold; (c) the Aztec Mine (probable source of the placer gold), discovered in 1868 to become the oldest and most productive gold mine in New Mexico, accounting for about 55 per cent of the early production—lode gold.

Dona Ana Co.

In Dona Ana County there is only one important gold-mining area, discovered in 1849, with a total gold production of about 13,500 ozs.

AREA, the Black Mt. and Texas Cr. districts, small area mines—free gold (near surface) and gold in sulfides in quartz.

ORGAN (district, including the N end of the Organ Mts. and extreme S end of the San Andreas Mts.; tot. prod., 11,435 gold ozs. as by-product of numerous regional copper-lead-silver-zinc mines.

Grant Co.

Several highly productive mineral districts in Grant County produced a total of 501,000 ozs. of placer and lode gold, with most gold production after 1912 coming as a by-product of base-metal refining.

AREA, W part of co.: (1) 16 mi. NE of LORDSBURG in Hidalgo Co., the Gold Hill district, discovered in 1884, area mines—lode gold in quartz; (2) a few mi. N, in S part of the Burro Mts., the Malone district, area base-metal mines—by-product gold.

CENTRAL (district, in E part of co., including SANTA RITA, HANOVER, and BAYARD, primarily base metals, with large-scale mining introduced in 1911; tot. prod., 1904–59, of 140,000 by-product gold ozs.): (1) in central part of district in the Santa Rita Basin, large-scale open-pit copper workings—by-product gold; (2) at BAYARD, area lead-silver mines —by-product gold; (3) at HANOVER, area lead-zinc mines (large-scale)— by-product gold.

DUNCAN, NE about 20 mi., in W part of co. about 4 mi. E of the Arizona line, the Steeple Rock district (tot. prod., 1880s–1955, of 34,050 gold ozs., mostly after 1932): (1) the Carlisle Mine (and others in the group), producer of $3,000,000 by 1897—lode gold; (2) the East Camp, area mines of rather recent development—by-product gold.

PINOS ALTOS (district, 8 mi. NE of SILVER CITY, in the Pinos Altos Mts. (tot. prod., 1860–1959, of 104,975 by-product gold ozs. from lead-silver ores and 42,647 ozs. of placer gold): (1) along Bear Cr. Gulch, its tributary Rich Gulch on the N, Whiskey Gulch on the E, and numerous other gulches in area of the old Gillette Mine (by-product gold), productive placers renewed annually by snow melt runoff; (2) some 30 regional mines (worked by 1862): (a) on E side of mts., principal producers—lode gold; (b) on W slope of mts., several good producers —lode gold.

SILVANIA (district), the Golden Eagle and Handcar mines—gold, in tetradymite.

Hidalgo Co.

Between 1922, when Hidalgo County was separated from Grant County, and 1952, the total gold production of 227,000 ozs. came as a by-product of copper-lead-silver ores.

HATCHITA, SW 12 mi., in central part of the Little Hatchet Mts.: (1) the Sylvanite district, discovered in the 1880s but gold not found until 1908, area mines—lode gold in quartz; (2) on W side of mts., all regional gulches and slopewash gravels, productive placers.

LORDSBURG: (1) S 2 mi. to rd. fork: (a) take W fork to ghost town of Shakespeare (the town "too mean to live"), area mines—lode gold; (b) SE, around cemetery about 1 mi., turn W to base of chain of low mts., many area pits and mine dumps—gold showings in copper minerals; (2) SW 3–8 mi., at N end of the Pyramid Mts., the Lordsburg district (discovered in 1870 but not well prospected until a stampede for dia-monds began prior to 1900 as the result of a fraudulent scheme involving salting one small area with diamonds; tot. gold prod., 1904–59, about 223,750 ozs.): (a) the Emerald Vein, especially the Eighty-five Mine (at 1,900 ft. the deepest in the state), producer of nine tenths of the tot. prod. in district)—lode gold; (b) S, to S end of district, the Pyramid camp (center of area containing some 85 old mines, e.g., the Atwood, Manner, Silver and Gold, etc., all primarily copper-silver)—by-product gold.

STEINS PASS (sta. on the Southern Pacific RR), N, and close to the Arizona line, the Kimball (Steins Pass) district: (1) numerous area mines and prospects—by-product gold; (2) the Federal Group of mines, most productive of district—lode gold.

Lincoln Co.

The most productive sources of gold in Lincoln County are lode mines discovered in the 1880s, with a total production through 1943 of 163,647 ozs.

Ancho, SE 7 mi., in central part of co. in the low Jicarilla Mts. at N end of the Sierra Blanca Range, the Jicarilla district: (1) area watercourse and slopewash gravels, especially along Ancho Cr., good placers (worked in 1850, but water shortage prevented large-scale operations); (2) area mines, probable sources of the placer gold—lode gold.

Carrizozo, NE 10–12 mi., around Baxter Mt. in the White Oaks Mts. (a N continuation of the Sierra Blanca Range), the White Oaks district, tot. prod., 1879–1940, of about 146,500 gold ozs.: (1) area mines, especially the North Homestake, South Homestake, and Old Abe mines, principal producers—lode gold; (2) regional watercourse and slopewash gravels around Baxter Mt., intermittently productive placers.

Nogal: (1) NE of Nogal Peak, in Dry Gulch, good placers discovered in 1865; (2) SW 6 mi., in the Sierra Blanca Range, the Nogal district, comprising early camps scattered over the mts. SE of Carrizozo of Vera Cruz, Parsons (Bonita), Schelerville (Church Mt.), and Alto (Cedar Cr.), tot. prod., 1868–1935, of about 12,850 ozs. of gold: (a) the American, Helen Rae, and nearby mines—gold by-product of copper-lead-zinc ores; (b) the Hopeful and Vera Cruz mines—by-product gold.

Otero Co.

Orogrande (SW cor. of co., about 50 mi. NNE of El Paso, Tex.), in the isolated Jarilla Hills, the Jarilla (Orogrande) district, tot. prod., 1905–18, of about 18,600 gold ozs.: (1) area mines—lode gold (with "3 ozs. of silver to every oz. of gold"); (2) SE slope of the Jarilla Hills just E of the Nannie Baird Mine (lode gold), all watercourse and slopewash gravels, placers (worked by dry-wash methods).

Rio Arriba Co.

Bromide, area copper-molybdenum mines—by-product gold.

El Rito, N 4 mi., in the Chama Basin (largely between El Rito Cr.

and Arroyo Seco), the El Rito district, 10 mi. long N–S by about 4 mi. wide (achieved dubious fame from a vastly overblown stock-selling scheme that boasted a $264,000,000,000 potential), area prospects in conglomerate and sandstone, with highest assays revealing $.30 per ton ($20.67 per oz. price).

HOPEWELL (about 15 mi. W of TRES PIEDRAS in Taos Co. and W extension of the Bromide district, about 25 mi. S of the Colorado line; tot. prod., 1881–1959, over $300,000 of lode and placer gold): (1) area mines and prospects in quartz—free gold, with pyrites; (2) in Eureka Gulch, placers produced over 90 per cent of district's total gold production: (a) immediately W of town, where valley narrows to a steep-sided channel, the Fairview placers, very rich—large nuggets; (b) 1 mi. farther down gulch, at jct. with a branch of Vallecitos Cr., the Lower Flat placers (gravels 35 ft. deep; water shortage put halt to hydraulicking efforts)—placer gold.

Sandoval Co.

BERNALILLO, E 8 mi., in N end of the Sandia Mts., the Placitas district: (1) area gravel-conglomerate deposits—placer gold ($1,013 produced in 1904, only recorded production in co.); (2) area lead-copper-silver mines—by-product gold.

COCHITI (Bland) district, tot. prod., 1870s–1940, about 41,500 gold ozs.: (1) area mines—lode gold; (2) the Albemarle Mine, discovered 1894, major producer—lode gold.

San Miguel Co.

PECOS, N, along the Pecos R. Canyon on old rd. 14 mi. to the Terrero store, turn E (1 mi. uphill), a long mine dump extending to the Willow Cr. campground, the Willow Cr. district (tot. prod., 1881–1940, of 178,961 gold ozs.), the Pecos (Hamilton or Cowles) Mine, base metals —by-product gold.

Santa Fe Co.

Total gold production of Santa Fe County, coming primarily from placer deposits, is estimated at between 150,000 and 200,000 ozs.

CERRILLOS: (1) area mines and prospects—gold, with silver; (2) on E side of the Ortiz Mts., between the San Pedro Mts. and the Cerrillos Hills, a large area known as the Old Placers (Ortiz, Dolores) district, tot. prod., 1828–1959, of 99,300 gold ozs.: (a) old town of Dolores, at mouth of Cunningham Canyon, along Dolores Gulch and Arroyo Viejo, gravel deposits to 100 ft. thick—placer gold; (b) S side of the Ortiz Mts., area watercourse and slopewash gravels—abundant placer gold; (c) several area mines and prospects (known in 1833), small producers—lode gold.

GOLDEN-SAN PEDRO, halfway between, in the San Pedro Mts., the New Placers (San Pedro) district, tot. prod., 1839–1946, of about 115,700 gold ozs.: (1) in foothills of the San Pedro Mts., all watercourse and slopewash gravels on the N, W, and S sides, profitable placers; (2) Tuerto Cr. (near GOLDEN) and all its tributaries, especially promising placer showings (in the gold-bearing "cement beds"); (3) in Lazarus Gulch, area gravels, especially productive placers, but water in very short supply; (4) numerous small lode-gold prospects in regional porphyry and sedimentary rocks—free gold.

Sierra Co.

One of the foremost mining counties in New Mexico, Sierra County produced 183,900 ozs. of gold from several districts between 1877 and 1959.

AREA, E side of the Rio Grande R., on SW slopes of the Caballos Mts., in secs. 16, 17, 21, and 22, (16S–4W), mainly along Trujillo Gulch, the Pittsburg (Shandon) district, tot. prod., about $175,000 in placer gold: (1) area watercourse and slopewash gravels, productive placers; (2) granite exposures in the mts., mines in gold-quartz veins—lode gold (source of the placer gold below).

HILLSBORO (25 mi. SW of TRUTH OR CONSEQUENCES, 17 mi. N of LAKE VALLEY, and 16 mi. W of the Rio Grande): (1) area watercourse, terrace, and slopewash gravels, rich placers worked intensively through 1905 (tot. prod. through 1959, about 149,000 gold ozs., including production from a few area lode-gold mines above the placers); (2) SE 6 mi. and E of the Animas Hills, the Hillsboro (Las Animas) district, an area of 18 sq. mi. of dissected alluvial wash; (a) all regional watercourse, bench, terrace, and slopewash gravels, very rich placers, especially (b) the Luxemburg Placers, at apex of the fan drained by Grayback, Hunkidori, and Greenhorn gulches (rich placers worked by dry-wash methods or by washing with water brought in by tank cars); (c) in Snake and Wicks gulches, productive placers; (3) in the Animas Hills, N of the Rio Percha, in SW part of a mineralized area, numerous area mines, especially the Rattlesnake Mine (located in 1877)—lode gold.

TIERRA BLANCA (district), area mines—lode gold, with silver.

Socorro Co.

Much of the 32,000 ozs. of gold produced from Socorro County's numerous mining districts came as a by-product of base-metal ores between 1882 and 1959.

COONEY (Mogollon) district, including Mill Canyon, Silver Mt., and Rosedale, area copper mines—by-product gold.

MAGDALENA (district, and most important mining area in co. for lead and zinc; tot. prod., less than 4,000 gold ozs.): (1) numerous area mines —by-product gold; (2) SE 3 mi., the ghost town of Kelly, area zinc mines —by-product gold; (3) SW 25 mi., in SW cor. of co., in N part of the San Mateo Mts., the Rosedale district (tot. prod., 1882–1940, of about 24,190 gold ozs.), area mines, especially the Rosedale Mine—lode gold only.

SAN MARCIAL, SW 30 mi. (about 20 mi. S of the Rosedale Mine), in S end of the San Mateo Mts., the Rhyolite district: (1) numerous small prospects along lode veins in a wide fault zone of rhyolite porphyry— gold showings; (2) N 3 mi., the San Jose (Nogal) district, along both sides of Springtime Cr. for several thousand feet, the Pankey vein (prominently revealed along the surface), numerous area claims and small mines—lode gold.

Taos Co.

AREA, the Rio Grande Valley, an auriferous region embracing several hundred sq. mi., with most promising gold prospecting beginning at the mouth of the Red R. in the N part of co. and extending along the Rio Grande to the S boundary of co. (and on into adjoining Rio Arriba Co.

as far as EMBUDO): (1) the Rio Grande: (a) all river bars, relatively productive placers; (b) deep-river gravels (depths unknown) contain placer gold, locally rich; (2) all regional bench and terrace gravels (capped in large areas by basalts interbedded with sediments): (a) It is thought that hydraulicking the bench gravels might be feasible in thicker deposits, and (b) that drift mining of "pay streaks" could be productive—placer gold.

NORTH CAROLINA

The first gold discovery in North Carolina was no mere showing but a solid, two-fisted lump of shining yellow metal weighing 17 pounds. Twelve-year-old Conrad Reed found it in Little Meadow Creek, a stream crossing his father's farm in Cabarrus County, while shooting fish with a bow and arrow. Because his father John Reed did not know what it was, he used the heavy nugget as a doorstop for three years, then sold it to a Fayetteville jeweler for the "whopping" price of $3.50. When he heard how much the jeweler later received for it, John Reed demanded and received an additional $3,000. Following this experience, Reed and three helpers returned to the creek and began finding more extraordinary gold nuggets, the largest of which, picked up in 1803, weighed 28 pounds. Altogether, the "Reed Mine" produced 153 pounds of nugget gold to bring about the first bona fide gold rush in American history.

After the first lode-gold mine was found in Long Creek in nearby Stanley County in 1825, so important did gold mining become that a federal mint was established at Charlotte in 1837, open today as the Mint Museum of Art. However, the first gold coins minted in America were stamped out by a private jeweler, Christian Bechtler, who produced coins in denominations of $1.00, $2.50, and $5.00 from 1831 to 1857. The Civil War forced the closure of the federal mint in 1861, after which it served North Carolina gold miners simply as an assay office.

From 1799 through 1965, with gold mining still viable today, North Carolina produced 1,170,587 ozs. of gold. There are approximately 660 inactive gold mines and prospects scattered throughout the Piedmont and mountain sections of the state, usually grouped into six geomorphic belts: the Eastern Carolina, Carolina Slate, Charlotte (Carolina Igneous), Kings Mountain, South Mountain, and Western Belt. Throughout these mineralized belts, the modern-day gold hunter will find literally hundreds of auriferous streams in the gravels of which to wield pick, shovel, and pan. There are no state laws pertaining specifically to prospecting for gold outside state parks. However, establishing a claim and performing any actual mining must be done in accordance with the North Carolina "Mining Act of 1971," a copy of which may be obtained at the office listed below.

Although regional placer deposits have been worked periodically throughout the years, much of the state's gold production has come from the lode deposits. Very many of the old mines, long since abandoned, have workings filled with water, surrounded by weathering dumps and dilapidated structures. Considerable danger may attend any effort to enter the old workings. In the more populated areas, housing developments have encroached over some mines, so that once-important gold producers have vanished entirely.

For more information and brochures, write: Director, Office of Earth Resources, Division of Mineral Resources, North Carolina Department of Natural and Economic Resources, Raleigh, North Carolina 27687.

Anson Co.

AREA, all regional stream gravels—placer gold.

ANSONVILLE, FAIRVIEW: (1) regional stream gravels—placer gold; (2) area old mines—lode gold, with pyrite, lead, and zinc.

Ashe Co.

AREA, S part of co., near headwaters of the New R., the Copper Knob (Gap Cr.) Mine—by-product gold.

JEFFERSON (E part of co.), E 8.2 mi. and 1 mi. N of Rte. 88, the Ore Knob Mine (tot. prod., 1855–1962, of 9,400 ozs.)—lode gold.

Buncombe Co.

AREA, along Cane Cr., all gravel bars and benches—placer gold.

Burke Co.

AREA: (1) very many auriferous stream and cr. gravel deposits scattered over co.—placer gold; (2) SW cor. of co., near Bee Bridge: (a) along Brindleton Cr., numerous lode mines and prospects—lode gold; (b) gravels of Hall and Silver crs.—placer gold; (3) at Brown Mt., area mines—lode gold, with platinum; (4) at Scott's Hill, area silver mines —by-product gold; (5) at Sugar Mt., area mines—minor lode gold.

BRIDGEWATER, area mines—lode gold, with manganese.

BRINDLETOWN, SE 8 mi., at S end of Richland Mt., on upper part of the First Broad R., placers unsuccessfully worked, 1953–54.

MORGANTON (in SW part of co.), SW 13 mi., in the South Mt. area, the Mills Property (Brindletown placers), tot. prod., 1828–1916, of $1,000,000—placer gold: (1) along Brindle Cr.: (a) originally, many placer operations; (b) several area saprolite mines, in quartz—lode gold; (2) N 13 mi., on Kingy Branch tributary of Upper Cr., the Brown Mountain Mine, in quartz—lode gold.

Burke, McDowell, and Rutherford Cos.

REGION. The Blue Ridge region, in which the South Mt. Belt, comprising the South Mt. Range, forms one of the most prominent eastern outliers of the Appalachian Mt. system, also constitutes the most important gold-mining area of North Carolina. The auriferous region embraces from 250 to 300 sq. mi., and panning for gold can be successful in practically all stream and watershed gravel deposits.

Cabarrus Co.

AREA (with overlap into S part of Rowan Co., q.v.), very many old mines predating the Civil War: (1) all regional stream and bench gravels —placer gold; (2) regional mines—lode gold.

CONCORD (E part of co.): (1) area old mines—gold, copper; (2) SE 6 mi.: (a) the Phoenix Mine—lode gold; (b) 1 mi. S of the mine, the Tucker (California) Mine—lode gold; (c) many other nearby mines and prospects—lode gold; (3) SE 10 mi., the Rocky River Mine—gold, with lead and zinc; (4) SE 11 mi., the Allen Furr Mine (23 mi. E of CHARLOTTE)—lode gold, with zinc; (4) SE 12 mi., the Nugget (Biggers) Mine

—coarse gold, lead; (5) S 13 mi., the Pioneer Mills Group of mines (not worked since the Civil War)—gold, molybdenum.

FAGGART: (1) E 1½ mi., the Barnhardt Mine—lode gold; (2) NE 3 mi., the Faggart Mine—lode gold.

GEORGEVILLE (SE part of co.), S 2 mi. and 2.8 mi. NW of LOCUST, the Reed Mine (original gold discovery in state in 1799, tot. prod. through 1930s of 50,000 gold ozs.): (1) along Little Meadow Cr., still occasionally worked—placer gold; (2) just above, the mine itself extending about 2,000 ft. along a ridge E of the cr.: (a) the Upper Hill workings, and (b) the Lower Hill workings—lode gold. In 1972 the Reed property was purchased by the state, with plans to develop the mine into a state historic site.

MOUNT PLEASANT: (1) SW 5 mi., the Harkey Mine—lode gold, with copper; (2) NW 8 mi., the Snyder Mine—lode gold.

TUCKER, N 3 mi., the Quaker City Mine—lode gold.

Caldwell Co.

AREA, around Grandmother's Mt., all stream gravels, placers.

COLLETTSVILLE, S, the Hercules Mine (12 mi. N of MORGANTON in Burke Co.)—lode gold.

HARTLAND, NW 1½ mi., adjoining mines of Miller and Scott Hill—lode gold.

Caswell Co.

AREA. The Carolina Igneous Belt (comprising Caswell Co. and parts of the adjoining cos.) is the second most important gold-mining region of North Carolina, varying from 15–30 mi. wide. Within the Caswell Co. portion, there are very many old workings: (1) all regional stream gravel deposits—placer gold; (2) associated mines (probable sources of the placer showings)—lode gold.

Caswell, Guilford, Davidson, Rowan, and Mecklenburg Cos.

REGION. These counties comprise the Carolina Igneous Belt. The most productive gold lode mines and placers occur in the counties other than Caswell Co.

Catawba Co.

AREA, Hooper's Quarry—gold, with calcite, graphite, pyrite.

CATAWBA, SE 5 mi., the Shuford Mine (and quarry), worked by dragline scraper, in numerous quartz seams—saprolite gold.

MAIDEN, many area mines and prospects—lode gold.

NEWTON: (1) area stream gravels—placer gold; (2) area mines—lode gold.

Cherokee Co.

AREA, along the Valley R.: (1) all regional sand and gravel bars, benches, old channels—placer gold; (2) all regional tributaries, in gravel deposits—placer gold.

MARBLE: (1) SE 1 mi., along the Valley R., in gravel deposits, placers; (2) NW, in the Snowbird Mts., the Parker Mine   lode gold.

MURPHY: (1) numerous area mines—lode gold, with lead and silver; (2) the No. 6 Mine—lode gold, lead.

UNAKA, E 1½ mi. on rd. to MURPHY, along Beaverdam Cr., numerous gravel deposits—placer gold.

VALLEYTOWN, all regional mts. extending from town 12–15 mi. to Vengeance Cr., along the lower slopes: (1) cr. and slopewash gravels, placers; (2) many area "diggings"—free gold.

Cleveland Co.

KINGS MT. (SW part of co.): (1) the Kings Mt. Mine and (2) 4 mi. E, the Crowder's Mt. Mine, rich producers—lode gold.

Davidson Co.

AREA: (1) all regional watercourse and bench gravels—placer gold possibilities; (2) many mines scattered throughout co.—lode and by-product gold.

LEXINGTON (heart of the Cid district, occupying about 125 sq. mi. in S part of co., extending from the Yadkin R. on the SW to about 1 mi. NE of CID; tot. prod., 1832–1907, of about 20,000 gold ozs.): (1) S and E, many abandoned mines—lode gold; (2) SE 5 mi., old camp of Silver Hill: (a) the Silver Hill Mine, silver—by-product gold; (b) S 1 mi., on

W side of Flat Swamp Cr. Valley, NW of the Silver Valley crossrds., the Silver Valley Mine—by-product gold; (c) W 5 mi., the David Beck's Mine—by-product gold; (3) E 6 mi., the Conrad Hill Mine (big producer in 1832 and first discovery site of gold in co., now inaccessible, with buildings in ruins)—lode gold; (4) SSW 8.9 mi. and 7 mi. NW of DENTON, the Silver Hill Mine, q.v. above.

THOMASVILLE: (1) SE 1½ mi., the Loftin Mine—lode gold; (2) SE 2 mi., the Lalor (Allen) Mine: (a) on dumps—gold showings, with pyrite; (b) ½ mi. W, the Eureka Mine—lode gold.

Davie Co.

AREA: (1) many old mines scattered throughout co., e.g., the Butler, County Line, Isaac, Allen, etc.—lode gold; (2) at Callahan Mt., several area mines—lode gold.

Franklin Co.

No early records were made of gold production, but modern estimates attribute several hundred thousand to one million dollars to placer workings and, more importantly, the Portis mine.

CENTERVILLE: (1) along Rte. 58, on rd. to INEZ, numerous old mine dumps—gold showings; (2) near Sandy Cr., on S side of first rd. out of town leading NW from Rte. 561, the Van Alston Prospect—lode gold.

DICKENS, E 2.4 mi., an old mine—lode gold.

JUSTICE: (1) NE 2.4 mi., and (2) ESE 1.6 mi., old mines—lode gold.

LOUISBURG, E 18 mi., in area streambeds—placer gold, occasional diamonds.

WOOD (P.O.), in NE cor. of co.: (1) just E by N 2.4 mi. and 0.4 mi. W of the Nash Co. line, the Portis placers (955 acres), sporadically worked, 1835–1935, tot. prod., about $9,000,000 (inadequate water prevented large-scale operations)—placer gold, with diamonds; (2) S about 1 mi., a group of mines straddling the Nash Co. line to N of Sandy Cr.—lode gold; (3) near Fishing Cr., the White House Tract (613 acres), in residual soils and gravel beds—free gold.

Gaston Co. (with overlap into Cleveland Co., q.v.)

ALEXIS: (1) at Clubb's Mt., area stream gravels, placers; (2) on Crowder's Mt., area stream gravels, productive placers.

CRAMERTON (S, in extreme SE cor. of co.): (1) the Oliver Mine (12 mi. SW of CHARLOTTE in Mecklenburg Co.), on W side of the Catawba R., with adjacent other mines—abundant lode gold, with lead (traditionally opened and operated for lead before 1776); (2) the McLean (Rumfeldt) Mine, 15–16 mi. SW of CHARLOTTE; (3) the Duffie Mine, and (4) the Rhodes Mine (18 mi. SW of CHARLOTTE)—all lode gold.

CROWDERS, the Kings Mt. area (cf. under Cleveland Co.).

DALLAS, NW 6 mi., the Long Cr. Mine—lode gold.

GASTONIA, numerous area mines in the Kings Mt. Belt area (though not actually in the belt itself)—lode gold.

Granville Co.

AREA: (1) all regional stream gravels—placer gold; (2) N half of co (with overlap into Person Co.), very many regional old copper-gold mines, e.g., the Royster (Blue Wing), Holloway, Mastodon, Buckeye, Pool, Gillis, Copper World, Yancy, etc.—lode gold, copper.

Guilford Co.

GREENSBORO (SW part of co., est tot prod, 1850s–80s, of $225,000 from several mines): (1) SW 5 mi., the Fisher Hill Mine—lode gold, with pyrite; (2) SE 6 mi.: (a) the Hodges Hill Mine—gold, copper; (b) the Twin Mine—lode gold; (3) SSW 6 mi., the Mills Hill Mine—free gold, with pyrite; (3) SW 8.1 mi. and 2.6 mi. ESE of JAMESTOWN, q.v., the Gardner Hill Mine: (a) the mine itself—lode gold, with copper; (b) area stream and bank gravel deposits, placers; (4) S 9.3 mi. and 8.1 mi. SE of JAMESTOWN, the North Carolina (Fentress) Mine on a vein traceable for 3 mi. along a quartz outcrop (before 1853 only lode gold was produced; afterward, only copper until 1865); (5) SW 10.6 mi. and 1.8 mi. S of JAMESTOWN, the Lindsay and Jacks Hill mines (both abandoned since the 1880s)—lode gold.

JAMESTOWN: (1) cf. also under GREENSBORO (2) S 2¼ mi., on N side of Rte. 29, the North State (McCullough) Mine—lode gold, with copper.

Halifax Co.

HOLLISTER, SE 4.2 mi. and 4.2 mi. SW of RINGWOOD, the H and H (House) Mine, 1940–57, lead, silver, zinc—by-product gold.

Henderson Co.

ASHEVILLE, SW 12 mi. and W of Boylston Cr. on S slope of Forge Mt., the Boylston Mine (worked in 1880s and again in 1935)—lode gold.

Jackson Co.

AREA, countywide streams, especially all along the slopes of the Blue Ridge, placers.

CASHIERS, area stream gravels—placer gold.

HOGBACK, area of Chimney Top Mts., regional stream gravels—placer gold.

Lincoln Co.

IRON STATION, NE 4 mi., the Graham Mine—lode gold, copper.

LINCOLNTON, nearby, the Hope Mine—lode gold.

Macon Co.

FRANKLIN, N 6 mi., in the Cowee Valley (take Rte. 23–141 N for 3 mi., turn NW on Rte. 28 about 4 mi. into signposted mining district) and up Cowee Cr. past the Cowee Baptist Church, many area mines for gemstones (fee charged to collectors at most mines), on dumps—gold-bearing minerals, often collectors' specimens.

McDowell Co.

AREA: (1) all regional stream gravels—placer gold; (2) Glade, North Muddy, and South Muddy crs., all gravels—placer gold, platinum; (3) at Hunt's Mt., area stream gravels—placer gold, platinum.

BRACKETTOWN, in the valley of the headwaters of South Muddy Cr., the Marion Bullion Co. Mine (copper, lead, zinc)—by-product gold, with platinum and diamonds.

DEMMING (SE part of co.), nearby: (1) the Sprouse Mine, 1885–1935, and other area mines—lode gold; (2) area stream gravels and slopewash deposits—placer gold.

DYSARTSVILLE: (1) numerous area Civil War era mines—lode gold; (2) all regional streams and benches, in gravels—placer gold; (3) South Muddy Cr. crossing of Rte. 26, in stream gravel deposits—placer gold, with diamonds.

GLENWOOD (S part of co.), S 5 mi., on the Second Broad R., the Vein Mt. Mine (a series of 13 parallel quartz veins crossing Vein Mt. in a ¼-mi.-wide belt): (1) extensive area placer operations prior to 1908; (2) the Nichols Vein, large and productive—lode gold, copper.

Mecklenburg Co.

CHARLOTTE: (1) within the city limits: (a) 1 mi. SW of intersection of Trade and Tryon Sts., the Rudisil Mine, operated 1829–1938, producing $1,000,000; and (b) 2,500 ft. N 25° E (presumably on opposite end of the same vein), the St. Catherine Mine, combined tot. prod. both mines, at least 60,000 ozs.—lode gold; (2) W 2½ mi., the Clark Mine —lode gold; (3) NW 5½ mi., between Rossel's Ferry and Beattie's Ford rds., the Capps Mine (famed for quality and amount of ore)—lode gold; (4) W 9 mi., the Stephen Wilson Mine—lode gold; (5) NW 11 mi.: (a) the Hopewell (Kerns) Mine, and (b) several other nearby mines—lode gold, with pyrite and copper.

Montgomery Co.

A total of around 50,000 gold ozs. were produced in Montgomery County, largely from two mines discovered in 1900 in the eastern part of the county.

AREA: (1) Beaver Dam Cr. jct. with the Yadkin R., NE 2 mi., the Beaver Dam Mine—free gold, with pyrite; (2) extreme NW cor. of co., many regional mines, e.g., the Bright, Ophir, Dry Hollow, Island Cr., Deep Flat, Spanish Oak Gap, Pear Tree Hill, Tom's Cr., Harbin's, Bunnell Mt., Dutchman's Cr., and Worth, all of which played an important part in the state's mining history—lode and by-product gold.

CANDOR: (1) W 2½ mi., the Montgomery Mine—lode gold; (2) W 3 mi., the Iola Mine—gold, with pyrite.

ELDORADO: (1) many area old mines—by-product gold (with silver, copper, zinc): (2) N 1½ mi.: (a) the Coggins (Appalachian) Mine—lode gold; (b) E, the Eldorado Mine—lode gold, with silver and pyrite; (3) SE 2 mi., on E side of the Uwharrie R., an old mine—lode gold; (4) E 3 mi., the Riggon Hill Mine—lode gold; (5) N 3 mi., near the Randolph Co. line, the Russel Mine—lode gold; (6) S 8 mi., the Moratock Mine —gold, copper.

OPHIR, W 2–3 mi., along the Uwharrie R.: (1) the Steele Mine (now nearly inaccessible, and 1832 structures in total ruins)—lode gold; (2) the Russel Mine (accessible from ELDORADO, q.v., tot. prod., about 15,000 ozs.)—lode gold.

STAR, W 3 mi., the Carter Mine (one of oldest in state)—lode gold.

TROY, NE 14 mi., the Black Ankle Mine—lode gold.

WADEVILLE, W, to within 3 mi. of jct. of Rtes. 24 and 27, the Sam Christian Mine—native gold (notable for large nuggets).

Moore Co.

AREA, numerous mines in co., such as the Bat Roost, Schields, and Grampusville—lode gold.

CARTHAGE: (1) NW 8 mi., the Bell Mine—lode gold; (2) NW 11 mi., the Burns (Alfred) Mine—lode gold: (a) N ¾ mi., the Cagle Mine, and (b) ¼ mi. W of the Cagle, the Clegg Mine—lode gold.

Nash Co.

ARGO, 5 mi. E of Ransom's Bridge, the Mann-Arrigton (Argo) Mine, also accessible from AVENTON, q.v.

AVENTON (NW cor. of co.): (1) NW 1.7 mi., the Arrington Mine— lode gold; (2) 1 mi. NW of the Arrington, the Portis Mine, on dumps —gold showings; (3) SW 2.4 mi.: (a) the Mann-Arrington (Argo) Mine, last worked in 1894 but with depth explorations in 1930s—lode gold; (b) 1 mi. W, 3 area mines along the Franklin Co. line—by-product gold; (4) many other regional mines in co., e.g., the Conyer's, Nick Arrington, Thomas, Kerney, Taylor, Mann, Davis, etc.—lode gold.

Nash and Franklin Cos.

REGION. The north portions of these two cos., plus the southern portions of Warren and Halifax cos., constitute the auriferous Eastern Carolina Belt, embracing about 300 sq. mi. Very many lode-gold mines are shown on quadrangle maps covering the region, and practically all streams, benches, and slopewash gravel deposits show placer gold.

Orange Co.

CHAPEL HILL, NW 12 mi., the Robeson Mine (major gold producer, now wholly abandoned), quartz veins—lode gold.

Person Co.

AREA, part of the Carolina Slate Belt (including sections of adjacent cos.), very many once-rich mines shown on quadrangle maps—lode gold, placers.

Polk Co.

AREA: (1) all regional stream, bench, and slopewash gravel deposits— placer gold; (2) very many regional mines in extensions of the South Mountains, e.g., the Double Branch, Red Spring, Splawn, and Smith mines, all important early producers—lode gold, copper, pyrite.

Randolph Co.

ASHEBORO: (1) W to near co. line and 18 mi. E by S from LEXINGTON in Davidson Co., the Jones (Keystone) Mine—native gold, with pyrite; (2) NE 10 mi., the Alfred Mine (open cuts)—gold, pyrite; (3) WNW 12 mi., on E side of the Uwharrie R., the Hoover Hill Mine (principal gold producer in co., tot. prod., 1848–1917, about 17,000 ozs.)—lode gold.

FARMER, also access to the Hoover Hill Mine, q.v. above.

Rowan Co.

Following the Reed discovery in 1799 in adjoining Cabarrus County, Rowan County was extensively prospected, and very many placer and lode-gold deposits were opened to exploitation. Between 1824 and 1915, a total of around 160,000 gold ozs. were produced.

GOLD HILL (district): (1) SW 0.6 mi.: (a) the Gold Hill Mine (Randolph shaft), tot. prod. through 1935, about $1,650,000 in lode gold; (b) such other area mines as the Randolph, Barnhardt, Barringer, Miller, Old Field, Honeycutt, Troutman, Union Copper, and Whitney, all major producers of the district—lode gold; (2) SW 1½ mi., the Mauney Mine—lode gold; (3) E: (a) 3 mi., the New Discovery Mine, and (b) ½ mi. farther E, the Dunns Mt. Mine (3½ mi. E of SALISBURY)—lode gold, with pyrite; (4) E 6 mi.: (a) on the Yadkin R., the Reimer Mine —lode gold; (b) 1½ mi. E of the Reimer, the Bullion Mine (little worked, more as traces of gold in a quartz outcrop)—free gold; (5) SE 9 mi.: (a) the Gold Knob Mine—lode gold; (b) 1 mi. farther S, the Dutch Cr. Mine—native gold, with pyrite.

Rutherford Co.

AREA: (1) many regional old mines, mostly quartz—lode gold; (2) at Sandy Level Church, area mines, on dumps—gold, platinum, rare diamonds; (3) the J. D. Twitty Placer Mine—raw gold, nuggets, occasional diamonds.

RUTHERFORDTON: (1) area mines, especially the Ellwood and the Leeds—lode gold; (2) N 5 mi., on divide between Cathey's Cr. and the Broad R., the Alta (Monarch, Idler) Mine, in 13 parallel quartz veins —lode gold.

Stanly Co.

ALBEMARLE: (1) NW 2½ mi., the Haithcock and Hern mines—lode gold; (2) E 4 mi., the Crawford (Ingram) Mine—placer gold; (3) NW 7 mi., the Parker Mine (long noted for its spectacularly large nuggets) —lode gold.

MISENHEIMER (4 mi. SE of GOLD HILL in Rowan Co., q.v.), the Barringer Mine, first gold discovery in co.—lode gold.

NEW LONDON: (1) in town, the Parker Mine (one of first mines worked in state, tot. prod. through 1935 est. above 10,000 ozs.)—lode gold: (a) colluvial area gravel deposits, productive early placers; (b) associated quartz veins and stringers iinterlacing the nearby country rock —native gold; (2) E 1 mi., the Crowell Mine—free gold in pyrite; (3)

W, in regional stream gravels along Mountain Cr. and its tributaries (overlapping into Cabarrus Co.)—placer gold, with diamonds.

Swain Co.

FONTANA VILLAGE (SW part of co.), NE 2½ mi., the Fontana Copper Mine, 1926–44, and 1,700 ft. deep—by-product gold.

Transylvania Co.

AREA, S part of co., mainly along Georgetown Cr. between lat. 35° 03' and 35°38' N, long. 82°50' and 83°00' W, in N drainage of the Toxaway R., the Fairfield Valley placers (most active in the 1800s, tot. prod. between 10,000 and 15,000 gold ozs.).

Union Co.

MONROE, W, including much of the W side of co.: (1) regional mines, e.g., the Lemmonds (Marion), New South, Crump, Fox Hill, Phifer, Black, Smart, Secrest, and Moore Hill (group 2 mi. S of Indian Trail) —free and lode gold, with pyrite; (2) N 14 mi., in extreme NW cor. of co., the Crowell Mine: (a) on dumps—gold showings; (b) ¾ mi. SW, the Long Mine, and (c) 3 mi. SE of the Long Mine, the Moore Mine —by-product gold, with copper, lead, and zinc.

POTTER'S STATION, N 1½ mi., the Bonnie Belle (Washington) Mine —free gold, with pyrite.

WAXHAW (in W-central part of co.): (1) area stream gravels, original placer-gold discoveries extensively worked; (2) NE 2.9 mi. and 3 mi. NW of MINERAL SPRINGS, the Howie Mine (largest mine in co., operated 1840–1942, tot. prod. estimated at 41,300 ozs. of lode gold through 1934; called the Condor Mine, 1940–42, it was the largest gold producer in North Carolina): (a) all area stream, bench, and slopewash gravels—placer gold; (b) on adjoining slopes, stringers in schist impregnated with fine-grained quartz bearing pyrite and gold.

Watauga Co.

BLOWING ROCK, area stream gravels—placer gold.

BOONE, area stream and bench gravel deposits, especially along Hardings Cr.—placer gold.

Wilkes Co.

DEEP GAP, E 6 mi., the Flint Knob Mine—gold, lead, silver.

TRAP HILL, area of Bryan's Gap, on E face of the Blue Ridge in a bold outcrop of quartz (traceable for nearly 3 mi.)—free gold, pyrite.

Yadkin Co.

YADKINVILLE, SE 8 mi: (1) the Dixon Mine, good producer, 1894–1914—lode gold; (2) neighboring Gross Mine, minor producer, 1913–14 —lode gold.

OREGON

On the Great Seal of the state of Oregon there is a miner's pick that symbolizes how gold mining was the original mainstay of the economy of the Oregon Territory. The truly great gold areas of Oregon are diagonally opposite each other: the far southwest corner, where gold was discovered in Jackson and Josephine counties in 1852; and the far northeast corner, where gold was discovered at Griffin Gulch, Baker County, in 1861.

Gold mining continued at a fairly high level throughout all the years until halted overnight by the World War II Administrative Order L-208. Although this order failed to achieve its goal of forcing miners to seek employment in strategically important base-metal mining, it dealt a crushing blow to gold mining in Oregon. By the end of the war, practically all experienced gold miners were gone forever, most of the productive mines were filled with water, and virtually none of the mines were put back into production.

With the beginnings of 1975, a strong renewed interest in prospecting for gold began sweeping Oregon's gold districts, based on the realization that since 5,796,680 ozs. of gold were produced between 1852 and 1965, there is undoubtedly much more to be found. During the Depression years of the 1930s, thousands of men were able to pan out a day's wages in many areas, and each season to the present finds hundreds more amateur and professional gold seekers at work in practically all the auriferous counties. Gold seekers, many of them armed with metal detectors, are invading scores of long-abandoned ghost mining towns and finding new gold in their environs. Most of Oregon is public domain open to the prospector, and almost the only restrictions placed upon him are the universal laws of courtesy and common decency. Beyond that about all that the gold hunter needs to achieve success is a smiling Lady Luck at his side.

For information and brochures, write: State Geologist, State of Oregon Department of Geology and Mineral Industries, 1069 State Office Building, Portland, Oregon 97201.

Baker Co.

In 1861 Henry Griffin's gold discovery in the gulch that bears his name brought intensive prospecting and great bonanza finds to Baker County. Extremely rich placer deposits and their sources soon found in nearby lode veins produced a total of 1,596,500 ozs. of gold through 1959, with mining continuing to the present. By 1975 numerous companies were formed to initiate a new era of prospecting old districts in depth with geophysical instrumentation.

AREA, along the W drainage of the Snake R., between lat. 44°21' and 44°44' N, long. 117°03' and 117°18' W, the Connor Cr. district (tot. prod. through 1959 of 97,000 gold ozs. from lode mines and 6,100 ozs. from placer deposits): (1) along Connor Cr., rich placers found in 1860s; (2) the Connor Cr. Mine—free gold, pyrite.

AUBURN, in Blue Canyon, rich early-day placers.

BAKER: (1) S a few mi., Griffin Gulch (original first discovery of gold), the Baker district, tot. prod., 1906–59, of 36,152 gold ozs., half from placers; (2) WSW 4–6 mi.: (a) W½ sec. 22, the Dale Mine—free gold; (b) in upper Washington Gulch, in secs. 20 and 29, the Stub (Kent) Mine—lode gold; (3) S end of Elkhorn Ridge, all area watercourse gravels—placer gold; (4) W 6–7 mi., in T. 9 S, R. 39 E: (a) Salmon and Marble crs., rich area placers, especially the Nelson placers; (b) on Salmon Cr., above the Nelson diggings, in SW¼ sec. 8, the Carpenter Hill Mine—lode gold; (c) in McChord Gulch, in NE¼ sec. 7, the Tom Paine-Old Soldier Group (Yellowstone), tot. prod. of $36,000—lode gold; (5) E 10 mi., near Virtue Flat, the Virtue district (tot. prod., 1862–1957, of 126,000 ozs. of lode and placer gold): (a) all area gulches leading up to the Virtue and White Swan lode-gold mines—placer gold abundant; (b) numerous other productive mines, e.g., the Brazos, Flagstaff, Hidden Treasure, Carroll B, Cliff, Cyclone, etc.—lode gold; (6) NW 10–15 mi., on N side of Elkhorn Ridge in upper drainage of Rock and Pine crs., the Rock Cr. district (tot. prod., 1880s–1958, of about 51,000 gold ozs.): (a) on N fork of Pine Cr., the Baisley-Elkhorn Mine (principal producer discovered in 1882, with 2-mi. underground workings)—lode gold; (b) 2 mi. W of the Baisley, in the Rock Cr. drainage, the Highland and Maxwell mines, major producers—lode gold; (c) the Chloride, Cub, and Western Union mines, minor producers—lode gold.

COPPERFIELD (E end of Rte. 86, and 67 mi. ENE of Baker), the Homestead district on the Snake R., the Iron Dyke Mine (copper), tot. prod., 1910–34, of 34,967 ozs.—by-product gold.

DURKEE (23 mi. SW of BAKER on U.S. 30), W and NW, between lat. 44°17′ and 44°43′ N, long. 117°10′ and 117°41′ W, the Burnt Cr. district (including WEATHERBY, GOLD HILL, DURKEE, CHICKEN CR., and PLEASANT VALLEY, tot. prod., 1880–1959, at least 50,000 ozs. of lode gold and 3,500 ozs. of placer gold): (1) all Burnt R. tributary streams and gulches, especially Shirttail Cr., bars, benches, and terraces, productive placers; (2) SE 6–12 mi., the Weatherby district, straddling U.S. 30 along the Burnt R.: (a) N of hwy., along Chicken and Sisley crs., important placers and area lode mines such as the Little Bonanza, Little Hill, Gleason, and Hallock mines on Chicken Cr.—lode gold; (b) S 4 mi., the Gold Ridge Mine, and (c) SE 4 mi., the Gold Hill Mine—lode gold; (3) SSW about 15 mi., to Rye Valley (ghost town), at heads of Basin Cr. and S fork of Dixie Cr., the Mormon Basin district (overlapping W into Malheur Co.), between lat. 44°22′ and 44°31′ N, long. 117° 23′ and 117°40′ W, tot. prod., 1863–1948, of about 117,500 ozs. of lode gold and 56,200 ozs. of placer gold: (a) the Rye Valley area, all stream, gulch, bench gravels, extensive placers; (b) many rich lode mines straddling both sides of the Baker-Malheur Co. line—lode gold.

GREENHORN (district, 50 mi. W of BAKER via ghost town of Whitney, in E part of the Greenhorn Mts., with overlap into Grant Co.), tot.

prod., 1877–1959, of 89,200 ozs. of lode gold and 10,382 ozs. of placer gold (most of it after 1930): (1) SE 1 mi.: (a) in sec. 10, (10S–34E), the Don Juan Mine, and (b) in E½ sec. 15, the Golden Eagle Mine (high-grade pockets)—native gold; (c) in NW¼ sec. 16, the Banner Mine and adjoining Banzette Mine—lode gold; (d) in NE¼ sec. 17, the Diadem Mine—lode gold; (2) all watercourse, bench, and terrace gravels surrounding Winterwille, Parkerville, and McNamee gulches, productive placers; (3) E 5 mi., in sec. 10, (10S–35E), the Bonanza Mine (principal producer, 1877–1910, on dumps [reworked large-scale, 1974–75])—gold, with copper, lead, and silver; (4) N and NE, each 2 mi., into Grant Co., many rich-producer lode mines along both sides of the boundary line in secs. 3, 10, and 11, e.g., the Golden Gate, Red Bird, Owl, Virginia, and other important mines shown on area quadrangle maps; (5) W 2½–3 mi., into Grant Co.: (a) in NE¼ sec. 7, the Bimetallic Mine, with (b) ½ mi. N in SE¼ sec. 6, the Intermountain Mine—lode gold; (c) in sec. 17, the Harrison Group of Windsor, Psyche, and Big Johnny mines, high-grade—lode gold.

HALFWAY (52 mi. E of BAKER on Rte. 86), NNW 12 mi., old ghost town of Cornucopia near head of Pine Cr., tot. prod., 1880–1941, of 272,777 ozs. of lode and placer gold: (1) along Pine Cr.: (a) within 4 mi. of Cornucopia, the Cornucopia Group of mines scattered in secs. 27 and 28 on both sides of cr.—lode gold; (b) all gravel deposits in Pine Cr. and its tributaries—placer gold; (2) NW 1–2 mi., the Whitman, Union-Companion, Last Chance, Wallingford, and Valley View mines, big producers—lode gold; (3) N 2½ mi., on E side of Pine Cr. near top of Simmons Mt., the Simmons Mine—lode gold; (4) NW 3 mi. and just under 1 mi. N of (and on opposite side of Bonanza Basin) the Lawrence Portal and the Queen of the West Mine (copper, lead, zinc)—byproduct gold.

HEREFORD (36 mi. SW of BAKER on Rte. 7), the Upper Burnt Cr. district, a large area between lat. 44°15' and 44°36' N, long. 117°35' and 118°20' W, including BRIDGEPORT, BULL RUN, UNITY, and HEREFORD; tot. prod., 1890s–1959, of about 9,300 ozs. of lode and placer gold: (1) along Burnt Cr.: (a) all gravel bars, benches, and terraces—placer gold; (b) all tributary streams, ravines, gulches—placer gold; (2) numerous regional mines—lode and by-product gold.

MEDICAL SPRINGS (district, 18 mi. NE of BAKER on Rte. 203): (1) area small mines and prospects—lode gold; (2) NE 4 mi., on divide between two tributaries of Big Cr., the Twin Baby Mine—lode gold; (3) E 10–20 mi., at S end of the Wallowa Mts. (accessible also N from KEATING), in upper drainage of Eagle Cr. and adjoining area on the Powder R. slope drained by Clover, Balm, and Goose crs., the Eagle Cr. district (tot. prod., 1860s–1951, of 87,782 ozs. of lode gold and 69 ozs. of placer gold: (a) area cr. gravels, minor placers; (b) near top of the Powder R.-Eagle Cr. divide, on branch of Goose Cr., the Sanger Mine (principal producer)—lode gold; (c) S on rough rd. (and 8 mi. NE of KEATING), on Balm Cr., the Mother Lode Mine (copper)—by-product gold.

RICHLAND (40 mi. E of BAKER on Rte. 86): (1) SE, along W drainage of the Snake R., between mouths of Burnt R. (near HUNTINGTON) and Powder R. (near RICHLAND), the Connor Cr. district, tot. prod., 1860s–1959, of about 97,000 ozs. of lode gold and 6,100 ozs. of placer gold: (a) 3 mi. up Connor Cr. from mouth, the Connor Cr. Mine (principal producer)—lode gold; (b) 7 mi. below HUNTINGTON, on steep slope above the Snake R., the Bay Horse Mine (silver)—by-product gold; (c) regional stream, bench, and terrace gravels, minor placers; (2) NW 10–13 mi., between Eagle Cr. and Powder R., the Sparta (a ghost town) district, tot. prod. through 1959 of 35,200 ozs. of lode gold and 7,700 ozs. of placer gold: (a) Eagle Cr. gravels, extensive placers; (b) ¼ mi. above the Powder R., in Maiden Gulch, the Maiden Mine—free gold, with pyrite; (c) WSW 2½ mi., the Crystal Palace Mine—lode gold; (d) in Sparta, the Del Monte Mine, and ½ mi. SW, the Gold Ridge (New Deal) Mine—lode gold; (e) many other area mines, e.g., the Union, Lone Star, Keystone, Minnie May, Rosebud, etc.—lode gold; (f) W 2 mi., the Gem Mine (most extensively developed in district)—lode gold.

SUMPTER (district, 26 mi. W of BAKER in the Elkhorn Mts., tot. prod., 1913–54, of 296,906 ozs. of placer gold): (1) entire Powder R. Valley (8 mi. long by 1 mi. wide), completely dredged placers; (a) N and NW, along Cracker Cr. and McCully fork, extensive placers; (b) all tributary streams, especially Buck and Mammoth gulches, rich placers; (c) all regional bench and terrace gravels along the Powder R. and its tributaries, worked in many places—abundant placer gold; (2) N 6 mi., in headwaters of the Powder R. in drainage area of Cracker Cr. and McCully fork, the Cracker Cr. district (tot. prod., 1907–59, of 189,389 ozs. of placer gold, but mostly lode gold): (a) at ghost town of Bourne, the North Pole-Columbia Mine (most productive in district, lode vein crosses Cracker Cr.) and many equivalent mines along the vein—lode gold; (b) up McCully fork, the Bald Mountain and Ibex mines, big producers—lode gold; (c) up all branch rds. (steep, rough, not maintained), many old mines between 5,000- and 8,000-ft. elevations—lode gold.

UNITY (2 mi. N of jct. of U.S. 26 with Rte. 7, in far E part of co.), SW 5–9 mi., on N flanks of Mine Ridge and Bullrun Mt., the Unity (Bull Run) district: (1) in sec. 2, (14S–36E), the Bull Run Mine—lode gold; (2) in sec. 3, the Orion Mine—free gold; (3) in sec. 1, the Record Mine (major producer, tot. $103,000)—free gold.

Clackamas Co.

AREA, near head of Ogle Cr. (tributary of the Molalla R. in far SE cor. of co.), just N of the Marion Co. line, the Ogle Mountain Mine and mill (in the North Santiam district of Marion Co.), tot. prod., 1903–19, of $10,000—free gold.

Coos Co.

BANDON: (1) area: (a) ocean beach black-sand deposits, rich placers (periodically worked and reworked)—placer gold, platinum; (b) E, regional elevated marine terraces (to 170 ft. high), many prospect and mining attempts in ancient black-sand deposits—placer gold; (2) N, along the beach all way to Cape Arago, in black-sand deposits—placer gold (very fine), platinum; (3) N 6 mi., on Cut Cr. (reached from the Seven Devils Rd.): (a) the Pioneer Mine (in ancient elevated beach terrace black-sand deposit, sluiced around 1915 and still occasionally reworked)—placer gold; (b) just N, the Eagle Mine (cf. the Pioneer Mine); (4) N 10 mi., in sec. 10, (27S–14W), the Seven Devils (Last Chance) Mine, worked for gold and platinum in early days and for chromium from 1942–43.

COOS BAY, W to ocean beaches, black-sand deposits both N and S of the mouth of the Coquille R.—placer gold, platinum.

MYRTLE POINT (including the Eden and Randolph districts): (1) regional streambed, -bench, and -terrace gravel deposits—placer gold (very fine); (2) S, along the S fork of the Coquille R., on Poverty Gulch: (a) in SE¼ sec. 29 and NE¼ sec. 32, (32S–12W), the Divelbiss (Coarse Gold) Mine—lode gold; (b) N of Poverty Gulch, on N slope of Salmon Mt., the Salmon-Mt. Mine (tot. prod., about $100,000 from 1885–98), an extensive hydraulic placer operation; (c) 6 mi. SE of Poverty Gulch, in E½ sec. 23, (33S–12W), the Nicoli Group of mines—gold, with pyrite; (3) near head of Eden Valley, on upper W fork of Cow Cr.: (a) in N½ sec. 10, (32S–10W), extensive placers; (b) down the W fork to near Sweat Cr., in sec. 36, (31S–10W), extensive placers.

POWERS, in secs. 19 and 30, (33S–20E), the Independence Mine—lode gold.

Crook Co.

PRINEVILLE, E 26 mi., in hills bordering Ochoco Cr., in sec. 30, (13S–20E), the Howard district, tot. prod. through 1923 of $79,885: (1) lower part of Scissors Cr. (above its jct. with Ochoco Cr. from the SW), numerous small placers, intermittently worked; (2) W bank of Ochoco Cr., the Ophir-Mayflower Mine (principal producer)—lode gold.

Curry Co.

AREA: (1) ocean beaches along entire co., but especially N and S of the mouths of the Chetco and Rogue rivers, in black-sand deposits—placer gold, platinum; (2) Rogue R. gravel bars, benches, terraces (from MARIAL, q.v., in the NE cor. of co. to its mouth at WEDDERBURN)—abundant placer gold.

AGNESS: (1) area Rogue R. gravel bars, benches and terraces, placers; (2) old mines and prospects along the lower Illinois R.—gold by-product of copper ores.

BROOKINGS, NE, to headwaters of the Chetco R.: (1) in T. 38 S, R. 10 W: (a) near center of sec. 26, the Frazier Mine—free gold, pyrite; (b) in SW¼NE¼ sec. 24, the Golden Eagle Mine, a hydraulic placer; (c) near head of Slide Cr., in NE¼ sec. 14, SE¼ sec. 11, and W½ sec.

12, the Golden Dream (Higgins) Mine—lode gold; (d) center sec. 23, the Peck (Robert E.) Mine, good producer—lode gold; (e) near edge of sec. 35, the Young Mine, rich pockets—free gold; (2) in sec. 10, (38S–9W), the Hoover Gulch (Williams and Adylott) Mine, worked intermittently since 1900—lode gold.

CHETCO, CORBIN, ECKLEY, MARIAL, OPHIR, PORT ORFORD, AND SELMA, all regional streams (gravel bars, benches, terraces), placers formerly worked and still yielding colors and nuggets to the seasonal gold panner.

GOLD BEACH, area ocean beach black-sand deposits (once extensively worked)—rich placer gold and platinum showings.

MARIAL: (1) SW 1 mi., in secs. 17 and 20, (33S–10W), the Mule Mountain Group of mines (11 lode claims, 1 placer, 1 millsite)—gold; (2) NE 2 mi., in NW¼ sec. 3 and NE¼ sec. 4, the Mammoth Mine —lode gold; (3) N 2½ mi., the Marigold (Tina H., Lucky Boy) Mine —lode gold (in quartz); (4) NNE 4½ mi., near Saddle Peaks (reached by rough trail), the Paradise Mine—free gold.

PORT ORFORD: (1) area beach black-sand deposits—placer gold and platinum (fine to very fine); (2) along S fork of the Sixes R., and at heads of Salmon and Johnson crs., numerous extensive placers.

SIXES: (1) W: (a) on inside of Cape Blanco, in sec. 4, (32S–15W), the Madden Mine—lode gold; (2) on N side of Cape Blanco, beach deposits of black sand on both sides of mouth of the Sixes R.—placer gold and platinum; (2) E on dirt rd. to near head of Rusty Cr.: (a) in SE¼ sec. 23, (32S–13W), the Big Ben (Rusty Butte) Mine—lode gold, with lead and silver; (b) in SE¼ sec. 22, the Combination Mine (high-grade pods) —lode gold.

Douglas Co.

AREA: (1) head of Last Chance Cr., in sec. 34, (32S–4W), the Puzzler Mine—lode gold; (2) on Quines Cr., in W½NW¼ sec. 1, (33S–5W), the Quartzmill Mine—lode gold.

GLENDALE, gravel bars, bench and terrace gravels along Cow Cr.: (1) numerous old hydraulic placers; (2) area mines, rather productive—lode gold.

MYRTLE CR., E and NE, in drainage basins of North and South

Myrtle crs.: (1) along tributaries of North Myrtle Cr., especially on Lee Cr. and Buck fork, extensive placers (tot. prod. in 1898 of $150,000); (2) near Letitia Cr. (tributary of South Myrtle Cr.), in NW¼ sec. 20, (29S–3W), the Chieftain and Continental mines (tot. prod., 1898–1930, of $100,000 for each mine)—lode gold.

Grant Co.

Embracing much of the southwestern part of the Blue Mountains of northeastern Oregon, Grant County was prospected right after the big gold strikes of adjoining Baker County, with rich discoveries made along Canyon Cr., at Granite, Greenhorn (cf. also under Baker Co.), North Fork, Quartzburg, and at Susanville, all ghost towns today. The total production of gold in Grant County approximates 470,600 ozs. of lode and placer gold through 1959.

BATES, NW 18 mi. down middle fork of the John Day R., the Susanville district (tot. prod., 1864–1959, of about 48,750 ozs. of lode and placer gold): (1) along Elk Cr. and N of the middle fork of the John Day R., extensive placer operations (main source of district gold production); (2) on S side of Elk Cr. about 2 mi. above its jct. with the middle fork, in S½ sec. 7, (20S–3⅜E), the Badger Mine, principal producer—lode gold; (3) other area mines include the Chattanooga in secs. 5 and 6; the Daisy, in NW¼ sec. 5; the Gem, in N½ sec. 5 (between forks of Elk Cr.); the Golden Gate (Poorman), in secs. 7 and 8 E of the Badger and N of Elk Cr.; the Homestake, on N side of Elk Cr. opposite the Badger; etc.; (4) N 8 mi., the Greenhorn district (cf. under GREENHORN in Baker Co.).

CANYON CITY: (1) the Canyon Cr. district (tot. prod., 1862–1959, of about 818,000 ozs. of lode and placer gold and continuing sporadically): (a) along the John Day R. and its tributaries, all bed and bench gravels, extensive placers; (b) area mines—lode gold; (2) SE 1½ mi., in SE¼ sec. 12, (14S–31E), the Golden West Mine (long abandoned)—lode gold; (3) SE 2 mi., on Canyon Mt., in N½ sec. 7, (14S–32E): (a) the Great Northern Mine (rich pockets periodically discovered)—free gold; (b) 1,000 ft. below, the Haight Mine, early good producer—lode gold; (4) SW about 4 mi., on NE slope of Miller Mt., in SE¼ sec. 22,

(14S–31E), the Miller Mt. Mine (operated till 1942)—lode gold; (5) ENE about 5 mi., near the Marysville School, in N½ sec. 33, (13S–32E), the Prairie diggings (surface cuts and shallow drifts along ½-mi. mineralized zone).

GRANITE (ghost town in far NE cor. of co. 14 mi. NW of SUMPTER in Baker Co., district tot. prod., 1862–1959, of about 160,000 ozs. of lode and placer gold): (1) along Granite Cr. and all its tributaries, especially Bull Run and Clear Cr.: (a) very extensive placers operated by bucketline dredges, (b) area mines—lode gold; (2) N, in drainage area of the N fork of the John Day R. and along Desolation Cr., very important early-day placers: (a) the Klopp Placer Mine, and (b) adjacent diggings in area of jct. of the N fork with Trail Cr.; (c) 6 mi. up the S fork from the Klopp mine, the French diggings, worked hydraulically; (d) 5 mi. below the Klopp mine, on the S fork, the Thornburg placers, hydraulic operations; (3) N and NE, in T. 8 S, R. 35½ E, many lode mines: (a) N 3 mi., the Ajax Mine (E½ sec. 22) and the Cougar Mine (NW¼ sec. 27, tot. prod., 1938–42, of 19,126 gold ozs.)—lode gold; (b) N 3½ mi., the Tillicum Mine (NE¼ sec. 23, on N side of Granite Cr.); (c) the Independence Mine (sec. 22); and (d) in adjoining Lucas Gulch, the Magnolia Mine—all free gold; (4) N 5 mi., in sec. 14, the Buffalo Mine (tot. prod. of 33,142 ozs.)—lode gold; (5) ½ mi. NE of the Buffalo, the Blue Ribbon Mine (2,000 ft. of workings)—lode gold; (6) NE 8 mi., in sec. 12, the Continental Mine, high-grade—lode gold; (6) NE 10 mi., in T. 8 S, R. 36 E: (a) on Onion Cr., in secs. 6 and 7, the La Belleview Mine (6,000 ft. of drifts, tot. prod. est. at $500,000, 1878–1941)—lode gold; (b) in secs. 18 and 19 (2 mi. E of the Buffalo Mine, q.v., the Monumental Mine (tot. prod., 1870–1928, of $100,000) —lode gold, with silver.

PRAIRIE CITY (12 mi. E of JOHN DAY on U.S. 26), N, mainly in drainage of Dixie Cr. tributary of the John Day R. and extending over divide into headwaters of Ruby Cr. tributary of the middle fork of the John Day R., the Quartzburg district (tot. prod., 1862–1959, of 45,100 ozs. of lode and placer gold): (1) along Dixie Cr., extensive placers extending for 5 mi. upstream from town; (2) around Dixie Butte, all regional watercourse, bench, and slopewash gravels—abundant placer gold; (3) N 5–7 mi., in T. 12 S, R. 33 E, many lode mines: (a) in NE¼ sec. 1, the Copperopolis Mine, small producer to 1906; (b) on W fork of Dixie Cr., in secs. 2 and 11, the Equity Group, operated 1878–1910 with prod. est. of $600,000; (c) in gulch W of the W fork 7 mi. above town, the Keystone Mine (last worked in 1933), and adjoining below, the Present Need Mine—lode gold (4–5 ozs. per ton of ore); (d) on E side of E fork of Dixie Cr., in NE¼ sec. 12, the Standard Mine, extensively developed—lode gold.

Harney Co.

AREA, extreme S part of co., in S part of the Steens Mts.-Pueblo Mts. Range: (1) area small prospects—gold showings; (2) in secs. 8 and 17, (40S–35E), the Farnham and Pueblo prospects—lode gold (a few small shipments of concentrates made).

BURNS, NE 20 mi., in area of Trout Cr. branch of the Silvies R., in sec. 4, (21S–32E), the Harney (Idol City-Trout Cr.) district: (1) creek bed, bench, and terrace gravels (mainly in valley fill), placers (tot. prod., 1891–1959, of about $50,000); (2) area prospects and small mines along a NW-trending shear zone for 1 mi.—lode gold.

Jackson Co.

Gold mining had its beginnings in Oregon with placer discoveries along Jackson Creek in this county in 1852, made by prospectors hurrying to the California Mother Lode mines. Through 1959, Jackson County produced some 450,000 gold ozs., most of which came from placer mines. Each summer sees very many amateur and professional gold hunters busily at work panning and sluicing along scores of auriferous streams and creeks.

AREA: (1) in NE¼ sec. 4, (33S–4W), near the Douglas Co. line, the Warner Mine—free gold; (2) head of Elk Cr., in sec. 29, (31S–2E), the Al Sarena (Buzzard) Mine, with 4,200 ft. of underground workings, tot. prod., 1897–1919, of $24,000—lode gold.

APPLEGATE (district, along the Applegate R.): (1) area diggings and hydraulic placers along r.—abundant colors, nuggets; (2) between lat. 42° 01' and 42°20' N, long. 123°00' and 123°15' W, the Upper Applegate district (very rich placers, tot. prod., 1853–1959, of about 210,000 gold ozs.): (a) along Forest Cr., first placer findings; (b) in Sterling, Humbug, and Thompson crs. (tributaries of the Applegate R.), extensive placers; (3) area mines: (a) in secs. 17 and 20, (40S–4W), the Steamboat Mine, major producer, tot. prod., 1869, of $350,000—lode gold, pyrite; (b) the Queen Anne and Sterling mines, minor producers—lode gold.

ASHLAND, W and S, in T. 39 S, R. 1 E and 1 W, the Ashland district, in watershed of Bear Cr. Valley, tot. prod., 1858–1959, of about 66,400 ozs. of lode and placer gold: (1) area placers; (a) along Ashland Cr., (b) along Bear and Anderson crs., (c) along Wagner Cr. and its tributaries, all productive placers; (d) in Arrastra Cr., Yankee and Horn gulches, productive placers; (2) W 2–3 mi., in T. 39 S, R. 1 W, area lode mines: (a) SW¼ sec. 6 and NW¼ sec. 7, the Ashland Mine, tot. prod., 1886–1942, of $1,500,000; (b) in sec. 13, the Burdic Group, i.e., the Ruth, Little Pittsburgh, and Growler mines, minor producers—lode gold; (c) NW¼ sec. 14, the Double Jack Mine, intermittently worked, 1870–1942; (d) NW¼ sec. 12 (1 mi. W of the Ashland Mine, q.v.), the Storty Hope Mine—lode gold, lead, pyrite; (e) NW¼ sec. 30, the Skyline Mine, worked intermittently from early 1900s to the present— lode gold; (3) E 8 mi., in N½ sec. 23, (39S–2E), the Barron Mine, tot. prod., 1917–31, of about $9,000—lode gold.

CENTRAL POINT, W, at Willow Springs on Willow Cr., extensive early placer workings.

GOLD HILL (district, tot. prod., 1853–1959, of about 80,000 gold ozs.): (1) area crs., e.g., Foots, Sam, Galls, Sardine, Evans, and Pleasant, all extensive early-day placers (many operated to 1942); (2) NE 2 mi., in SW¼NE¼ sec. 14, (36S–3W), the Gold Hill Pocket near top of hill (most famous of all gold-pocket discoveries, a mass of nearly pure gold in a very few cu. ft. of earth worth more than $700,000 when found in 1857); (3) S 2 mi., in SW¼ sec. 28, (36S–3W), the Braden Mine, good producer 1885–1916—lode gold; (4) at the mouth of China Gulch 2½ mi. S of the Gold Hill Pocket: (a) the Roaring Gimlet Pocket, very rich —native gold; (b) several area small additional pockets immediately to E (altogether valued at $40,000)—native gold; (5) NE 3 mi., in S¼ sec. 2, (36S–3W), the Sylvanite Mine, major producer 1880s–1940—lode gold; (6) S 5 mi., on Kane Cr. in sec. 17, (37S–3W), the Revenue Pocket ($100,000)—massive native gold; (7) NW 6 mi., W of left fork of Sardine Cr., in NW cor. sec. 29, (35S–3W), the Lucky Bart Mine, tot. prod., 1890–1916, of about $200,000 (worked intermittently since)—lode gold.

JACKSONVILLE (Medford district, in the Bear Cr. Valley, tot. min. prod., 1851–1959, at least 26,000 ozs. of placer and lode gold): (1) along Jackson Cr., extensive placers; (2) S, along Sterling Cr., largest hydraulic placers in district; (3) Jackson Cr. reservoir, W 800 ft., in S½ sec. 25, (37S–3W), the Town Mine, major producer—lode gold; (4) W 2 mi., in NW¼ sec. 36, the Opp Mine (tot. prod., 1880s–1941, of over $100,000); (5) numerous other lode mines, especially where the Jacksonville district overlaps into Josephine Co., q.v.

PHOENIX: (1) along Bear Cr., extensive early placers; (2) NW, the Forty-nine diggings (a group of highly productive placers, 1860–1900, and intermittently worked to the present).

Jefferson Co.

ASHWOOD (26 mi. E of MADRAS), district in the Horse Heaven mercury mining area of rugged, rolling hills: (1) NE 3 mi., the Oregon King Mine (tot. prod., 1935–50, of 2,419 gold ozs.), by-product of copper-lead-silver ores; (2) farther E, around Axhandle Butte, numerous prospects—gold showings.

Josephine Co.

One of the leading gold-producing counties in Oregon, Josephine County produced an estimated 1,235,000 ozs. of lode and placer gold between 1852 and 1959.

AREA: (1) almost all streams draining from the Klamath Mts., in gravel deposits of beds, benches, and adacent slopes—placer gold (colors, nuggets) and platinum; (2) Josephine Cr. and its tributaries (Canyon Cr. and Fiddler Gulch), among most important and productive placer localities; (3) W side of co., between lat. 42°13′ and 42°29′ N, long. 123° 38′ and 124°05′ W, the Illinois district (tot. prod., 1852–1953, between 5,000 and 10,000 ozs. of placer gold): (a) along the Illinois R., downstream from mouth of Josephine Cr., very productive placers; (b) Sixmile Cr., Hoover Gulch, Rancherie, and Oak Flat, all bars, and benches, rich placers; (c) upper Briggs Cr. Valley, in sec. 7, (36S–8W), the Barr Mine, a rich placer operation; (d) along lower Briggs Cr., in area of Red Dog and Soldier crs., rich placers; (e) NW side of Briggs Cr., in sec. 24, (36S–9W), the Elkhorn placers, very productive; (f) Secret and Onion crs. (both tributaries to Briggs Cr.), productive early placers; (4) in T. 38 S, R. 9 W, the Illinois-Chetco district (with overlap into Curry Co.): (a) SW¼ sec. 7, the Becca and Morning Group, minor production— lode gold; (b) in secs. 5 and 8, the Calumet Mine (between forks of Rancherie Cr.), a prominent quartz ledge—free gold showings; (c) SW¼ sec. 14, the Gold Ridge (Pocket Knoll) Mine, numerous shallow cuts active in 1900; (d) in sec. 10, the Hoover Gulch Mine—lode gold.

CAVE JCT.: (1) area cr. and terrace gravels: (a) placer workings, and (b) area lode-gold mines; (2) S several mi., in mts., small mines—lode gold.

GALICE (district, including Mt. Reuben, tot. prod., around 268,000 gold ozs.). (1) area. (a) local placer operations include the Ankeny, Courtney, Carnegie, California-Oregon, and Last Chance; (b) hillside just W of the Galice Range (¼–½ mi. wide, extending 4 mi. to SW), patches of gravel on benches about 500 ft. above present streams as dissected by tributaries of Galice Cr.—placer gold; (c) the "high-bench gravels" along both sides of the Rogue R. (gold-bearing but not much worked to date); (2) downstream: (a) the Dean and Dean, and Rocky Gulch placer mines; (b) in Hellgate Canyon, the Hellgate placers, productive; (3) along Galice Cr., many rich placers (discovered in 1854 and later worked by Chinese); (4) W 1 mi., the Old Channel Placer Mine (largest hydraulic mine in state, discovered in 1860); (5) NE (and 21 mi. SW of GLENDALE in Douglas Co.), in secs. 22, 23, 26, and 27, (33S–8W), the Benton Mine, near Mt. Reuben (located in 1893 and largest underground mine in Oregon; closed in 1942); (6) the Almeda, Gold Bug, Oriole, Black Bear, and Robertson (Bunker Hill) mines, important producers—lode gold.

GRANTS PASS (district, occupying the Rogue Valley SE of mouth of Jumpoff Joe Cr., except the Applegate Valley, and including the Jumpoff Joe Cr., Rogue R., Winona, and Merlin camps; tot. prod., about 22,000 ozs. of lode and placer gold): (1) near town: (a) the Dry diggings, (b) along Picket and Jumpoff Joe crs., and (c) all Rogue R. gravel bars and bench deposits—very productive early placers; (2) NE 9 mi., on Louse Cr., the Granite Hill Mine (flooded out in 1908 after $75,000 prod.); (3) N 10 mi. to the Hugo interchange on I-5, then E 10 mi., at head of Jumpoff Joe Cr.: (a) the Daisy (Hammersly) Mine, tot. prod., 1910–41, of about $250,000; (b) such other mines as the Anaconda, Cloudy Day, Copper Queen, Dorothea (Marshall), Forget-Me-Not, Gold Note, Hayden (Little Dandy), Horseshoe Lode, Ida, and many more in the whole district; (4) NE about 18 mi. and 5 mi. E of I-5 at the Grave Cr. bridge, in NE part of co. from WINONA to King Mt., the Greenback-Tri County district (a group of lode-gold mines along adjacent boundaries of Douglas and Jackson cos.): (a) on Tom East Cr. 1½ mi. N of the old camp of Placer, the Greenback Mine, tot. prod., 1897–1912, of $3,500,000—lode gold; (b) along Grave Cr. and tributary Coyote and Wolf crs; extensive placers, especially dredging on S side of Grave Cr. upstream from LELAND (largest operations in co. history); (c) along Tom East Cr., the Columbia placers (tot. prod. through 1941 of $400,000 and still productive to panners); (d) on upper Jumpoff Joe Cr., just below Brass Nail Gulch, the Cook and Howland placers (active in 1930s); (e) near head of Jumpoff Joe Cr., in Bummer Gulch, the Sexton placers, productive; (f) on Jack Cr. and nearby Horse Cr., placers worked before 1910; (g) along Louse Cr., the Forest Queen, Granite Hill, and Red Jacket placers, operated productively into the 1960s.

HOLLAND, S 1½ mi. along Althouse Cr., in stream gravel deposits, and benches—gold colors, nuggets.

O'BRIAN (in far SW cor. of co. on U.S. 199), E to TAKILMA-WALDO, the Waldo district (tot. prod., 1853–1931, of 194,000 gold ozs.): (1) area: (a) along Althouse and Sucker crs; extensive early placers, including the Llano de Oro (Esterly), Deep Gravel, Platerica, and Leonard placers, very rich, worked by thousands of miners in the 1850–60s; (b) in Allen and Scotch gulches, gravel deposits, productive placers; (c) at head of Allen Gulch, in secs. 33 and 34, (40S–8W), the High Gravel Mine, extensively hydraulicked in 1917; (d) in NW¼ sec. 36, the Queen of

Bronze Mine (primarily copper but producer of about $120,000 in by-product gold, over 7,000 ft. of underground workings); (2) E fork of the Illinois R., near Takilma, rather extensive placer operations; (3) near headwaters of the Illinois R. at Waldo, the famed "Sailors' Diggings" (rich placers discovered by sailors in 1852 en route from Coos Bay to the Jacksonville mines. The sailors dug a 41-mi. ditch to bring water for sluicing and hydraulicking. Placering continued into 1942, with intermittent activity to the present. This area is noted for its large nuggets, one found weighing 15 lb.); (4) W of Waldo, in Fry Culch, the Bailey Mine, a productive placer dragline operation.

Selma, at head of Soldier Cr. tributary of Briggs Cr., the Eureka Mine, opened in 1901—lode gold.

Williams (in SE part of co., at lat. 42°07′ N, between long. 123°15′ and 123°36′ W, the Lower Applegate district, tot. prod., 1904–59, recorded as 4,180 gold ozs. but probably totaling over 10,000 ozs. of lode and placer gold, 1852–1870s): (1) area: (a) along Williams Cr., numerous placers worked after 1852; (b) along Slate and Oscar crs. and Missouri Flat, area gravels—abundant placer gold; (c) the Porcupine Mine, early producer—lode gold; (2) NW 3 mi.: (a) in N½ sec. 21, (38S–5W), the Humdinger Mine, major producer in 1900—lode gold, with pyrite; (b) in SE¼ sec. 16, the Oregon Bonanza Mine, operated continuously 1936–40, quite productive—lode gold.

Lake Co.

New Pine Cr. (15 mi. S of Lakeview on U.S. 395), E a few mi. in hills, several small mines in the High Grade district (an overlap from Modoc Co. in Calif.)—lode gold.

Plush, N 10 mi., in T. 35 S, R. 23 E, the Lost Cabin (Coyote Hill, Camp Loftus) district, discovered in 1906, many small, shallow prospect pits—gold showings.

Lane Co.

Gold mining began in Lane County in 1858 and continued until World War II stopped all gold mining, with a total production through 1959 of around 50,000 gold ozs.

BLUE R. (well E of SPRINGFIELD on U.S. 126, district overlapping into Linn Co., tot. prod., 1887–1959, of 10,200 ozs. of lode gold): (1) N, in T. 15 S, R. 4 E; (a) in secs. 32 and 33, the Lucky Boy Mine, 14 patented claims lapping into secs. 4 and 5, (16S–4E), tot. prod., 1887–1942, exceeding $200,000; (b) other area mines include the Rowena, Tate, Treasure, and Union—lode gold; (2) cf. Linn Co. for additional lode-gold mines in district.

COTTAGE GROVE, SE 35 mi., on divide between the Umpqua and Willamette rivers, between lat. 43°35′ N and long. 122°35′ to 122°45′ W, the Bohemia district (largest and most productive gold district in the W Cascade Mts., tot. prod., 1858–1959, of 38,637 ozs. of lode and placer gold): (1) along Sharps, Martin, and Steamboat crs., gravel bars and benches—placer gold; (2) in T. 23 S, R. 1 E: (a) in N½ sec. 13, the Champion (Evening Star) Mine, with 18,000 ft. of underground workings—by-product gold from copper-zinc ores; (b) 1 mi. W of the Champion, in N½ sec. 14, the Music Mine (14 claims) and (c) on Sharps Cr. in NE¼ sec. 15, the Lower Music Mine (together a tot. prod., 1891–1949, of $280,000); (d) in N½ sec. 11, the Crystal (Lizzie Bullock) Mine—by-product gold; (e) other area mines, e.g., the El Capitan (President) Group, Grizzly Group, Leroy Group, Shotgun (Carlisle), and Star—all lode gold; (3) in T. 23 S, R. 2 E: (a) on Grizzly Mt., in SW¼ sec. 7, the Helena Mine (tot. prod., 1896–1950, of $250,000); (b) in NE cor. of co., the Mayflower Mine on Horseheaven Cr.—lode gold; (c) other area mines include the Noonday, Oregon-Colorado, etc. —lode gold.

OAKRIDGE, N 11–12 mi., between Fall and Christy crs. in T. 19 S, R. 4 E, the Fall Cr. district (low production): (1) numerous prospects in secs. 13, 18, and 19 (most caved in or difficult to reach); (2) in E½ sec. 13, the Golden Eagle (Jumbo, Highland) Prospect (1,300 ft. of underground workings)—lode gold; (3) in E½ sec. 18, the Ironside Mine, worked intermittently for many years—lode gold (no definite veins).

Lincoln Co.

NEWPORT, along coast from Yaquina Bay to TOLEDO, in all black-sand deposits—placer gold and platinum (very fine).

Linn Co.

Foster, NE 27 mi. via the Green Peter dam and reservoir rd., to drainage area of Quartzville Cr. (major tributary of the Middle Santiam R.), in T. 11 and 12 S, R. 4 E, the Quartzville district (tot. prod., 1863–1959, of 8,557 gold ozs.): (1) along Quartzville Cr., gravel bars and benches, placers (this area is a popular summer gold-panning hobby locality—abundant gold colors and occasional nuggets); (2) in T. 11 S, R. 4 E: (a) along Dry Gulch on White Bull Mt., in secs. 21, 22, 23, 26, and 27, the Lawler Mine (tot. prod., 1861–98, of $100,000); (b) near head of Dry Gulch, in sec. 23, the Albany Mine (tot. prod., 1888–1920s, of $50,000); (c) other area productive mines, e.g., the Bob and Betty (Smith and McCleary), Lucille, Munro (Mayflower) Group, Red Heifer (Silver Signal), Riverside Group, Savage (Vandalia) Group, and the Tillicum and Cumtillie (Golden Fleece) claims—lode gold; (3) in T. 12 S, R. 4 E: (a) in SW¼ sec. 1 and NW¼ sec. 12, the Paymaster Claim; (b) in NW¼ sec. 11, the Galena Mine—both lode gold.

Sweet Home, SE along the Calapooya R. to end of rd. and just N of the Lane Co. line, the Blue R. district (cf. also under Blue R. in Lane Co.): (1) in E½ sec. 28, (15S–4E), the Cinderella Mine, 1902–67—lode gold; (2) in W½ sec. 28, the Great Northern Mine (tot. prod. since 1900, about $25,000); (3) in secs. 31 and 32, the Poorman Group, minor production to 1916—lode gold.

Malheur Co.

Between 1904 and 1959 Malheur County produced a total of 13,522 ozs. of lode gold and 13,860 ozs. of placer gold.

Area. For the Mormon Basin district, which straddles the Baker-Malheur Co. line, see localities listed under Durkee in Baker Co.

Ironside (47 mi. NW of Vale on U.S. 26), NE, to old camp of Malheur City, the Malheur district (10 mi. WSW of the Mormon Basin in N part of co., with overlap into Baker Co.), most active in 1875 after completion of the Eldorado Ditch, tot. prod. through 1959, about 9,600 ozs. of lode gold from numerous area mines (some surprisingly deep).

Marion Co.

MEHAMA (22 mi. SE of SALEM on Rte. 22), E 23 mi. via the Little North Santiam rd., in T. 8 S, R. 4 and 5 E, the North Santiam district (with overlap into S part of Clackamas Co., q.v.), tot. prod., 1877–1947, of only 454 ozs. of by-product gold from such mines as the Black Eagle, Capital, Crown Mine, Silver King Group, Ruth (Amalgamated), and Santiam Copper—very minor by-product gold.

Union Co.

STARKEY, SE, to N end of the Elkhorn Range and near N boundary of the Bald Mt. batholith, near head of the Grand Ronde R., the Camp Parson district (20 air mi. N of SUMPTER in Baker Co.): (1) at head of Tanners Gulch, the Camp Carson placers, extensively hydraulicked; (2) area quartz prospects (little developed), lode gold; (3) the Orofino Mine, on dumps—gold showings.

Wheeler Co.

ANTONE, just E, the Spanish Gulch district: (1) along Rock and Birch crs., bed and terrace gravels—placer gold; (2) area small mines in quartz exposures—gold and silver in galena.

PENNSYLVANIA

Only a single gold-producing locality exists in Pennsylvania. For information on other minerals, write: Director, Bureau of Topographic and Geological Survey, Pennsylvania State Planning Board, Harrisburg, Pennsylvania 17120.

Lebanon Co.

CORNWALL, S, and just NW of a rd. at Big Hill (on old Rte. 322), extensive mine dumps—gold traces. The nearby Cornwall Iron Mine, opened in 1742 for iron with small amounts of copper, has produced 37,459 ozs. of by-product gold since 1908, most of it after 1937.

SOUTH CAROLINA

The gold-bearing formations of South Carolina are similar in origin to those of the Slate Belt in North Carolina. Placer gold was first discovered in 1827 at the Haile Mine which, a few years later, became the largest gold producer in the southeastern states. Since that early date, South Carolina has produced a total of 318,801 ozs. of both placer and lode gold until the mines were ordered closed in 1942. However, with the greatly increased world price of gold and the opening to Americans everywhere for unlimited possession of gold, high expectations throughout South Carolina exist for reopening many mines and for greatly renewed prospecting in all of the state's auriferous districts.

For more information, write: Director, The Division of Geology, South Carolina State Development Board, P.O. Box 927, Columbia, South Carolina 29202.

Abbeville Co.

AREA. A long gold belt, known as the Carolina Slate Belt, extends across several counties in a northeast–southwest direction as a continuation of the same geologic formation in North Carolina: (1) very many streams, creeks, benches, and terraces throughout the belt—placer gold; (2) many regional mines "upcountry" from the placer gravels and probable sources of the placer gold—lode gold.

ABBEVILLE: (1) area mines—lode gold; (2) S 20° E 5.8 mi., W of Long Cane Cr., the Jones Mine—free gold in pyrites; (3) S 5° E 7.3 mi., the Lyon Prospect (several quartz veins cutting phyllite)—free gold; (4) S 10° W 8 mi., the Calais and Douglas Mine—lode gold; (5) S 9 mi. and ½ mi. E of Beula Cross Rds., a mine—lode gold.

LOWNDSVILLE, S 46° W 3 mi., the Cook Prospect (small quartz lenses, mill site)—lode gold.

Anderson Co.

EASLEY, S 9° E 6.4 mi., and 300 ft. N of Three and Twenty Cr., the Henderson Prospect (shallow pit in quartzose vein)—lode gold.

Cherokee Co.

AREA (reached from SMYRNA in York Co.): (1) north: (a) 0.7 mi., and W of Caanan Church, the Wyatt (Bradley?) Mine—free gold; (b) 1.7 mi., the Dixon Mine (worked in 1938)—lode gold; (c) 1.8 mi., the Wallace Mine (worked in 1914 and during 1930s)—lode gold; (2) west: (a) 2 mi., the Southern Gold (Terry, Terry & Horn) Mine, early-day good producer and reworked in 1930s, on standby in 1970—lode gold; (b) 3 mi., on both sides of Beech Cr. in Cherokee and York cos., the Bar Kat (McGill) Mine, rich pyrite vein—free and lode gold.

COWPENS: (1) NE 3 mi.: (a) the Love Springs (Old Palmer) Mine— lode gold; (b) along nearby cr., placers worked long ago; (c) S of the branch cr., several prospect pits—gold showings; (2) SE 3.8 mi.: (a) the Hammet Mine (one of oldest gold mines in state)—lode gold; (b) area stream gravels (first gold discoveries)—placer gold.

GAFFNEY: (1) S 25°E 3 mi., the Lockhart Mine—lode gold; (2) SE 10½ mi., S of Flint Hill and SW of Smith's Ford: (a) the Love (Old Wilkey, Kennedy) Mine, large, rich veins—lode gold; (b) adjoining cr. gravels—placer gold; (3) SE 11 mi. (1 mi. W of Smith's Ford on the Broad R.), the old Darwin Mine—lode gold; (4) S 12 mi., just N of the Pacolet R. and 1–2 mi. above the SC-18 bridge, the Nuckols and Norris, and Nott Hill mines: (a) in the mines—lode gold; (b) around top of hill, in loose chunks of quartz—free gold; (5) SE 12 mi., near the 12-mi. post, on Flint Hill, area mines and prospects—gold showings.

KINGS CREEK (STA.): (1) area: (a) W of the Caanan Church, the Bolin, Wyatt, and other nearby mines—lode gold; (b) E of Rte. 97 (S of Rte. 5), the Dixon, Eutis, Southern Gold, Wallace, and other mines and prospects (cf. also under AREA), primarily copper and lead—by-product gold: (2) SE 0.8 mi., the Eustis (Lowe) Mine—lode gold; (3) SW 2 mi., the Bar Kat Mine—lode gold.

Chesterfield Co.

Between 1828, when the first placer gold was found, and 1939, Chesterfield County produced 21,840 recorded ozs. of gold.

AREA, numerous old mines scattered all over the co.—gold, pyrite.

JEFFERSON: (1) area including the JEFFERSON-PAGELAND uplands, between Fork Cr. on E and Little Fork Cr. on W: (a) the flat part of the Coastal Plain near White Plains Church, and (b) the flanking stream valleys, slopes, and hillsides—placer gold; (c) S of town, along Fork Cr. and along Gold Dust branch N of Rte. 109, placers; (2) on the S flood plain of Nugget Cr., between Rtes. 39 and 40, many small placers (worked by tenant farmers in the 1930s); (3) N 54° W 2 mi.: (a) the Leach Mine—minor lode gold, and (b) the adjoining Gregory Placer Mine, productive; (4) S 10° E 2.3 mi., the Kirkley Mine, worked 1939–40—lode gold; (5) W 3 mi. to Brewer Knob, on the Lynches R. adjacent to the co. line in NW part of co.: (a) the Brewer Mine, including the Hartman, Topaz, and Brewer pits (practically the total co. gold production attributed to this mine); (b) the Tanyard Placer (in the sediments of the Coastal Plain; (c) all regional stream and bench gravel deposits (numerous well-worked placer operations first developed in 1828 and extensively hydraulicked during the 1880s)—abundant placer gold obtainable today by pan, sluice, and rocker.

PAGELAND, the Oro Mine (worked in 1941)—lode gold

RUBY: (1) N 87° W 3.9 mi., the Hendrix Prospect (layers of brown "clay slate" impregnated with limonite cubes)—lode gold; (2) N 70° W 7 mi., the Edgeworth and Brewer Mine, fairly rich—lode gold.

Edgefield Co.

EDGEFIELD: (1) N 12 mi., on E side of Sleepy Cr. 2½ mi. N of U.S. 378, the Landrum and Quattlebaum mines (separated by small valley running NW into Sleepy Cr., worked in 1856 through early 1930s)—lode gold; (2) large surrounding area: (a) mines and prospects—lode gold; (b) regional stream, bench, and slopewash gravels—placer gold.

Greenville Co.

GREENVILLE: (1) N 11 mi. and 0.6 mi. S of Locust Hill, the Westmoreland Placer, very productive stream gravels; (2) N 15° E 15 mi.: (a) the Wild Cat Mine, smoky quartz vein—lode gold; (b) area stream and hillside gravels, early-day placer workings.

GREER: (1) area mines—lode gold; (2) N 6½ mi., on W side of the

Middle Tyger R.: (a) the Briggs Prospect—lode gold; (b) adjoining, the McBee (Carson) Placer ("remarkable for producing more gold than any mine in the northwestern part of the state"); (2) 1 mi. above the McBee Placer: (a) the Cureton Mine (pyritic quartz vein with irregularly high values)—lode and free gold; (b) on opposite side of the Tyger R. from the McBee Placer, several area mines—lode gold, pyrite.

PRINCETON, NW 3 mi., on W side of Mountain Cr., 0.4 mi. below the Greenville rd. bridge, the Desota Prospect—lode gold.

Greenwood Co.

TROY, area, the Young Mine—lode gold.

VERDERY, SE 2½ mi. and 0.9 mi. E of railroad, the Bradley Prospect, productive—free gold.

Kershaw Co.

CAMDEN, NW 9 mi. and 1 mi. NE of Getty's bridge over Samneys Cr., the Lamar Mine (reopened in 1960)—lode gold.

Lancaster Co.

Southern Lancaster County contains the largest gold mine in all the southeastern states, the Haile Mine, which is credited with most of the total 278,080 gold ozs. produced in the county between 1829 and 1942.

AREA, access from JEFFERSON in Chesterfield Co.: (1) S 76° W 4½ mi., the Ingram Mine—lode gold; (2) N 79° E 6 mi., the Stevens Mine, productive in 1840s, on dumps—pannable gold; (3) N 8 mi., the Funderburk Mine, worked in 1897 and again in 1941—lode gold; (4) N 69° W 9 mi., the Knight's Prospect (minor operation in the 1850s)—lode gold.

KERSHAW (on the Kershaw Co. line): (1) N 51° E 3.8 mi.: (a) the Haile Gold Mine (large-scale, opened in 1829 and finally closed in 1942)—lode gold; (b) all area stream and bench gravels surrounding the mine —placer gold (original discoveries leading to finding the Haile lode); (c) adjoining the Haile, the Brassington Mine—lode gold, pyrite; (d) on the SW extension of the Haile veins, the Clyburn (Cay) Mine—lode gold; (e) all area saprolite and gulch gravel deposits—placer gold; (2) N 7 mi., the Phiffer Prospect (in selvages close to vein)—lode gold; (3) N 5° E 8 mi., on Flat Cr. tributary of the Lynches R., the Blackmon Gold Mine —gold, pyrite. This mine was a good producer, 1850s–1910, located ½ mi. N of jct. of Rtes. 86 and 167.

LANCASTER: (1) N 79° E 6 mi., the Stevens Mine (cf. under AREA); (2) N 77° E 9 mi., the Belk Mine, productive 1830s–40s—lode gold; (3) E 9½ mi., the Johnson (Strand) Mine, in quartz—lode gold (abandoned in 1855); (4) E 10 mi., the Stroud Prospect (quartz vein exposed for more than 1 mi. near coarse basaltic dike—minor gold showings.

OSCEOLA: (1) N 3 mi., the Izell (Ezell?) Mine, productive of fine cabinet ore specimens (not worked after 1850s)—native gold; (2) NW 4 mi., the Hagin Mine, massive pyrite vein—minor lode gold.

Laurens Co.

LAURENS, S 35° W 8 mi., the Raeburn (Rabon) Cr. Prospect, quartz —lode gold.

WATERLOO, W 7 mi., on W side of rd. on high ridge between Saluda and Reedy rivers, the Mt. Olive Prospect (a little-prospected quartz ledge)—gold showings.

McCormick Co.

Almost all of McCormick County's recorded production of 43,700 ozs. of gold have come from a single mine, the Dorn Mine, in the central part of the county.

BEULA CROSSROADS: (1) SE ½ mi. (and 9 mi. S of ABBEVILLE in Abbeville Co.), the Neill (Neel) Mine—lode gold; (2) S 2 mi., the Link Mine, quartz seam—lode gold (in pyrolusite).

McCORMICK: (1) in town, the Dorn Mine (tot. prod., 1852–80, of $1,000,000, second most important mine in state next to the Haile Mine in Lancaster Co.; reopened for short period in 1932—lode gold, with copper, lead, and zinc; (2) S 2½ mi., the Self Mine (short, rich vein) —free gold; (3) S 3 mi., the Butler Prospect—gold, with pyrite; (4) SW 3 mi., the Jennings Prospect, pyritic quartz vein—lode gold; (5) SW 5½ mi., the Searles Prospect—gold, in pyrite.

Newberry Co.

PROSPERITY, SW 4 mi., adjoining the Bush R., the Lester(?) Prospect, shallow pit in 1908 in brown quartz—gold showings (minor).

Oconee Co.

AREA, bed, bar, and bench gravels between the Toxaway and White Water rivers, the White Water-Toxaway placers (discovered in 1850s), quite productive.

ADAMS CROSSING, SW 3 mi.: (1) the Cochran (Lawton) Placer, productive; (2) area nearby mines in schists cut by quartz—lode gold.

CHERRY, SW 1½ mi., the Pickens Prospect—lode gold.

KEOWEE STA., N 0.2 mi.: (1) the Sloan Placer (worked before the Civil War); (2) adjacent exposure of quartz stringers in mica schist (numerous small mines and prospects)—free gold.

PULASKI, N 4 mi. (10 mi. N 40° W of FORT MADISON), the Cox Prospect, quartz vein cutting slates 1 mi. S of the Cox house—lode gold (low-grade).

SENECA, S 5½ mi., the Sitton Prospect (quartz stringers in sericite schist)—lode gold, small nuggets.

WALHALLA: (1) N 11½ mi., area of old mines—gold, in pyrite; including (a) the Jesse Lay Mine—lode gold, and (b) nearby cr. bed gravels—placer gold; (2) W 14 mi., on E scarp of the Chatooga R. 1 mi. W of Rogues Ford, the Henckel Mine—lode gold; (3) N 15 mi., on middle fork of Cheohee Cr.: (a) the Kuhnman (Old Cheohee) Mine—lode gold, and (b) nearby cr. and bench gravels—placer gold.

Pickens Co.

CALHOUN, N 1 mi.: (1) the Calhoun Placer (beds of small creeks on S side of the main stream), worked by slaves in the 1850s; (2) adjoining quartz outcrops in mica schists—lode gold.

Saluda Co.

NEWBERRY, S 12 mi., in fork of Big Cr. and Little Saluda R., the Culbreath Mine—lode gold (distributed through abundant pyrite).

Saluda, NNE 5½ mi., E of Hwy. 19 and ½ mi. NE of jct. of Rtes. 31 and 43, the Yarborough Mine (re-examined in 1960)—lode gold.

Spartanburg Co.

Area: (1) co.-wide stream and bench gravels—placer gold, zircons, rare diamonds; (2) 7 mi. N of greer in Greenville Co. (straddling the co. line), along E bank of the Middle Tyger R.: (a) the Wolf and Tyger Placer, and (b) along Wolf (Wolf Swamp?) Cr. along the co. line, rich placers—both extensively hydraulicked.

Spartanburg, all area stream and bench gravels—placer gold, zircons.

Union Co.

Glenn Springs: (1) E 3 mi., the Nott Mine (productive in 1844, reworked in 1932 and for several years afterward)—lode gold; (2) S 77° E 3.8 mi., the Mud (Harman) Mine—lode gold, copper; (3) S 60° E 6 mi., the West Mine: (a) in upper levels—placer gold; (b) in depth (deepest mine at 115 ft. in state in 1844) lode gold.

Union, N 72° W 8 mi.: (1) the Bogan Mine—lode gold; (2) immediate area streambed gravels—placer gold.

West Springs: (1) SE 1.2 mi., the Ophir (Thompson, Fair Forest) Mine, tot. prod., 1844–1904, of $100,000; (2) on Fair Forest Cr., the Nott Mine (cf. under Glenn Springs).

York Co.

Area, many well-known old mines in co. (check regional quadrangle maps)—lode and by-product gold.

Bethany, N 30° W 1.7 mi., the Patterson Mine (worked quite extensively before 1912)—lode gold, pyrite.

Blacksburg, E, to Kings Mt. Battleground, then E 2 mi., the Ferguson Mine—lode gold, pyrite.

Clover, SW 5.7 mi., on Bullock's Cr. tributary of the Broad R., an old mine—free gold, with pyrite and magnetite in slate.

Fort Mill, N 30° W 4.3 mi., the Clawson (Sutton) Mine—abundant free gold with pyrite (cubes to ½ in. on edge).

HICKORY GROVE: (1) N 1 mi., the Wylie Mine—lode gold; (2) SE 1½ mi., the Thunderhead Prospect (placers discovered in 1961 along a cr. on the Ollie Proctor property)—panning gold to nugget size; (3) W 1½ mi., just N of Rte. 46–816, the Mercer Mine (several pits)—free gold; (4) SE 2 mi., alongside Rte. 211, the Thunderbird Mine—free gold, pyrite; (5) W 2.2 mi., on Smith's Ford rd., at head of a branch of Guin Moore's Cr., on S side of rd., the Magnolia Mine—lode gold, with malachite: (a) 0.4 mi. W, the Parker No. 1 Mine, and (b) 0.8 mi. SW, the Parker No. 2 Mine—lode gold; (6) SW 3.2 mi., the Brown Mine (worked before 1900, and by 1905 had 1,600 ft. of underground workings)—lode gold; (7) W 4 mi., the Allison Prospect—gold in chalcopyrite; (8) NW 4.6 mi. and 1 mi. E of Smith's Ford, the Arrowwood Mine —minor lode gold.

HICKORY GROVE-SMYRNA, area of quartz vein outcrops extending NE between both towns, some 50 lode-gold mines and prospects along both sides of Rtes. 5, 97, and 211; the mine dumps reveal a reported 28 varieties of collectible minerals and gemstones, including native gold nuggets and galena.

KINGS CREEK: (1) S 2½ mi. and N of Wolf Cr., the McGill (Bar Kat) Mine, worked in 1895 and again in 1930s—lode gold; (2) SW 3 mi., on Wolf Cr. tributary of the Broad R., the Carroll and Ross Mine—free gold, pyrite; (3) S 5 mi., the Bolin Prospect—lode gold.

OLD LONDON STA., SE ½ mi., the Bradley (Wyatt?) Mine—minor lode gold.

SMYRNA: (1) N 0.7 mi. and E of Caanan Church, the Dickey (Allison) Mine, pit and tunnels worked before 1904 and again in the 1930s—lode gold; (2) NE 1 mi.: (a) the Horn Mine, and (b) numerous other area mines, e.g., the Hardin, La Peire, Love No. 2, McCarter, etc.—lode gold; (3) W 1 mi., the Martin Mine—lode gold (some record-size nuggets); (4) SW 1.2 mi., the Cassady Mine—lode gold; (5) E 2 mi., the Castles and Scoggins Mine—gold, pyrite; (6) W 3 mi.: (a) the Carroll and Ross (Wolf Cr.) Mine, productive—lode gold; (b) ½ mi. S, the Darwin Mine—coarse gold; (7) W 3½ mi., the Hull Mine—lode gold, lead; (8) NE 3½ mi. and N of Hwy. SC-55, the Falkner Mine (worked in 1930s with further prospecting in 1960s)—gold showings; (9) SW 4.4 mi.: (a) the Dorothy Mine, 1860–1942, productive, and (b) adjoining on SW, the Schlegelmilch Mine (similarly productive)—lode gold; (c) ½ mi. E, an old prospect—gold, copper; (10) N 25° E 6½ mi.: (a) the Ferguson Mine (productive before 1904, reworked in 1930s, reprospected in 1960s)—lode gold; (b) 1.3 mi. NE, the Ellis Mine—lode gold, with pyrite and chalcopyrite.

YORK: (1) NE 3½ mi., the very old Wallace Mine—lode gold; (2) NE 6 mi., the Wilson (Big Wilson) Mine, opened in 1840s—lode gold, with native copper nuggets and chalcopyrite, (3) NE 10 mi. and 2 mi. NW of Nannys Mt., the Barnett Mine—lode gold; (4) NE 13½ mi., the Campbell Mine (numerous pits)—gold showings, pyrite cubes.

SOUTH DAKOTA

In 1874 when General George A. Custer's expedition was reconnoitering the Black Hills, two miners named Ross and McKay attached to the expedition discovered gold in gravel bars of French Creek near present-day CUSTER in Custer County. Despite Sioux Indian opposition to Americans invading their domain, a literal gold rush ensued almost immediately and, in 1876, the Sioux were forced to cede the Black Hills to the United States. Subsequently immensely rich placer deposits poured out a golden wealth, and even richer lode-gold mines came into production, so that through 1965 a total of 31,207,892 ozs. of gold were recorded, with intensive gold mining continuing to the present.

First discovered in 1875 in pre-Cambrian outcrops in the LEAD district of what is now Lawrence County, a number of lode mines were later combined into the Homestake Mining Co. The Homestake Mine became America's largest and richest gold producer, and it still ranks first in the nation's annual production of gold. Thus South Dakota is ranked third among America's gold-producing states, with Lawrence County easily ranking first among all counties in the United States.

For more information, write: State Geologist, State Geological Survey, University of South Dakota, Vermillion, South Dakota 57069.

Black Hills Region

Placer gold is found in the gravel bars of all the present streams in the Black Hills and in the various terraces that border their valleys. Few of these deposits have failed to yield gold in paying quantities, and many have produced handsomely. Many of the high-level gravel deposits on the east side of the Black Hills contain considerable placer gold.*

Custer Co.

CUSTER: (1) all regional stream, bench, and terrace gravel deposits—placer gold; (2) E 6 mi., along French Cr. (original 1874 gold discovery), all bed, bench, and terrace gravels—placer gold. The auriferous deposits extend all the way to FAIRBURN.

Lawrence Co.

Locale of the Homestake Mine, Lawrence County produced a total of 26,386,000 ozs. of gold between 1875 and 1959, a record unmatched for any similar area on the North American continent.

DEADWOOD (district, in E-central part of co., including old camps on Deadwood, Two Bit, Strawberry, and Elk crs.; tot. prod. through 1959 of 284,000 ozs. of lode and placer gold): (1) along Deadwood, Whitewood, Gold Run, Bobtail, and Blacktail gulches, very rich placer workings, still productive to the amateur; (2) along Spearfish Cr., many productive placers; (3) in Spring Gulch, the Hidden Treasure Mine, major gold producer—lode gold; (4) on W edge of town on U.S. 14 Alt., the Broken Butte Mine—lode gold, with pyrite and sphalerite; (5) E 3½ mi., the Mascot Mine, discovered in 1892, earliest records of gold production in district—lode gold; (6) SE 3–4 mi., in Strawberry Gulch, the Oro Fino Mine (discovered in 1893)—lode gold; (7) SE 8 mi., near Roubaix on Elk Cr., the Cloverleaf Mine (tot. prod. through closing year of 1937 of 43,885 ozs.)—lode gold.

LEAD (district, including Yellow Cr. in the central part of the co.): (1) S side of town on U.S. 14 Alt. and U.S. 85: (a) the Homestake Gold Mine (largest gold producer in North America, with tot. prod., 1875–1959, of 24,450,000 ozs. of lode gold and an average annual output to the present of more than 500,000 ozs., with silver a substantial by-

---

*N. H. Darton and Sidney Paige, *Central Black Hills of South Dakota* (Washington, D.C.: U.S. Geological Survey, 1925), Folio 219, p. 29.

product. The Homestake embraces 654 mining claims covering 5,639 acres. Visitor tours are available June through August, except Sundays and holidays.); (b) other area smaller lode mines with lesser production —lode gold; (2) NW 1–3 mi., in Blacktail and Sheeptail gulches and False Bottom Cr., the Garden (Maitland) district: (a) the Maitland Mine, discovered in 1902 (tot. prod. through 1959 of 137,000 ozs. of lode gold); (b) other area smaller mines bring tot. prod. of district to 176,000 ozs. of lode gold; (3) W about 3 mi., in the Bald Mt. area: (a) on W slope of Bald Mt., the Trojan Mine—lode gold; (b) the Squaw Creek district (including the Ragged Top, Elk Mt., and Carbonate areas, tot. prod. through 1959 of about 75,800 ozs. of lode gold); (c) W and S of Ragged Top Mt., area mines—lode gold; (d) in Squaw Cr. and Annie Cr. areas, numerous mines (active primarily before 1914)—lode gold; (e) all regional stream, bench, and terrace gravels—placer gold.

SAVOY, the Ragged Top Mountain district (5 mi. long, lying immediately W of Spearfish Cr. in the large bend of Spearfish Canyon), first area mines opened ½ mi. N of Ragged Top Mt. 2 mi. NE of town, at DACY—lode gold.

TINTON: (1) the Nigger Hill district, many area productive mines—lode gold; (2) along Bear, Potato, Nigger, Poplar, and Mallory gulches, richly productive placers; (3) along Sand and Beaver crs., many placer workings.

Pennington Co.

Lying immediately south of Lawrence County in the southern part of the Black Hills, Pennington County produced an estimated 128,000 ozs. of gold, largely from lode mines but with minor amounts from placer deposits.

AREA, along Spring, Castle, and Rapid crs. and Palmer Gulch, placer gravels (discovered in 1875 and still productive to the gold panner).

HILL CITY (district, with numerous mines in a belt extending from 6 mi. SW of town to 5 mi. NE; tot. prod., 1875–1938, estimated at 35,400 ozs. of lode gold): (1) area mines in and about town—lode gold; (2) near head of Spring Cr., several mines—lode gold.

KEYSTONE (district, in W part of co., extending from 3½ mi. NW of town to 1½ mi. SE; est. tot. prod., 1875–1959, of 85,000 lode-gold ozs., plus unrecorded amounts of placer gold): (1) area all along Battle Cr., productive placers; (2) the Keystone-Holy Terror Mine (est. tot. prod. of 76,000 ozs.)—lode gold; (3) S 4 mi. and 2 mi. E, the Spokane district: (a) area mines—by-product gold, with silver; (b) area streambed gravels and bench deposits—placer gold showings.

MYSTIC, SW, all along Castle Cr. (bars and terrace deposits made up of quartz gravels), important placer localities in (1) Crooked Gulch, (2) Hoodoo Gulch, and (3) at Chinese Hill—placer gold, still obtainable by panning, sluicing, or with a rocker.

MYSTIC-PLACERVILLE-BIG BEND-RAPID CITY, all along Rapid Cr. for 40 mi., ranging from the creek bed to the heights above, very productive placer deposits, especially: (1) at Neilsen's and (2) near PLACERVILLE: (a) the Placerville Bar, (b) Swede Bar, and (c) Stockade Bar, all highly productive—placer gold.

ROCHFORD (district, including a belt of mines 9 mi. long, extending from 4 mi. SW of town at MYERSVILLE to 5 mi. N 30° W, at NAHANT in Lawrence Co.): (1) in the "Hornblende Belt," many mines—lode gold, lead, silver; (2) in the "Iron Quartz-Tremolite Belt," many mines —by-product gold.

ROCKERVILLE, center of an area of several sq. mi. of "high terrace" gravel deposits, very productive—placer gold.

SHERIDAN, along Spring Cr., many productive placer workings.

SILVER CITY: (1) W, numerous mines (quartz veins in schist)—free gold, with lead-antimony minerals; (2) N 2 mi., along Rapid Cr., numerous area mines—lode gold, with copper and silver.

TENNESSEE

The principal gold occurrences in Tennessee lie in a narrow belt of auriferous stream gravels about 50 miles long in southern Blount, Monroe, and Polk counties along the North Carolina state line. Most of the 23,800 gold ozs. produced in Tennessee between the first discoveries in 1831 and 1959, of which 14,872 ozs. were a by-product of the Ducktown copper ores in Polk County, came from placer deposits in Monroe County.

For more information, maps, and guides, write: Director, Department of Conservation, Division of Geology, State Office Building, Nashville, Tennessee 37202.

Blount Co.

MARYVILLE, area of Montvale Springs: (1) regional stream gravels—placer gold; (2) E of the Chilhowee Mts., regional stream gravels—placer gold.

Monroe Co.

AREA, SE part of co., along Coker Cr., numerous placer deposits and mines with first discoveries of gold made in 1827 (tot. prod. by 1854 about $46,023 and, through 1962, of 9,000 gold ozs.): (1) present stream gravels, along with bench and "high terrace" deposits—abundant placer gold: (a) area alluvial cones, rather extensively mined—placer gold; (b) numerous area small mines (low-grade ores) in fissure veins, not commercially valuable but of interest to the occasional gold hunter—gold showings; (2) along Whippoorwill Branch (of the Tellico R.): (a) stream and bench gravels—placer gold; (b) several area mines—lode gold.

Polk Co.

COPPERHILL-DUCKTOWN (district, embracing many copper mines scattered over many sq. mi., major gold-producing area of Tennessee as a by-product of copper refining), on many mine dumps—gold showings in quartz and copper minerals.

TEXAS

The quantity of gold produced in Texas has been quite insignificant, the total production through 1942 being only 8,277 ozs., nearly all of which came as a by-product of silver ores from the Presidio Mine in Presidio County and the copper-silver ores of the Hazel Mine in Culberson County. There is no public land in Texas. All mineral rights belong to the landowners; therefore, without exception, royalties must be paid by the gold discoverer or operator of any mine, either to the private landowner or to the state of Texas, if such operations are contemplated on state lands.

For more information, write: Director, Bureau of Economic Geology, University of Texas at Austin, Austin, Texas 78712.

Bastrop Co.

AREA, all along Gazley Cr., in gravel deposits of Eocene age—reportedly gold colors obtainable by panning.

Bastrop, Caldwell, and Gonzales Cos.

REGION. All exposures of Eocene sediments appear to contain finely disseminated gold particles. The gold hunter must necessarily be referred to a geologic map of these counties in order for him to trace out the auriferous formations.

Blanco Co.

AREA, all along Walnut Cr. just S of the Llano Co. line, in Cambrian conglomerates worked briefly in the 1890s—placer gold.

Brewster Co.

AREA, in the Quitman Mts., scattered lode mines—by-product gold.

ALTUDA: (1) in the Altuda Mts., area lead-silver mines—by-product gold; (2) near town, the Bird Mine—gold traces.

SOLITARIO (district), area lead mines—gold traces with galena.

Culberson and Hudspeth Cos.

ALLAMOORE, S, in the Van Horn Mts., area small mines and prospects —by-product gold.

VAN HORN: (1) area, the Van Horn-Allamoore district: (a) small mines and prospects—by-product gold; (b) the Sancho Panza, Black Shaft, and Pecos mines (all copper, lead, zinc)—minor by-product gold; (2) vicinity (including the Lobo RR sta.): (a) area of the Carrizo Mts., (b) SE flanks of the Diablo (Sierra Diablo) Plateau, (c) the W escarpment of the Wylie Mts., and (d) NE of Eagle Mt. (a small area), all gravel deposits, little investigated or worked—placer gold potential; (3)

E of the Sierra Diablo Plateau: (a) numerous copper-lead-zinc mines—minor by-product gold; (b) the Hazel Mine (old and best-developed property and principal copper-silver producer)—minor by-product gold; (4) on S flank of the Van Horn Mts., near co. line, the Plata Verde Mine —by-product gold.

Howard and Taylor Cos.

REGION: (1) all watercourse and terrace gravel deposits—placer gold potential; (2) exposures of pre-Cambrian metamorphic formations (cf. a geologic map), in quartz veinlets and pegmatites—minor gold showings.

Hudspeth Co.

AREA, the Quitman Mts.: (1) numerous small mines and prospects—by-product gold; (2) the Bonanza and Alice Ray mines (both lead-zinc) —by-product gold; (3) in the Eagle Mts., area lead-zinc mines—by-product gold.

Irion Co.

MERTZON, area exposures of iron-stained, decomposed limestone—assays show presence of minor amounts of gold.

Irion, Uvalde, and Williamson Cos.

REGION, all limestone exposures—minor gold showings (very finely disseminated).

Llano Co.

AREA: (1) along Crabapple, Coal, and Sandy crs., in bed and bench gravels—reportedly placer gold (colors); (2) the Kiam Prospect, mostly molybdenum—minor gold showings.

KINGSLAND: (1) drainage basins of the Llano R. to the E of the W line of Mason Co., all regional watercourse beds and benches, potential good for placer discoveries; (2) along Sandy Cr.: (a) the drainage area (385 sq. mi.), watercourse sand and gravel deposits—potential placer gold. This basin includes large areas of metamorphic rocks (good possibilities for substantial gold occurrences), and the basin has shown the most extensive and notable gold showings in Texas; (b) the upper basin of Sandy Cr., between RILEY and Cedar Mts., in stretch between Potato Hill and Click Gap, in sand and gravel deposits along bottom of gorge at base of loose wash deposits—probable placer gold; (c) near mouth of Sandy Cr., in bed gravels of the narrows—possible placer gold; (3) just S of the Gillespie Co. line, a narrows in the course of Cole Cr., in base gravel and sand deposits, potential placers; (4) above all three narrows: (a) in extensive areas of flattish land, scattered over the flatlands—gold showings (some of which might be commercially profitable as placer deposits); (b) all regional main crs. and tributaries, gravels and sands just above bedrock, potentially profitable placers; (5) below Enchanted Rock, in the lower courses of Crabapple and Cole crs., and in the main Sandy Cr., in gravels immediately above bedrock, potential placers.

LLANO, NE 5 mi., and N of the LLANO-LONE GROVE rd., the Heath Mine (primarily copper, intermittently worked in soils above pre-Cambrian schists, in quartz stringers)—by-product gold (very small amounts).

Llano and Mason Cos.

THE LLANO UPLIFT. In central Texas and confined to Llano and Mason Cos., the Llano Uplift contains gold-bearing consolidated bedrocks bearing auriferous quartz veinlets and stringers in metamorphosed schists, quartzites, phyllites, slates, and marbles. The auriferous exposures occur in a narrow strip along the eastern two thirds of the N line of Gillespie County and in a small area in the center of the west part of Burnet County—all told, not more than 1,000 sq. mi. of rocks that might contain any appreciable amount of gold.

Since the Llano Uplift has never been adequately prospected, and since erosion of gullies and larger watercourses, floods, landslides, and grass and brush fires constantly expose new bedrock outcrops, there is the possibility that one or more of these processes may disclose a commercially valuable gold-bearing vein. The Llano Uplift would, therefore, appear to be the most logical area in Texas for amateur and professional gold hunters to prospect "virgin" territory.

Mason Co.

AREA, in N part of co., the Mason Mts., in basal gravel and sand deposits of Cretaceous age (especially where they fill ancient valleys and channels in the underlying rocks) are possible natural concentrators of placer gold worth prospecting.

Presidio Co.

SHAFTER (district): (1) the Presidio Mine, 1880–1942 (producer of about nine tenths of all gold recorded in Texas, principally as a by-product of silver ores bearing lead and zinc); (2) other area small mines, including the Chinati and Montezuma, intermittently worked for lead and silver—minor by-product gold.

Williamson Co.

GEORGETOWN, N 20 mi., in weathered outcrops of iron-stained lime-stone (assays made in 1883 showed presence of good values)—finely disseminated gold.

UTAH

Utah's great concentrations of mineral wealth lie in a central north–south zone of transitional mountains within a few minutes' to a few hours' drive of SALT LAKE CITY. On the west the Oquirrh Range is the site of the world's largest open-pit copper mine, in Bingham Canyon, and as early as 1863, gold placers were found in the same canyon. Subsequent to the development of copper mining, its by-product gold became an important addition to the state's economy.

Between 1863 and 1965, Utah produced 17,765,000 gold ozs. to place sixth among the nation's gold-producing states. While small amounts of gold occur in all of the mineral deposits of Utah, with the exception of iron and uranium-vanadium localities, most of the gold was recovered as a by-product of base-metal ores.

Commercial placer operations in Utah have been generally unsuccessful or discouraging. However, in many parts of the state there are extensive areas of stream and terrace gravels that have not been thoroughly prospected for gold. Some are along stream courses draining the major mining districts. In many areas of Utah, lack of water imposes a serious handicap for usually accepted placer mining methods.

For more information, write: Director, Utah Geological and Mineralogical Survey, University of Utah, Salt Lake City, Utah 84101.

Beaver Co.

Between 1860 and 1950 a total of 55,850 gold ozs. were recorded for Beaver County.

MILFORD: (1) E, in the Mineral Mts., area copper mines—by-product gold; (2) W approx. 23 mi. on Rte. 21, old ghost town of Frisco in the San Francisco Mts., the San Francisco district (tot. prod. to 1917 of 21,822 gold ozs.): (a) many area mines—by-product gold; (b) the Horn Silver Mine (discovered in 1876, producer of 60 per cent of co. gold output and for 10 yrs. the most successful gold-silver mine in Utah); (c) the Cactus Mine, minor producer—lode gold.

Box Elder Co.

PARK VALLEY (district), area mines—by-product gold.

Cache Co.

NEWTON, area mines—by-product gold (minor).

Garfield Co.

AREA, the White Canyon district, all regional watercourse, bench, and terrace gravel deposits—placer gold.

Grand Co.

AREA, Wilson Mesa, regional mines and claims—lode gold.

Iron Co.

Most of the 12,760 gold ozs. produced in Iron County came from the Stateline district in the extreme western part of the county near the Nevada boundary.

UVADA, area along the Utah-Nevada line at lat. 38° N, the Stateline (Gold Springs) district: (1) the Gold Springs camp, area mines, veins in latite—by-product gold; (2) N, in the Stateline area, several mines, veins in rhyolite—lode and by-product gold.

Juab Co.

FISH SPRINGS, area lead-silver mines, rich—by-product gold.

TINTIC (district, including the East Tintic district adjoining in Utah Co., one of three most important mining districts in Utah, tot. prod., 1869–1959, of 2,648,000 gold ozs., of which about 35 per cent came from the East Tintic): (1) a great many regional base-metal mines (copper, lead, zinc, and silver), on dumps—gold showings; (2) principal mines include the Chief, Centennial, Eureka, Mammoth, Gemini, Eureka Hill, Iron Blossom, Tintic Standard, North Lilly, and Eureka Lilly—by-product gold; (3) all watercourse and bench gravel deposits below the big mines, potential placers.

WEST TINTIC, area lead-copper-silver mines—by-product gold.

Kane Co.

KANAB, E 15–20 mi. on dirt rd. to the Paria R. and ghost settlement of Paria: (1) area of the Vermilion Cliffs, a few small mines—gold, silver; (2) in the regional Shinerump clays, finely disseminated gold. Paria was the site of an abortive gold rush in 1911, when considerable machinery was freighted in from MARYSVALE to hydraulic the clays; remains of an assay office still stand, but the gold content runs about 5 cents a cu. yd.

Piute Co.

Between 1868 and 1959 a total of about 240,000 gold ozs. were produced in Piute County, principally from lode mines in the Tushar Range in the west part of the county.

MARYSVALE: (1) SW 6 mi., in the Tushar Range, the Mt. Baldy (Ohio) district, tot. prod., 1868–1958, of 77,500 gold ozs.: (a) in Pine Gulch, bar and bench gravels, placers; (b) the Ohio camp, many area mines—minor gold, with silver; (c) the Mt. Baldy camp, many area mines and claims—lode gold; (d) the Deertrail Mine (principal producer, primarily lead-silver)—lode and by-product gold; (2) WNW 10 mi., in NW part of co., the Gold Mountain (Kimberly) district, with some mines in adjoining Sevier Co. (tot. prod., 1889–1959, about 159,000 gold ozs.): (a) the Annie Laurie Mine, most productive in district—lode gold; (b) the Sevier Mine, second most productive—lode gold; (c) other area mines, claims, and prospects—gold showings.

Salt Lake Co.

Most of the 10,651,000 ozs. of gold produced in Salt Lake County have come as a by-product of the great Bingham Canyon copper ores 10 miles southwest of SALT LAKE CITY in the Oquirrh Range.

ALTA (district SE of SALT LAKE CITY, elev., 10,000 ft. in the Wasatch Mts., now a noted ski resort), many old mines—lode gold, with lead and silver.

BINGHAM (West Mountain) district, on E slope of the Oquirrh Range, leading mining district in Utah and fourth-largest American producer of gold, discovered in 1863: (1) Bingham Canyon: (a) entire length and including Bear Gulch to its head, in cr. gravels to bedrock, bench and rim gravel deposits, rich placers; (b) all tributary crs. and gulches, productive placers; (2) all neighboring canyons: (a) in channel fillings covering the main bedrock under present streams, and (b) in gravel deposits on earlier streambeds (now left as isolated remnants upon the canyon walls above present streams), and (c) all associated bench deposits, present and ancient, productive placers; (3) in Middle Bingham Canyon: (a) on Argonaut, Dixon, Cherikino, and upper Clays ground, productive placers in bench deposits; (b) also rich placers in bottom and rim gravel deposits; (4) in Carr fork, the Gardella Pit, rich placer workings; (5) in Lower Bingham Canyon: (a) the St. Louis, Lashbrook, and Schenk placers, very productive; (b) rim deposits (best developed), in the Old Channel, Clays, and Mayberry placer workings, rich; (c) West Mountain and Bingham placers, workings in deep gravels; (6) along Damphool Gulch: (a) several bench levels, productive placers; (b) in adjacent gulch, the Clays Bar, very productive placers—coarse angular gold, nuggets to considerable size; (c) at mouth of Damphool Gulch, rich placer deposits (deep, reached by shafts and stoping along bedrock); (7) the West Jordan Claim (in Bingham Canyon, original discovery site for the placer gold which, in toto for the district, is estimated between 600,000 and 1,000,000 ozs.); (8) the Highland Boy Mine (an original gold prospect that led to discovery of pyritic copper ores)—by-product gold; (9) the huge modern Kennecott Copper Corp. open-pit mine, largest in the world; not worth a gold-hunter's attention but well worth visiting, tot. prod. through 1959 of more than 10,610,000 ozs. of by-product gold.

BRIGHTON (now a ski resort), many area old mines—gold, with copper and silver.

COPPERTON, area copper mines—by-product gold.

SALT LAKE CITY: (1) due E, on mountainside a short distance above the Pioneer Monument, a large cement slab seals the entrance to a tunnel which, rumor has it, is an undeveloped, extremely rich lode-gold mine owned by the Church of Jesus Christ of Latter-Day Saints (Mormon) as a "reserve" source of wealth should the Church ever require it; (2) SE 20 mi., on W slope of the Wasatch Range, the Cottonwood district (immediately N of the American Fork district in Utah Co. and just SW of the Park City district in Summit Co.), tot. prod., 1865–1959, of about 30,275 gold ozs.: (a) Little Cottonwood camp, several area mines opened in 1866 for lead and silver, especially the Emma Mine—by-product gold; (b) Big Cottonwood camp, discovered in 1868, more than 1,000 claims located by 1880—minor by-product gold; (c) the Old Emma and Sells mines (bismuth, tungsten)—by-product gold.

San Juan Co.

AREA, the Blue Mts., regional mines and prospects—gold showings.

BLUFF (scene of the "Bluff excitement" of 1892, when 1,200 prospectors arrived but went away empty-handed), regional placer gold is in all watercourse, bench, and terrace gravel deposits but is so extremely fine that recovering it is difficult: (1) mouth of Montezuma Cr.; (2) 3 mi. below the mouth of Nakai Canyon at Zana Camp (about 20 mi. above jct. of the San Juan R. with the Colorado R.); (3) at Spencer Camp, in the Great Bend of the San Juan 6 mi. below Zana Camp; (4) 4 mi. below Nakai Canyon—placer gold (too fine for recovery, but much heavy equipment freighted into these localities for hydraulic efforts).

Summit and Wasatch Cos.

These two counties adjoin one another along the east slope of the Wasatch Range, with the major mining district straddling the boundary line.

PARK CITY (Summit Co. district, 25 mi. SE of SALT LAKE CITY, encompassing the Uinta district in the SW cor. of Summit Co. and the Snake Cr. and Blue Ledge districts in the NW cor. of Wasatch Co., tot. prod., 1869–1959, of 790,000 ozs. of lode and by-product gold): (1) many area large mines in and about town (lead-silver ores)—by-product gold; (2) the Ontario Vein (discovered in 1872, a "bonanza" production)—lode gold; (3) in W part of district: (a) the Treasure Hill, lead and silver—by-product gold; (b) the Silver King Mine, extensive workings—by-product gold; (4) the United Park City Mining Co. and New Park Mining Co. mines, major producers—by-product gold.

Tooele Co.

Gold is the chief mineral commodity of Tooele County, which spans the Great Salt Desert from the Oquirrh Range to the Nevada line. Between 1858 and 1959 Tooele County produced 1,257,000 ozs. of gold.

GOLD HILL (in SW part of co.), the Clifton (Gold Hill) district near the Utah-Nevada line, tot. prod., 1892–1944, of about 26,000 gold ozs.: (1) numerous area mines, especially the Cane Springs, Alvarado, and Gold Hill mines (all major producers and active to 1940)—by-product gold; (2) the Deep Cr. area, small copper-lead-silver mines, some tungsten—by-product gold.

MERCUR (no longer on maps, but in 1902 about 12,000 residents, reached from Rte. 36 S of TOOELE a short distance S of Rte. 73 turnoff E to OPHIR, q.v.), the Camp Floyd (Mercur) district, 55 mi. SW of SALT LAKE CITY, tot. prod., 1870–1945, of 1,115,000 gold ozs. (with mercury and silver as by-products): (1) the Mercur, Delamar, Geyser-Marion, Sacramento, Sunshine, Overland, Daisy, and La Cigale mines—disseminated gold. Mercur, totally destroyed by fire, was Utah's only all-gold mining city and was considered to be one of the world's greatest gold camps, surpassed in its heyday only by Goldfield, Nev. Here the cyanide process was first introduced into N America (the enormous settling basins are spectacular, as are the sky-scraping tailing piles within the townsite), and the first long-distance power transmission line (from PROVO) ever developed to that time was brought in to run the giant mills; (2) S 4 mi., in Sunshine Canyon, the ghost camp of Sunshine, and (3) on W extremity of the Oquirrh Range, the old camp of West Dip, many area mines—lode gold, with lead and pyrite.

OPHIR (between the Camp Floyd district on the S and the Bingham district on the N, reached S from TOOELE via Rtes. 36 and 73), the Ophir-Rush Valley district, tot. prod., 1864–1959, of about 104,000 ozs. of by-product gold: (1) on W edge of town, the Ophir Mine (lead-silver) —by-product gold (with excellent pyrite crystals on dump); (2) Rush Valley, area mines, 1913–75—by-product gold.

STOCKTON (7 mi. S of TOOELE, in the Ophir district), just E, the Honorine Mine, major producer—by-product gold.

WILLOW SPRINGS (district, in far SW cor. of co., in the S part of the Deep Creek Range, tot. prod., 1891–1950, of 11,650 gold ozs.), area mines and prospects—by-product gold.

Utah Co.

AREA, cf. TINTIC in Juab Co., the East Tintic district, many large and productive mines—by-product gold.

AMERICAN FORK (district, in NE part of co., in the Wasatch Range about 5 mi. SE of the Cottonwood district, tot. prod., 1870–1959, of about 45,000 ozs. of by-product gold): (1) the Miller Mine, chief producer of lead and silver, 1870–1936—by-product gold; (2) numerous other area mines, mostly idle since 1880—by-product gold.

SANTAQUIN, SILVER LAKE (district), numerous regional lode mines for lead and silver—by-product gold.

Wasatch Co.

HEBER CITY: (1) co. area: (a) along North Fork and (b) Snake Cr., numerous mines—lode and by-product gold; (2) Grey Head Mt., NE (a region of about 100 sq. mi. lying about 50 mi. W of DUCHESNE in

Duchesne Co. via U.S. 40: (a) Indian Lake, (b) Avintequin, and (c) Sams Canyon, many mines and prospects—lode and by-product gold.

VERMONT

Vermont contains relatively few gem and mineral localities of interest to the gold hunter. An extremely scenic state of 9,564 square miles, including its lakes and waterways, all parts of Vermont are readily accessible to the motorist over fine paved roads and highways. Probably far more gemstone and mineral localities may exist than have been found to date, since the basal formation of the state's bedrock is granite. However, the rock and mineral collecting pastime has not been notable among Vermont residents in past years.

For general information, write: Director, Vermont Department of Libraries, Geological Publications, Montpelier, Vermont 05602.

Bennington Co.

READSBORO, a few area base-metal mines—by-product gold (on dumps, minor gold showings).

Windsor Co.

BRIDGEWATER, area old mines and prospects (primarily for lead)—minor gold showings.

VIRGINIA

Virginia was one of the first gold-producing states among the thirteen original colonies. The earliest printed reference to gold anywhere in the colonies was made in 1782, when Thomas Jefferson reported on a 4-pound rock that contained 17 pennyweights of gold. The rock had been picked up on the north side of the Rappahannock River about 4 miles below the falls.

Virginia's auriferous area lies east of the Blue Ridge Mountains in a belt 15 to 25 miles wide and 200 miles long. The state's most productive years for gold mining were from 1830 to 1856, with all gold mining halted by the Civil War. Despite a greatly reduced mining effort following that conflict, and a short-lived renewed flurry of prospecting and mining during the 1930s, the total recorded gold production in Virginia amounted only to 167,558 troy ozs.

For more information, write: State Geologist, Department of Conservation and Economic Development, Division of Mineral Resources, Natural Resources Building, Box 3667, Charlottesville, Virginia 22903.

Buckingham Co.

ARVONIA, GOLD HILL, JOHNSTON, AND NEW CANTON, many regional old mines (originally opened for lode gold and later producing copper) —lode gold.

DILLWYN: (1) S side of town at the historical marker, area hillsides, several old mines—lode gold; (2) NW 1 mi.: (a) the Buckingham Mine (the Eldridge Mine until 1853)—lode gold; (b) adjoining on the SW, the London and Virginia Mine (open cuts and drifts in a 10-mi.-long mineralized shear zone, in quartzite)—free gold, with some silver, lead, and zinc; (3) SW 4 mi., the Booker Mine (extensively worked before the Civil War)—lode gold.

Carroll Co.

WOODLAWN, area exposures of cellular quartz veins, numerous prospects in hornblende-mica schists—minor gold, with pyrite.

Culpeper Co

WILDERNESS, W 3 mi., the Culpeper Mine—lode gold.

Fauquier Co.

MORRISVILLE, NE 2.3 mi., on 594-acre tract along a dirt rd. off Rte. 634, about 1.6 mi. E of its jct. with Rte. 806, the Franklin Mine (one of most notable lode-gold mines in Virginia, 1837–77, when it caved in) —lode gold.

Floyd Co.

AREA: (1) SE side of Laurel Ridge, along Laurel Cr., the Laurel Creek Prospect (placers discovered in 1879): (a) all streambed gravels, and (b) collateral bench gravel deposits, productive—placer gold; (2) 2 mi. S of the mouth of Brush Cr., the Luster McAlexander Prospect, in stream alluvium, productive placers—gold, as grains of 5–80 mg.

FLOYD, in bed of Black Run (tributary of the Little R.), the Black Run Prospect, quartz veins 3–4 ft. wide in mica schist—gold, with lead and molybdenum.

Floyd and Montgomery Cos.

REGION, W side of the Blue Ridge, at SE base of Pilot Mt., along Brush and Laurel crs., numerous productive placers worked from 1830s.

Fluvanna and Goochland Cos.

REGION. A gold belt traverses the boundary between Fluvanna and Goochland cos., crossing the James R. at Bremo Bluffs into Rockingham Co. This belt was extensively prospected and mined between 1830 and 1860, with many mines developed, the chief one being the Tellurium (cf. under TABSCOTT in Goochland Co.), as well as the Bowles, Payne, Page, Hughes, Moss, Fisher, Busby, Taugus, Gilmore, etc.—lode gold.

Goochland Co.

AREA: (1) the James R. district, the Collins Mine—lode gold; (2) along Byrd Cr., near the Fluvanna Co. line, the Ruth Mine, a placer operation worked by dragline, 1935–42.

COLUMBIA, NE 3¼ mi., on E side of Big Byrd Cr.: (1) the Bertha and Edith Mine, opened in 1877: (a) the Oak Hill vein and (b) the Maple Branch vein—lode gold; (2) Maple and Camp branch tributaries of Big Byrd Cr., bed, bench, and terrace gravel deposits, productive placers (worked before the Civil War and again in the 1930s–40s).

LANTANA, NE 1 mi., in the valley of Little Byrd Cr., along a tributary 4 mi. S of TABSCOTT: (1) along creek bed gravels, productive placers; (2) on N side of cr., 30–40 ft. above present channel, terrace gravel deposits, rich placers (operated by dragline, 1934–36, area now a swampy marsh covered by dense vegetation).

TABSCOTT: (1) SW 1½ mi., extending ¼ mi. along the SE side of Rte. 605, the Moss Mine (tot. prod., 1825–1938, estimated between $20,000 and $25,000)—lode gold; (2) SW 2½ mi., on both sides of a USFS fire rd. approx. 0.55 mi. NW of its jct. with Rte. 605 (overlapping into Fluvanna Co.), the Tellurium Mine, discovered in 1832 and site of perhaps the first stamp mill erected in the United States (50-lb. wooden stamps with iron shoes): (a) the "Little" and "Middle" veins in garnetiferous quartz-sericite, and (b) the "Big Sandstone" vein (a 3-ft.-wide ledge of quartzite with auriferous stringers)—lode gold.

Halifax Co.

VIRGILINA (district, extending about 18 mi. N–S into Person and Granville cos. of N.C., very productive, with most gold mines located on a low, flat-topped ridge): (1) area, many old mines, on dumps—gold showings, with copper minerals; (2) NE 3 mi., the Poole and Harris Prospect, many crosscuts and shallow pits—lode gold; (3) NE 4½ mi., and just N of the abandoned Red Bank store: (a) the Red Bank (Gold Bank) Mine—lode gold, with specular hematite; (b) ¼ mi. NE, the Luce and Howard Mine (operated until 1906)—lode gold; (4) N 6 mi., two old abandoned mines—gold showings.

Louisa Co.

MINERAL, N 5 mi., upstream from Contrary Cr., a large mining area, with the outlying mines quite productive—lode gold, with copper minerals on dumps.

PENDLETON, S 1 mi., follow dirt rd. beginning just E of a farm pond to W, to dead end on the McPherson farm, take trail about 1 mi. into forest to an old mine—free gold (obtained by crushing ore detritus and panning).

Montgomery Co.

AREA, NW of Laurel Ridge and SE of Pilot Mt., along Brush Cr. and its tributaries: (1) the Brush Cr. Prospect, discovered in 1879 and intermittently worked through the 1920s, streambed and bench gravel deposits—placer gold; (2) all regional stream and bench gravels in an area 20 mi. long and 4 mi. wide, productive placers.

Orange Co.

AREA: (1) on E side of Rte. 667 approx. 0.45 mi. NE of jct. with Hwy. 3, the Wilderness Mine (opened in 1911, with development work in

1923)—lode gold; (2) along a woods rd. off W side of Rte. 667 approx. 2.4 mi. NNE of its jct. with Hwy. 3, the Melville (Rapidan) Mine, discovered before 1885 and worked until 1938—lode gold.

FREDERICKSBURG, W 18 mi., in NE part of co., along a woods rd. off W side of Rte. 667 (extending approx. 1.7 mi. NNE of its jct. with Hwy. 3): (1) the Vaucluse Mine (tot. prod. through 1959 of 50,000 gold ozs.) —lode gold; (2) area streambed and bench gravels, placered extensively in 1832; (3) area gold-bearing ledges (several area mines worked through the 1930s—lode gold; (4) 1 mi. NE of the Vaucluse, the Melville Mine, open pits in quartz lenses—lode and free gold.

Patrick Co.

PATRICK SPRINGS, NE 4½ mi., near Polebridge Cr., area pyritiferous quartz vein exposures, many prospects—gold, pyrite.

Spotsylvania Co.

AREA, in W part of co., 1½ mi. NW of Shady Grove Church, the Whitehall Mine (tot. prod., 1848–84, of $1,800,000)—lode gold (one 3-ft.-sq. pocket yielded $160,000 in pure gold at the $20-per-oz. price).

CHANCELLORSVILLE, WNW, into extreme NW cor. of co., near the Rappahannock R; oldest mine in co.—lode gold.

Stafford Co.

HARTWOOD: (1) in area, the Eagle Mine (12 mi. NW of FREDERICKS-BURG in Spotsylvania Co.), extensively worked until 1894—lode gold; (2) 2 mi. farther NW, the Rappahannock Mine, operated before and after the Civil War—lode gold.

MORRISVILLE, GOLDVEIN (on U.S. 17 NW of FREDERICKSBURG), very many old mines (make local inquiry), on dumps—gold showings.

WASHINGTON

Washington has not been a major gold-producing state, even though between 1860 and 1959 a total of 2,671,026 gold ozs. were recovered from very many and widely scattered lode and placer mines. Annual gold production has been steady for more than a century, and today Washington is the only state in America where gold recovery is increasing.

Early-day miners, following the first discovery of gold in the Yakima River in 1853, certainly failed to recover all the gold in the many stream placers where they wielded pan, rocker, sluice, and large-scale hydraulic and dredging equipment. Recently, an old placer operation at LIBERTY, Kittitas County, was reworked along a small part of the once extensively placered channel, with a new production of twenty pounds of gold nuggets being recovered.

Commercial placer gold prospecting and mining has continued from the earliest days right to the present, and each summer finds very many amateur gold hunters and scuba divers busily hunting new gold in the old districts, especially along the Columbia and Snake River bars, where a new supply of fine placer gold is deposited by every spring runoff of snow-melt waters. Considerable unexploited virgin placer deposits are known, and mining claims may still be taken up and worked at a profit. Some of the as yet undiscovered auriferous deposits were simply overlooked by the prospectors of the past; other deposits are only just now becoming accessible over recently built roads and trails.

For a complete description of every lode and placer mine, claim (patented or unpatented), or prospect in Washington, as well as for every other commercially mined metal, request from the office listed below Vols. 1 (text) and 2 (maps) of Bulletin 37, *Inventory of Washington Minerals*, Part II (1956), by Marshall T. Hunting, 498 pp.; price, $4.50.

For information and brochures, write: Supervisor, State Department of Natural Resources, Geology and Earth Resources Division, Olympia, Washington 98504.

Asotin Co.

CLARKSTON: (1) area sand and gravel exposures, and (2) the Snake R. bars: (a) the Clarkston placers, and (b) the Snake R. placers, all productive.

Benton Co.

AREA: (1) all Columbia R. low-water sand- and gravel bars—placer gold; (2) N bank of the Columbia R.: (a) in sec. 1, (5N–28E), and (b) in sec. 6, (5N–29E), the Berrian I. (Goody) Placer (four separate operations), productive.

ALDERDALE, E 4 mi., in deserted channel of the Columbia R. (covering several sq. mi.), the Artesian Coulee Placer in sec. 6, (4N–24E), productive.

BURBANK, NE, along the Snake R. above its jct. with the Columbia R., low-water sand- and gravel bars—placer gold.

PATTERSON, on Blalock I. in the Columbia R.: (1) the Blalock I. Placer (3 operations), and (2) the Gone Busted Placer (2 operations, with dry-wash plant operated 1938–40)—placer gold.

PROSSER, E 3 mi., on S side of hwy. in SW¼ sec. 33, (9N–25E), the Prosser Mine—lode gold.

Chelan Co.

Between 1860 and 1959 a total of 1,565,425 gold ozs. were produced in Chelan County. There are 139 lode mines and 24 placer workings, many still in operation.

AREA: (1) on headwaters of Roaring (Raging) Cr., 10 mi. W of mouth of Chikamin Cr., the Allen Magnesium Mine—lode gold; (2) from mouth of Nigger Cr.: (a) upstream for 2 mi., the Nigger (Negro) Cr. placers; (b) in sec. 9, (22N–17E), the Amigo Mine—lode gold; (3) (a) along Railroad Cr., in secs. 16 and 17, (23N–17E), the Railroad Cr. placers; (b) in sec. 36, at mouth of Ruby Cr., the Ruby Cr. placers; (4) in sec. 19, (27N–18E), at mouth of Deep Cr., the Deep Cr. placers; (5) Horseshoe Basin (accessible from Lake Chelan by rd. and trail), in N½ sec. 29, (35N–15E), some 13 area lead-silver mines—by-product gold.

BLEWETT (Peshastin) district, in S part of co. at lat. 47°25′ N, long. 120°40′ W (tot. prod., 1860–1959, of about 850,900 gold ozs.), all listed mines in T. 22 N, R. 17 E: (1) S a few hundred ft., in SW¼ sec. 1, the Black and White (Diamond Dick) Mine—free gold, pyrite; (2) W 1,200 ft., in SE¼ sec. 2, the Peshastin (Blewett, La Rica) Mine—free gold, with copper and lead minerals; (3) N ¼ mi., on E side of Peshastin Cr.: (a) in SW¼ sec. 1, the Blue Bell (I.K.L.) Mine—free gold, pyrite; (b) many other nearby mines—free gold; (4) along Peshastin Cr., many early mines, e.g.: (a) 1 mi. above Nigger Cr., in sec. 1, the Bloom Placer, and (b) the Cook, Crawford, Solita, and (c) on upper reaches of cr., the Peshastin Cr. placer workings, productive; (5) in the Culver Cr. area: (a) on S side, in SW¼ sec. 2, the Alta Vista (Pole Pick No. 2) Mine—lode gold, pyrite; (b) from head of gulch W to and across Nigger Cr., in NE¼ sec. 3, the Blinn Mine; in SW¼ sec. 2, the Bobtail (Wye) Mine; on N side of gulch ½ mi. above its jct. with Peshastin Cr., the Golden Eagle (Lucky King) Mine; and many other mines—lode and free gold; (6) in the adjoining Chiwawa district, very many other mines—lode gold.

ENTIAT, in E-central part of co., between lat. 47°40′ and 48°00′ N, long. 120°10′ and 120°45′ W (largest district in state, with 790 sq. mi., tot. prod. through 1930 of about 10,000 gold ozs.): (1) N, along the Entiat R., the Entiat placers; (2) NW, along the Mad R., the Mad R. placers; (3) in Crum Canyon tributary of the Entiat R., in N½ sec. 36, (26N–20E): (a) the Rex (Rogers) Mine, major producer—lode gold, silver; (b) 1½ mi. E of the Rex, in SW¼ sec. 25, the Pangborn Mine —lode gold, silver; (4) about 3 mi. up the Entiat R. from its mouth, near top of ridge N of the r., (25N–20E), the Cook-Galbraith Mine—free gold; (5) WNW 10 mi., in (26N–20E): (a) the Savage Mine (4 claims) —free gold, silver; (b) 2 mi. farther up, the Sunset Mine—free gold.

HOLDEN, in NW part of co., on E slope of the Cascade Range between lat. 48°07′ and 48°19′ N, long. 120°30′ and 120°45′ W, the Chelan Lake (Holden, Railroad Cr.) district, tot. prod., 1890s–1951, of 514,525 gold ozs.: (1) the Holden (Howe Sound, Irene) Mine, in secs. 18 and 19, (31N–17E), 13 patented and 78 unpatented claims (largest gold producer in state from copper ores)—by-product gold; (2) in cirque basin SW of Hart Lake in NE¼ sec. 8, (31N–16E), the Crown Pt. (Aurelia Crown, Crown Power) Mine—lode gold, molybdenum.

STEHEKIN (district): (1) many area mines—lode gold; (2) on Company Cr., the Well-Known Mine; (3) at head of Agnes Cr., the Goericke Mine; (4) on Maple Cr., the Lulu Mine; and (5) on Flat Cr., the Sunset Mine—all lode gold.

WENATCHEE (district, in SE cor. of co., at lat. 47°22′ N, long. 120° 20′ W, tot. prod., 1855–1959, of about 190,000 gold ozs.): (1) on W side of Squillchuck Cr., near center sec. 22, (22N–20E), the Golden

King (Wenatchee, Squillchuck, Gold King) Mine, tenth largest lode-gold mine in America—lode gold, with silver and pyrite; (2) in sec. 3, (22N–20E), the Wenatchee Placer Mine, productive.

Clallam Co.

ARCH-A-WAT (W of NEAH BAY, on the Pacific O.), S along beach to: (1) between Portage Head and Pt. of Arches, in secs. 18, 19, and 30, (32N–15W), the Shi Shi Beach (Lovelace) Placer; rd. from NEAH BAY reaches a place on the cliff above the beach, productive placers—gold, platinum, iridium, zircons; (2) 2 mi. N of mouth of Ozette R. in sec. 12, (31N–16W), the Ozette Beach Placer (access trail from L. Ozette) —gold, platinum; (3) in T. 30 N, R. 15 W: (a) at mouth of Little Wink Cr., in S½SW¼ sec. 1, the Little Wink (Japanese, Sand Point) Placer; (b) near small stream in NW¼NW¼ sec. 19, the Main and Bartnes Placer—fine placer gold, platinum, etc.; (c) at high-tide level in SW¼SW¼ sec. 18, the Morrow Placer—flour gold; (d) 2 mi. S of Sand Pt., in SW¼ sec. 18 (reached by hiking along beach from Ozette R. mouth or N from LA PUSH, q.v.), the Yellow Banks Placer—gold, platinum.

LA PUSH: (1) N 4 mi. along beach, at Johnson Point (Cape Johnson), in NW¼ sec. 5, (28N–15W), the Johnson Point Placer: (a) in beach black sands, and (b) adjacent bench black-sand deposits—placer gold, platinum; (2) N 10 mi. along beach, near mouth of Cedar Cr.: (a) in E½ sec. 18, (29N–15W), the Cedar Cr. (Starbuck) Placer—fine gold, platinum; (b) in sec. 19, black-sand beach deposits—gold, platinum; (3) N on beach to 125 ft. NW of mouth of Big Wink Cr., in NW¼NW¼NE¼ sec. 12, (30N–16W), the Morgan (Big Wink Cr.) Placer—gold, platinum.

PORT ANGELES: (1) S, between Ennis Cr. and head of White Cr., on Melick farm, probably in sec. 23, (30N–6W), the Port Angeles Mine —lode gold, silver; (2) SE 6 mi. and 3½ mi. by trail, on Cowan Cr. tributary of Little R., in sec. 6, (29N–6W), the Angeles Star (Winter, Gregory-Savage) Mine—lode gold, with copper and zinc.

Clark Co.

BRUSH PRAIRIE, area black-sand deposits—placer gold.

CAMAS, Columbia R. low-water bars—placer gold.

MOULTON: (1) E 4 mi., on E fork of the Lewis R., in approx. sec. 21, (4N–4E): (a) the McMunn Placer—gold, platinum; (b) terrace gravel deposits above stream—placer gold only; (2) E 5 mi., on the Lewis R. rd. 1 mi. above the McMunn claim, the Lewis R. Placer (access rd. up E fork of the Lewis R.), in stream black-sand deposits.

Columbia Co.

MARENCO, SSE, in SW¼NE¼ sec. 3, (9N–41E), near N end of the Hoffeditz farm, the Hoffeditz Mine (100–200 ft. above river)—lode gold.

Cowlitz Co.

WOODLAND, in sec. 6, (5N–1E), the Green Mountain Mine (small open cut)—gold, copper, lead, silver, and zinc.

Douglas Co.

COLUMBIA RIVER (far SW cor. of co.), the Columbia River Placer.

STEVENSON'S FERRY (a Columbia R. landmark), the Trouble Placer, locally rich occurrences—placer gold.

Ferry Co.

Between 1896 and 1958 Ferry County produced a minimum of 839,000 gold ozs., more than 99 per cent of which came from the Republic district, with about 6,000 ozs. from the Danville district and Columbia River placer operations. From 1904 to 1938 Ferry County was the leading gold producer in Washington. There are 164 lode mines and 35 placer workings within the county.

AENEAS, S to Lyman Lake, the Crown Point Mine—lode gold.

BELCHER (13 mi. NE of REPUBLIC): (1) NE 1 mi.: (a) the Belcher (Blue Bell-Belcher) Mine—lode gold, with copper and iron; (b) 1 mi. W, in sec. 31, (38N–34E), the Winnipeg (Yaki) Mine—lode gold, copper, iron; (2) near center of E½ sec. 18, the Pin Money Mine—gold, silver, copper, nickel, cobalt; (3) in NE¼ sec. 6, the Hawkeye Mine—gold, copper, silver, iron.

COVADA: (1) near mouth of Stray Dog Cr. in sec. 12, (31N–36E), the Blue Bar Placer in the Columbia R., annually renewed—flour gold; (2) on Columbia R. 2 mi. above Blue Bar, near Turtle Rapids, the Turtle Rapids Placer (terrace 60 ft. above high water extending several mi., paystreak several in. thick), extensively worked by Chinese; (3) due E, in sec. 6, (31N–37E), on a Columbia R. bar, the Thompson Placer.

CURLEW (NE of REPUBLIC), near N¼ cor. sec. 6, (39N–34E), the Panama Mine—lode gold, with silver, copper, lead, zinc.

DANVILLE: (1) on Big Goosmus Cr., the Goosmus Cr. Placer (very productive along narrow cr. bottom); (2) S 3 mi.: (a) the Danville Mine, rich—lode gold; (b) on E side of Kettle Cr., in SW¼ sec. 16, (40N–34E), the Morning Star Mine, 10 claims—free gold, with pyrite, copper, tungsten.

GEROME (P.O.), on N bank of the Columbia R., in sec. 4, (29N–36E), the Wilmont Bar Placer (2 terraces, 20 and 100 ft. above r.)—abundant placer gold, cerium, thorium.

GRAND COULEE DAM, up Columbia R. 6 mi., on N side, in sec. 17, (28N–32E), the Plum Bar Placer, extensively worked 1938–39.

HUNTERS, down Columbia R. 2 mi., on W bank, in sec. 23, (30N–36E), the Rogers Bar Placer, 1,500 acres and 3½ mi. long: (1) rich deposits at 30, 75, and 100 ft. above r.; (2) low-water river bars—placer gold, platinum.

KELLER: (1) E, to Columbia R. by dirt rd., near mouth of Ninemile Cr., probably in sec. 16, (29N–35E), the Ninemile Placer, terrace deposits 30 and 100 ft. above r., paystreak in each 1–3 ft. thick (old Chinese workings for ¾ mi. along r. and 200 ft. back)—flake gold; (2) SE, to Columbia R.: (a) on r. 3 mi. above mouth of the Spokane R., in sec. 11, (25N–35E), the Threemile Placer—flour gold and large, thin flakes; (b) at mouth of Sixmile Cr. (6 mi. above mouth of the Spokane R.), in sec. 34, the Sixmile Placer—flake and flour gold.

METEOR, N 3 mi., in NW¼ sec. 5, (32N–36E), the Gold Twenty Mine—lode gold, with lead, silver, zinc.

NESPELEM, along the Columbia R., between mouths of Nespelem and Kettle rivers to the S and E of Grand Coulee Dam, in bars and terrace gravels—placer gold.

PEACH: (1) opposite, on N side of the Columbia R., in secs. 15 and 16, (27N–35E), the Kirby Bar Placer; (2) between town and mouth of the Spokane R., large is. in the Columbia R., the Peach Is. Placer, a productive gravel deposit.

REPUBLIC (Eureka) district, 25 mi. S of Canada, most consistent gold producer in Washington, tot. prod., 1896–1959, of 836,393 gold ozs.: (1) SW ½ mi., the Republic (Blaine Republic) Mine, 13 patented claims—lode gold, silver, selenium; (2) ½ mi. out, in S½SW¼ sec. 34, (37N–32E), the Morning Glory (Old Gold) Mine lode gold, silver; (3) SW 1 mi., near base of E slope of Copper Mt., near center sec. 12, (36N–32E): (a) the Princess Maude (Southern Republic) Mine—lode gold, silver, pyrite; (b) in sec. 4, the Alva Stout Placer; (4) NW 1 mi., near W line of NW¼ sec. 35, (37N–32E): (a) the Insurgent Mine, and (b) adjoining on the E, the Lone Pine Mine—lode gold, silver; (c) the Last Chance Mine—lode gold, silver; (d) on E side of Eureka Gulch, near SW ¼ sec. 35, the Quilp (Imperator, Eureka) Mine—lode gold, silver; (5) NW 2 mi.: (a) on W side of Eureka Gulch, on line between secs. 34 and 27, (37N–32E), the Ben-Hur Mines—lode gold, silver; (b) in area, the Day (Aurum) Mine with 92 patented claims—lode gold, silver; (c) at head of Eureka Gulch, in W½SE¼ sec. 27, the Knob Hill Mine (13 patented claims, in 1960 the third most productive gold mine in America)—lode gold, silver.

Grant Co.

ELECTRIC CITY, E ½ mi., at foot of E wall of Grand Coolee, in sec. 14, (28N–30E): (1) the Electric City (Big Four, Daniels, Black-Rosauer) Mine—by-product gold (from bismuth, beryllium, molybdenum, silver); (2) in SW cor. sec. 14, in floor of Grande Coolee, the Hope No. 1 Mine—lode gold, silver.

GEORGE, SE, to Columbia R., in sec. 22, (18N–23E), gravel bars and terraces along r., placers.

PRIEST RAPIDS, far SW cor. of co., near toll bridge jct. of Rtes. 24 and

243, in secs. 10 and 11, (13N–24E), at Chinaman's Bar, area river gravels—placer gold.

Grays Harbor Co.

MOCLIPS: (1) area of Cow Pt., ocean beach black sands, placers; (2) in sec. 8, (20N–12W), on beach, the Moclips Placer.

OYHUT, area ocean beach black sands, the Oyhut Placer.

Jefferson Co.

KALALOCH, N to Ruby Beach, on the tombolo between Abbey I. and Ruby Beach, in E½NE¼ sec. 31, (26N–13W), in black-sand deposits, the Ruby Beach Placer—very fine-grained gold.

QUINAULT, NE 22½ mi. on dirt rd., take trail 1½ mi., on Rustler Cr. in sec. 31, (25N–7W), the Rustler Cr. Mine, 5 claims, quartz vein in slate—lode gold (good values).

King Co.

There are 99 lode mines and 4 placer operations in King County.

BARING (N-central part of co.): (1) W to the Tolt R., in SE¼ sec. 29, (26N–8E), the Tolt R. Placer (in top 18 in. of river bar gravels)— fine placer gold; (2) out 3½ mi. by rd. and trail, in sec. 15, (26N–10E), the Yellow Jacket Mine, 10 claims—lode gold, silver; (3) SW, the Climax Claims—by-product gold (with silver in chalcopyrite).

GREENWATER, in (19N–8E), the White River Mine in sec. 6, stamp mill in 1890—lode gold.

GROTTO (NE cor. of co.): (1) in secs. 29, 30, and 31, (26N–11E), the Money Cr. district: (a) the Aaron Mine—lode gold; (b) in secs. 20 and 29, in bed and bench gravels of Money Cr., the Money Cr. Placer; (2) near headwaters of Money Cr., in SW¼ sec. 34, (26N–10E), the Miller R. district: (a) the Apex (Bondholders Syndicate) Mine—gold, copper, lead, silver; (b) adjoining the Apex, in sec. 33, the Damon and Pythias Mine—lode gold, with lead and copper; (c) in sec. 30, (26N–11E), the Bergeson (Normandie) Mine—gold in arsenopyrite; (d) in secs. 31 and 32 (2 mi. from Money Cr.), the Kimball Mine—lode gold, with antimony and silver; (3) 8 mi. up Miller R. from its mouth by rd. and 2 mi. by trail, in sec. 13, (25N–10E), the Coney Basin Mine, 14 claims

—by-product gold from copper-lead-silver-zinc ores.

NORTH BEND: (1) N 23–24 mi. along the N fork of the Snoqualmi R., to (25N–10E): (a) ½ mi. up Illinois Cr., in secs. 8 and 9, the Beaverdale Mine—lode gold; (b) the Rushing Mine, 6 claims—lode gold; (c) in the Buena Vista district, in secs. 7 and 18, the Lennox Mine (45 claims, bunkhouses, mess hall, etc.), base metal by product gold; (d) 2 mi. by trail from the N fork and 3½ mi. above mouth of the Lennox R., in secs. 9 and 10, the Lucky Strike Mine—lode gold, silver, copper; (e) near head of the N fork of the Snoqualmi R., in NW¼ sec. 4, the Monte Carlo Mine, 13 claims—lode gold, silver; (f) E on dirt rd. to the Taylor R. district, in sec. 16, (24N–10E), the Rainy (Western States Copper) Mine—by-product gold; (2) near Snoqualmi Pass, on the S fork of the Snoqualmi R., in secs. 7 and 8, (22N–11E), the Carmack Mine, 5 claims—gold, copper, lead, silver.

SNOQUALMI: (1) WNW, along the Snoqualmi R., in bars and bench gravels, the Snoqualmi R. Placer; (2) NNE, along the Raging R , in gravel bars, the Raging R Placer.

Kittitas Co.

There are 47 lode mines and 27 placer workings in Kittitas County. With a total gold production of 37,095 ozs., only the Swauk district produced more than 10,000 ozs.

CLE ELUM: (1) along the Cle Elum R.: (a) near town, the Cle Elum Placer, productive; (b) from headwaters to a place about halfway to its mouth, the Cle Elum R. Placer; (2) on Mammoth Mt.: (a) the Broncho Mine—lode gold; (b) E of Eagle Mt., the Mammoth Mine—by-product gold; (3) on Fortune Cr.: (a) the Queen of the Hills Mine—lode gold; (b) the Ruby Mine (primarily antimony, copper, lead, silver)—by-product gold; (4) in T. 23 N, R. 14 E: (a) contiguous to Fish L., in Stevenson Gulch, the Silver Bell Mine—lode gold, silver; (b) in sec. 12, the Silver Cr. Mine—lode gold, silver; (c) on Hawkins Mt., near the SW cor., sec. 24, the Cle Elum Mine and, nearby, the Cascade Mine—lode gold, silver, some lead; (d) in NW¼ sec. 25, the Maud O. Mine—free gold, in pyrite; (e) in S½ sec. 24 and NE¼ sec. 25, the Ida Elmore Mine —lode gold and silver.

LIBERTY: (1) area, the Black Jack and Sunflower placers, productive;

(2) in sec. 1, (20N–17E): (a) in SW¼, the Old Bigney Placer, and (b) in the N½, the Ewell (Flag Mountain) Mine, stamp mill—lode gold; (3) along Williams Cr.: (a) near town, and (b) at jct. with Swauk Cr., the Williams Cr. Placer (good pay gravel within 3–4 ft. of bedrock and 70–80 ft. below present stream level); (c) ¾ mi. up Williams Cr., in sec. 2, the Ollie Jordan Mine (rich, erratic pockets)—free and wire gold; (4) the Cougar Gulch area in (21N–17E): (a) 4 mi. up gulch, in SE¼SW¼ sec. 26, the Cascade Chief (Morrison, First of August, Gladstone) Mine —free gold; (b) 5½ mi. up gulch (1½ mi. by trail), in sec. 30, the Wall St. Mine, 3 claims—lode gold; (c) all bed and bench bars in Cougar Gulch, especially below the mines—placer gold (abundant).

RONALD: (1) out 20 mi. to just N of Camp Cr., on line between secs. 25 and 26, (23N–14E), the Camp Cr. Mine, 4 claims of lead-silver-zinc ores—by-product gold; (2) out 24 mi. by rd. and trail: (a) on E side of the Cle Elum R., in secs. 26 and 27, (24N–14E), the Aurora (Lynch, Paramount) Mine, 16 claims—free gold, with copper and silver; (b) W of the Aurora, on Eagle Mt., the American Eagle Mine and, nearby, the Boss Mine—lode gold, silver.

THORP, on Red Mt., the Thorp Mine—lode gold, silver.

VIRDEN, N, between lat. 47°14' and 47°16' N, long. 120°28' and 120° 42' W, the Swauk district (tot. prod., 1868–1959, of 7,141 ozs. of lode gold and 4,972 ozs. of placer gold): (1) area, near the Blewett Pass hwy., the Gold Reef Mine—lode gold, silver; (2) in T. 20 N, R. 17 E: (a) at jct. of Boulder and Williams crs. in sec. 1, the Boulder Cr. Placer (pay dirt at bedrock, productive); (b) on E fork of Williams Cr., the Gold Leaf Mine—free gold (some in perfect octohedral crystals), silver; (c) on Snowshoe Ridge, in sec. 2, the Clarence Jordan Mine, 3 claims—free gold, silver; (d) in NE¼ sec. 3, the Cedar Cr. Placer; (e) in secs. 3 and 10, the Swauk Mining and Dredging Placer, very productive, large-scale; (f) in sec. 10, the Bryant (Deer Gulch) Placer; in NE¼, the Burcham Placer; and in the SE¼, the Dennett Placer, all productive; (g) along Baker Cr., above its jct. with Swauk Cr., the Baker Cr. Placer; (3) in NE¼SW¼ sec. 13, (21N–17E): (a) the Golden Fleece (Mercer, T-Bone) Mine, 3 claims, mill—free gold, silver, pyrite; (b) in sec. 33, the

Bear Cat Placer, productive; (c) in NE¼ sec. 15, the Sylvanite Mine —free gold; (4) near head of Swauk Cr., in sec. 8, (21N–18E), the Zerwekh (Big Z) Mine—free gold.

Lewis Co.

LONGMIRE, 1½ mi. above, on W slope of Eagle Mt., near NW cor. sec. 27, (15N–8E), the Eagle Peak Mine (copper)—by-product gold.

MINERAL, SE: (1) on Mineral Cr., the Waterfall Mine—lode gold, with lead and silver; (2) on S fork of Mineral Cr., the Tacoma Mine— lode gold.

Lincoln Co.

AREA, along the Columbia R., from Grand Coulee Dam to LINCOLN: (1) practically all: (a) low-water river bar gravels, and (b) all upper-terrace gravel deposits—placer gold (usually very fine); (2) on E side of the Columbia R., in secs. 12 and 13, (27N–35E), the China Bar Placer, 250 acres (low bar has paystreak in upper portions); (3) in SE¼NE¼ sec. 8, (28N–33E): (a) the Clark Placer, productive and extensively worked in late 1930s; (b) opposite the mouth of the Sanpoil R., in E½ sec. 8, the Keller Ferry (Angle) Placer—fine gold; (c) in S½ sec. 9, the Winkelman Bar Placer—fine gold (rather abundant).

CRESTON FERRY, in sec. 2, (27N–34E), the Creston Ferry Placer (in river bench, partly worked before 1910).

DAVENPORT: (1) a few mi. above SANPOIL: (a) the Hell Gate Bar, and (b) Peach Bar—fine placer gold; (2) NE 10 mi. (9 mi. SE of Pitney Butte), in Mill Canyon, the Iron Crown Mine, 5 claims—lode gold, with copper; (3) N 20 mi., in steep valley on NW side of Pitney Butte, in SW¼NW¼ sec. 32, (28N–37E), the Fouress (Pitney Butte) Mine, 5 claims of copper-lead-silver-zinc ores—by-product gold.

GRAND COULEE DAM (in Ferry Co.), E 15 mi., the Latta and Phillips Placer, terrace gravel deposits—placer gold.

PLUM (E of Grand Coulee), ½ mi. below ferry at Swawilla, in sec. 7, (28N–33E), the Barnell Placer (in 1938 produced from $200 to $400 a week).

Okanogan Co.

Okanogan County is one of Washington's largest counties, and it contains several extensive mining districts. Placer gold was discovered along the Similkameen River in 1859, and lode mines were opened up near Conconully in 1871. The total gold production from 270 lode mines and 32 placer operations through 1959 is estimated at between 85,000 and 90,000 ozs.

BRIDGEPORT, E, on N side of the Columbia R., in NW¼ sec. 10, (29N–25E), on a terrace 100 ft. above the r., the Shotwell Placer (covering a large area)—fine gold.

CARLTON, NE 3½ mi. up Leecher Cr., in NW¼ sec. 23, (32N–22E), the Minnie Mine, 4 claims—lode gold, with silver and zinc in pyrite.

CHESAW, in the Myers Cr. district, (40N–30E), tot. prod., 1888–1959, of about 9,500 gold ozs.: (1) area: (a) near center of NW¼ sec. 21, the Eagle Mine—lode gold; (b) center of W½ sec. 35, the Crystal Butte (Mother Lode) Mine, 6 claims—by-product gold from lead-silver-zinc ores; (2) N ½ mi., in SW¼ sec. 16, the Reco Mine—lode gold, copper, silver; (3) E 1 mi., in secs. 21 and 22 (or 16), the Chesaw Mine, open cuts—lode gold, silver; (4) S 1 mi., on Mary Ann Cr., the Mary Ann Cr. Placer, 14 claims—gold (from grass roots to bedrock); (5) NW 4 mi., in SW¼SE¼ sec. 11, (40N–29E), the Poland China (Molson, Overtop) Mine, 11 claims—lode gold, lead, silver; (6) on Deadman Cr., the Deadman Cr. Placer: (a) along creek bed, in gravel deposits, and (b) to 250 ft. on benches above stream—placer gold (abundant).

COLVILLE INDIAN AGENCY, W, along the Columbia R., near Katar: (1) the Gold Bar Placer, productive; (2) in sec. 11, (30N–28E), the Murray Placer, intermittent producer; (3) along the Nespelem R., the Nespelem R. Placer, productive.

CONCONULLY: (1) 1 mi. below, on 'the Salmon R., the Ballard Placer; (2) on N side of Sinlahekin Cr., near NE cor. sec. 20, (37N–25E), the Okanogan Copper Mine (access only by trail), a group of claims—lode gold, copper.

GILBERT (end of the Twisp R. rd. NW from TWISP), just N, in sec. 11, (34N–18E), the Pay Day Mine, 5 claims—lode gold, copper, silver.

LOOMIS: (1) N ½ mi., in NE¼NE¼ sec. 1, (38N–25E), the Palmer Mt. Tunnel, 56 base-metal claims—by-product gold; (2) take rd. to old P.O. of Golden, W ½ mi., in NE¼ sec. 10, (39N–26E), the Triune (Crescent) Mine, 5 claims—lode gold, copper, lead, silver; (3) N 3 mi. on Palmer Lake rd., in SW¼ sec. 19, (39N–26E), the Pinnacle Mine, 6 claims—free gold, with pyrite, silver, base metals; (4) N, to Palmer Lake: (a) on Palmer Mt., in NE¼ sec. 36, (39N–25E), the Black Bear (Alice, Summit, War Eagle) Mine—by-product gold from copper-silver ores; (b) on W slope of mt., the Bunker Hill Mine, rich—lode gold; and near SW cor. sec. 19, (39N–26E), the Leadville Mine, 13 claims—free gold (with copper, lead, silver); (c) ½ mi. E, in N½ sec. 6, the Empire Mine—by-product gold; (5) between Palmer Mt. and Wannacut L., in NE¼ sec. 22, (39N–26E), the Rainbow Mine—free gold (with copper, lead, silver); (6) in SW part of sec. 33, (40N–26E), the Bullfrog Mine —lode gold, silver; (7) W 3 mi., in sec. 5, (38N–25E), the Gold Hill Mine, 86 claims—by-product gold; (8) 8 mi out, the Chopaka area: (a) in SW¼ sec. 32, (40N–25E), the Gold Crown Mine, 3 claims—lode gold, copper, silver; (b) other area mines—gold showings.

MALOTT, E 3 mi., in S½NW¼ sec. 22, (32N–25E), the Rustler Mine—lode gold, molybdenum.

MAZAMA: (1) NW 3 mi., in SE¼ sec. 14, (36N–19E), the Mazama Queen (Continental) Mine, 3 claims, mill—by-product gold; (2) in T. 36 N, R. 20 E: (a) ½ mi. N of town, in N½ sec. 30, the Mazama Pride (Hotchkiss) Mine, 9 claims—lode gold, silver, copper; (b) 1 mi. NE of town, on cliff in SE¼ sec. 30, the American Flag (Oriental and Central) Mine, reached 1½ mi. by rd. and ½ mi. by trail, 3 claims (copper-silver-zinc)—by-product gold; (c) near NW cor. sec. 30, the Gold Key Mine —lode gold, copper; (d) 5 mi. from town on the Goat Cr. rd., the Imperial (Crown Pt.) Mine in NW¼ sec. 16 and the Rosalind Mine in N½ sec. 17—lode gold, copper, silver.

METHOW (Squaw Cr.) district, 9 mi. above mouth of the Methow R., tot. prod., 1887–1959 (mainly 1932–59), of 16,482 gold ozs.: (1) on NE slope of Hunter Mt., in E½ secs. 12 and 13 and W½ secs. 7 and 18, (30N–22E), the Methow (London, New London) Mine, 4 claims E of mt. on W side of the Methow R. and 3 claims E of r.—lode gold (with copper, silver, tungsten); (2) ¼ mi. E of the Methow R., in SW¼NW¼ sec. 18, (30N–23E), the Roosevelt Mine, 75 ft. above r. —lode gold, copper, tungsten; (3) on S fork of Gold Cr., in sec. 25, (31N–21E), the Bolinger Mine, 5 claims, major producer—lode gold, silver.

NIGHTHAWK (NW of OROVILLE), 2 mi. S of Canada at base of Mt. Chopaka, near SE cor. sec. 7, (40N–25E), the Golden Zone Mine, 5 claims—by-product gold.

OROVILLE-NIGHTHAWK (Palmer Mt.) district, between lat. 48°50′ and 49°00′ N, long. 119°28′ and 119°42′ W, tot. prod., 1859–1959, of 49,000–50,000 gold ozs. (half placer): (1) between town and NIGHTHAWK, along the Similkameen R.; (a) in river bars and terrace gravels, the Similkameen Placer—flake, shot, and nugget gold; (b) at Similkameen Falls, placers of same name, productive; (2) N 3 mi., on N side of Similkameen Cr.; (a) in SE¼NW¼ sec. 19, (40N–27E), the Okanogan Free Gold (Owasco, Allison) Mine, 5 claims—lode gold, silver; (b) in SE¼ sec. 23, the Chicago Mine (several veins over 2 ft. wide, bearing also lead and silver)—lode gold; (c) in SW¼ sec. 25, the Combination Mine—lode gold, silver; (3) many other lode-gold mines in the T. and R. area; (4) SW, to Wannacut L. (and district): (a) 200 yds. from E shore of lake, in sec. 14, (39N–26E), the Anaconda Mine—lode gold, copper, silver; (b) on rd. E of the War Eagle property, the Golden Crown Mine—lode gold (on dump, assays run to $105 a ton).

PARK CITY, N 1½ mi., on Strawberry Cr. in S½ sec. 35, (34N–31E), the Crounse (Strawberry Cr.) Placer, productive.

PATEROS (in the Squaw Cr. district, cf. under METHOW): (1) in SW¼ sec. 20, (30N–23E), the Friday (Tom Hall) Mine, 5 claims, access by paved rd. and cable car across Methow Cr.—lode gold, silver; (2) near head of Squaw Cr., on N side of valley near NE cor. sec. 17, (30N–22E): (a) the Chelan (Pennington) Mine, 7 claims—lode gold (with tungsten); (b) many other nearby lode-gold mines; (4) NW 5 mi., on E side of the Methow R., the Sullivan (Pateros) Mine, 5 claims—lode gold, copper, silver.

TWISP: (1) out 4 mi., in S½ sec. 13 and N½ sec. 24, (33N–21E), the Twisp View Mine, 2 claims (copper, lead, silver, zinc)—by-product gold;

(2) in secs. 25, 26, 35, and 36 (5 mi. SW of town), the Alder Mine, 17 claims, copper-silver-zinc ores—by-product gold; (3) S½ mi., in SE¼ sec. 18, (33N–22E), the Rattlesnake Mine, base metal and silver— by-product gold; (4) on lower W slope of Pole Pick Hill, in SE¼ sec. 18, (33N–23E), the Red Shirt Mine—lode gold, copper, silver.

WAUCONDA, near E co. line and 20 mi. S of Canada, the Cascade (Wauconda) district, tot. prod., 1901–57, between 10,000 and 15,000 gold ozs.: (1) area, in SE¼ sec. 7, (37N–31E), the Wauconda Mine, 32 claims (22 near town and 10 out 3 mi.), 5 ledges 152 ft. wide, with quartz veins to 40 ft. wide in argillite—lode gold; (2) NE 10 mi. to BODIE on Toroda Cr., in SW¼ sec. 3, (38N–31E) and sec. 34, (39N–31E), the Bodie (Northern Gold) Mine, major producer—lode gold, silver.

WINTHROP, out 27 mi., on Isabella Ridge in NW¼SE¼ sec. 15, (38N–21E), the Mountain Beaver Mine, 3 claims—lode gold, with bismuth, copper, silver.

Pacific Co.

FT. CANBY, in black-sand beach deposits at mouth of the Columbia R.—placer gold, platinum.

ILWACO, just S, on island at mouth of the Columbia R., the Sand I. Placer, productive—gold, platinum.

NASELLE, at mouth of the Naselle R., on beach along Chetlo Harbor peninsula ½ mi. from U.S. 101, the Chetlo Harbor Mine, a vein 3 ft. wide extending for 300 ft. along beach—lode gold.

OCEAN PK., area beaches, in black-sand deposits—placer gold and platinum (very fine).

Pend Oreille Co.

There are 38 lode mines and 5 placer claims listed in Pend Oreille County.

AREA: (1) near SE cor. sec. 19, (39N–43E), in a draw near Schultz's cabin, the Schultz Placer; (2) on the Pend Oreille R., ½ mi. W of the Slate Cr. rd., in N-center sec. 26, (40N–43E), the Harvey Bar Placer; (3) on E side of the Pend Oreille R. just below mouth of the Z Canyon gorge, 2½ mi. W of Crescent L., the Schierding Placer (worked by dragline dredge)—placer gold (flat colors, rounded edges; nuggets).

METALINE FALLS: (1) E, on Sullivan Cr., the Sullivan (O'Sullivan) Cr. Placer—nuggets to 2 ozs.: (2) in sec. 2, (39N–45E), the Deemer Mine, 5 claims, open cuts (mostly caved in now)—lode gold.

NEWPORT: (1) out 4 mi., in Pine Canyon, the Farmer Jones Mine, mill—lode gold; (2) in T. 30 N, R. 43 E: (a) on main hwy., in sec. 24, the Nevell Mine—lode gold, silver; (b) in NE¼ sec. 24, the Wold Mine —lode gold; (3) in sec. 14, (30N–43E), 1 mi. from rd., the Hansen Mine (copper, lead, zinc)—by-product gold; (4) in sec. 22, (30N–44E), the Sunrise Mine—lode gold, silver; (5) in sec. 24, (31N–44E), the Gilbert Mine, in quartz—free gold; (6) in T. 34 N, R. 44 E: (a) at E end of Brown's L., in NW¼NE¼ sec. 24, the Brown's L. Placer—occasional gold colors; (b) in SE¼NW¼ sec. 26, the Isabella Mine—lode gold (with copper, silver).

Pierce Co.

AREA, far E part of co., adjacent to the Yakima Co. line: (1) near head of Morse Cr., the Morse Cr. Placer; (2) the Summit district, in (17N–10E): (a) in sec. 12, the Ogren Placer; (b) in sec. 25, the Silver Cr. Mine (reached by trail from end of rd. up Silver Cr.), base metals—lode gold; and in the NE¼, the Silver Cr. Placer (coarse gold, productive); (3) in sec. 25, the Washington Cascade (New Deal) Mine (reached 7 mi. by rd. from the Chinook Pass hwy.)—lode gold, silver, copper; (4) center sec. 35, the Current Mine, 9 claims—lode gold, silver; (5) on S end of Crystal Mt., at head of Silver Cr., the Campbell Mine (reached 3 mi. by trail from end of the Morse Cr. rd.), 5 claims—lode gold.

CLAY CITY, 1 mi. out, in sec. 30, (17N–5E), the Seigmund Ranch Mine—free gold.

Skagit Co.

AREA: (1) along the Skagit R., almost entire length through co., in stream gravel bars and adjacent benches and terraces—placer gold possibilities; (2) near headwaters of Ruby Cr., the Ruby Cr. Placer (first gold discovery W of the Cascades).

ANACORTES: (1) in city limits, in N½ sec. 26, (35N–1E), the Stephens

(Fidalgo) Mine, copper-silver—by-product gold; (2) area beach black-sand deposits, the Anacortes placers; (3) SW 3½ mi. from town center (but within city limits), the Anacopper Mine—by-product gold; (4) out 4½ mi., in secs. 1 and 2, (34N–1E), the Fidalgo I. Mine (copper, silver) —by-product gold; (5) SW 6 mi., the Matrix Mine (copper)—by-product gold.

FIDALGO (I.), area beaches along entire shoreline, in black-sand concentrations—placer gold, chromite.

LYMAN, area of Day Cr., the Day Cr. Placer, productive.

MARBLEMOUNT, area of Johnsburg: (1) on S side of the N fork of the Cascade R., in W½ sec. 25 and E½ sec. 26, (35N–13E), the Midas Mine, 5 claims—minor gold (with lead, silver, zinc); (2) SE of the Midas, in SE¼ sec. 25 and NW¼ sec. 36, the Soldier Boy Mine, 5 claims—lode gold (with copper, silver, zinc).

MT. VERNON: (1) SE 4½ mi., on Devils Mt., in S½ sec. 4, (33N–4E), the Mt. Vernon (Devils Mt. Pacific) Mine, 2,100 acres overlapping into adjoining sections, in nickel-chromite ores—free gold; (2) E, near summit of Bald Mt., in sec. 17, (34N–6E), the Bald Mt. Mine, copper primarily—by-product gold.

Skamania Co.

There are 27 lode mines and 2 placer operations listed as gold producers in Skamania County.

AREA: (1) on McCoy Cr., in (10N–8E). (a) in sec. 15, the Hudson and Meyers Placer; (b) probably in sec. 10, the Perry Mine—lode gold, arsenopyrite; (c) along both sides of Camp Cr. in sec. 10, the Primary Gold Mine (access 4 mi. by trail from end of the Niggerhead rd.), 18 claims—free gold (with silver and platinum)—and the adjoining Bruhn Mine—free gold; (d) in creek bars and bench gravels, the Camp Cr. Placer, quite productive; (2) on N side of Quartz Cr., in SW¼SE¼ sec. 7, (8N–8E), the Plamondon Mine, access 3½ mi. by trail from end of rd., 29 claims, in chalcedonic quartz—lode gold, silver.

SPIRIT LAKE: (1) NW, in extreme NW cor. of co., on Black Mt.: (a) in secs. 2, 3, 10, and 11, (10N–5E), the Independence Mine, 6 claims (copper, lead, silver)—minor lode gold; (b) opposite, the Minnie Lee

Mine—lode gold (with copper, silver); (c) near Grizzly Cr., probably in sec. 20, the Grizzly Cr. Mine (access by trail)—lode gold, with lead and silver; (2) NE, in T. 10 N, R. 6 E: (a) in secs. 8 and 17, the Germania Mine, 12 claims—lode gold, copper; (b) near W¼ cor. sec. 16, the Golconda Mine (access ½ mi. up the Ryan L. trail from the Green R. trail)—lode gold, with copper and zinc.

STEVENSON, NW, in E½ sec. 25, (4N–5E), the Texas Gulch Placer, productive.

Snohomish Co.

Snohomish County contains 236 lode-gold mines and 42 placer-gold operations, although the total production of gold from 1903 through 1956 is only 9,595 ozs., most of it from the Monte Cristo and Silverton districts.

DARRINGTON: (1) near town, on Deer Cr., the Deer Cr. Placer, productive; (2) S, on White Horse Mt.: (a) in sec. 14, (32N–8E), the Calf Moose Placer—fine to coarse gold, nuggets; (b) in sec. 13, the Cow Moose Placer, productive; (3) near town, on W side of the Sauk R., in NW¼ sec. 22, (32N–9E), the Darrington Placer, well worked; (4) in (32N–9E): (a) on Jumbo Mt., in SE¼ sec. 27, the Queen Anne Mine (base metals)—by-product gold; (b) on E side of Gold Mt., in NW¼ sec. 27, the Elwell-Darrington Mine (and adjacent Burns Prospect)— lode gold, copper; (c) in NE¼NE¼ sec. 29 (5 mi. from town), the Sunrise (North Star, Oldfield) Mine, 4 claims—lode gold; (d) on N side of mt., in sec. 32, the Sloman Mine—lode gold, copper, silver; (5) in sec. 24, (32N–9E) and sec. 19, (32N–10E), the Blue Bird Mine, several claims (base metals and silver)—by-product gold; (6) on W side of Gold Mt., in NW¼ sec. 18, (32N–10E), the Burns Mine, 7 claims—lode gold, with copper, silver; (7) area 7 mi. from Bedel and 7 mi. from Barlow Pass, the Perm Mine (lead, silver, zinc)—lode gold.

INDEX: (1) W ½ mi., on N side of N fork of the Skykomish R., in NE¼ sec. 19, (27N–10E), the Line a Little Placer (2 claims), productive; (2) out 9 mi., on N fork of the Skykomish R., in secs. 19 and 20, (28N–11E), the Bonnie Belle Placer; (3) on W slope of Iron Mt., possibly in sec. 29, (28N–11E), the Black Hawk Mine, 4 claims—lode gold.

Monte Cristo (district, in S-central part of co. which, with the
Index, Silverton, Silver Cr., and Sultan contiguous districts, embraces
territory 20 mi. long by 10 mi. wide, extending from the center of the
co. to the King Co. line): (1) in (28N–11E): (a) the area along Silver Cr.,
in SE¼ sec. 3, the Remonille Mine, 3 claims—lode gold; (b) on W side
of Silver Cr., in sec. 6, the Red Cloud Mine, 3 claims—lode gold, silver,
lead; (c) on W side of cr., in NE¼ sec. 19, the Michigan Mine (and
many others in area)—lode gold, silver, etc.; (2) NE¼ mi., in sec. 22,
(29N–11E), the Rainy (Ben Lomond) Mine, 3 claims—lode gold, silver;
(a) in sec. 22 (1 mi. E of town), the Keystone Mine—lode gold, copper,
lead, silver; (b) nearby, in sec. 26, the Monte Cristo (Mystery, Pride)
Mine, 14 claims, primarily copper, lead, silver, zinc—by-product gold;
(3) ½ mi. NE of Goat L., in SE¼ sec. 2, the Glory of the Mt. Mine,
7 claims—lode gold, silver; (4) W 2½ mi., on E slope of ridge dividing
the Sauk, Sultan, and Stilaguamish watersheds (overlooking Crater L.),
the Del Campo Mine, 3 claims—lode gold, copper, silver; (5) in sec. 33,
(29N–11E): (a) on N side of Silver Cr., ¼ mi. E of Red Gulch (in W½
sec. 32), the Mineral Center (Bonanza, Edison, Louise, Washington-
Iowa) Mine—lode gold; (b) 2 mi. from town on the Mineral City rd.,
the Good Hope Mine, 10 claims—lode gold and silver (with lead and
copper); (c) on E side of E fork of Silver Cr., the Minnehaha Mine—
lode gold, silver, pyrite; (6) in the Wilmon Peak area: (a) on NW side,
in sec. 27, (29N–11E), overlooking Monte Cristo, the Justice (Golden
Chord) Mine—lode gold, silver; (b) on W slope, in NE¼ sec. 27, the
Peabody (Sidney) Mine, 7 claims (lead, silver, zinc)—by-product gold;
(7) W side of Weden Cr., the Northwest Consolidated Mine—lode
gold, with copper and silver; (8) very many other mines on area quadrangle map.

Silverton: (1) just W, on Marten Cr., in NW¼ sec. 11, (30N–9E),
the Silver Coin Mine, base metal—by-product gold; (2) SE½ mi., at
mouth of Silver Gulch: (a) in NE¼ sec. 19, the Copper Independent
(Independent) Mine—lode gold, silver; (b) 1,000 ft. below the outlet of
Copper L., in SW¼ sec. 33, the Cornucopia Mine, 4 claims—lode gold,
with lead and silver; (3) NE 2 mi., in secs. 8 and 9, (30N–10E), the
Double Eagle Placer, productive; (4) on S side of Stilaguamish R., at foot
of Huckleberry Ridge, in secs. 17–19, the Eclipse (Dahl) Mine, 51

claims and 5 mi. underground workings—lode gold, with copper, mercury, silver.

SULTAN: (1) at W end of town, on the Sultan R., area of (28N–8E): (a) the Sultan R. Placer, probably in sec. 31: (b) along the Sultan R. to its mouth, all stream and bench gravels—placer gold; (2) on the Sultan R., just below mouth of Sultan Canyon: (a) in sec. 17, the Aristo placers; (b) in secs. 8 and 17, the Bertha D Placer, and (c) in N½ secs. 4 and 5, the Black Hawk Placer; (3) in NE¼ sec. 5 and SE¼SE¼ sec. 32, (29N–8E), the Sultan Canyon (McCloud) Placer; (4) in secs. 8 and 17, (28N–8E): (a) the Merry Mixer Placer, (b) the Mom and Pop Placer, and (c) the Pork Barrel Placer; (5) on S side of r., in NW¼ sec. 35, (29N–8E), the Great Northern Mine, 3 claims—lode gold, copper, silver; (6) 26 mi. from town, in Sultan Basin, ½ mi. from rd. near center of N½ sec. 27, (29N–10E), the Calumet Mine—lode gold, copper, silver, zinc.

Stevens Co.

Since 1902 there have been 204 lode-gold mines and 21 placer operations developed within Stevens County. Most of the total production of 52,145 gold ozs. through 1959 came from lode mines.

AREA: (1) along the Columbia R. entire length of co. line, very many placer workings; (2) bench and terrace gravels to well above river level —placer gold.

ALADDIN, W, on the J. W. Scott ranch, in sec. 6, (37N–41E), the Scott Mine—lode gold, silver.

ARDEN, out 2½ mi., in SE¼NE¼ sec. 34, (35N–39E), the Rocky L. Mine, 8 claims—lode gold (minor) in base-metal ores.

BLUECREEK: (1) NW1½ mi., in NE¼SE¼ sec. 26, (33N–39E), the Krug (Hartford) Mine, 10 claims, primarily copper—minor lode gold; (2) SW 2½ mi., in N½NE¼ sec. 2, (32N–39E), the Liberty Copper Mine, 6 claims—lode gold, some silver.

BOSSBURG: (1) S 5 mi., in sec. 22, (37N–38E), the Sandoz Placer, productive; (2) between town and MARCUS, on E side of the Columbia R., in secs. 16 and 21, (37N–38E), the Valbush Bar Placer.

BOYDS, out 2 mi., in NW¼ sec. 3, (37N–37E), the Napoleon Mine (second major Orient district, q.v. under ORIENT, producer)—by-product gold (from iron and copper).

CEDONIA, NE 10½ mi. and ¼ mi. N of the ADDY–BISSELL rd., in SW cor. sec. 19, (32N–38E), the Columbia Tungsten (Black Horse, Stockwell) Mine, 15 claims, primarily tungsten—lode gold and silver (with molybdenum and zinc).

CHEWELAH: (1) E 1½ mi., in N part sec. 7, (32N–41E), the Chewelah Standard (Nellie S.) Mine, 5 claims, mainly copper-silver-molybdenum—minor by-product gold; (2) out 6 mi., near center N¼ sec. 32, (33N–41E), the Chinto (Banner) Mine—minor lode gold; (3) W 18 mi.: (a) the Belle of the Mountain Mine—lode gold; (b) adjoining on S, the Blue Belle Mine—lode gold; (4) very many other mines in the Kaniksu Nat'l. Forest—lode and by-product gold.

DAISY, out 2½ mi., along the Columbia R., the Collins Placer.

EVANS: (1) near town, in secs. 29 and 32, (39N–39E), the Nobles Placer; (2) out 1½ mi. and 1,000 ft. E of Hwy. 22, in sec. 15 and 22, (37N–38E), the Gold Bar Mine, 8 claims (lead, silver, zinc)—lode gold.

KETTLE FALLS: (1) E 1 mi.: (a) the Gold Hill (Colonel Fish) Mine, some lead and silver—lode gold; (b) in SW¼ sec. 16, (36N–38E), the City View Mine, copper-lead-silver—lode gold; (2) NE 2 mi., in S½SE¼ sec. 7, the Sunday (Sunday Morning Star, Golden Hope) Mine —lode gold, silver; (3)(a) in SW¼SE¼ sec. 9, on top of hill, the Gold Reef (Benvenue, Golden Reef) Mine, 4 claims—lode gold, copper, silver; (b) in SW¼ sec. 4, the Gold Ledge Mine (near rd.)—lode gold, silver; (4) out 6 mi.: (a) below town, in sec. 29, (35N–37E), the Holsten Placer; (b) S and 1 mi. off Hwy. 22, near center N½ sec. 11, the Ark (Silver Queen) Mine, 19 claims of silver and base metals—minor lode gold.

LOON LAKE, N 6 mi., near center sec. 33, (31N–41E), the Loon Lake Copper Mine—minor lode gold, silver.

NORTHPORT: (1) area: (a) N, along the Columbia R. to the Canadian line (camp on W bank of reservoir just N of town), the Evans Placer, a lease of 5 mi. along the E shore—placer gold abundant; (b) on Onion Cr., the Alice May Mine—lode gold; (2) W 1 mi., on Sheep Cr., in sec. 25, (40N–39E), the St. Crispin Mine—lode gold, with copper, lead, silver; (3) SE 3 mi., on N side of Fish Cr., in SW¼ sec. 16, (39N–40E), the Northport Mine—lode gold; (4) near rd. in NW¼SE¼ sec. 7, (40N–40E), the Elvick Mine—lode gold, lead.

ORIENT (Pierre L.) district, in far NW cor. of co., between lat. 48° 50' and 48°57' N, long. 118°05' and 118°10' W, formed in 1902 (with 86 mines currently listed in T. 39 and 40 N, R. 36½–37 E), tot. prod. through 1942 of 45,057 gold ozs.: (1) E 3 mi. via the Lookout rd., to First Thought Mt.: (a) in secs. 7 and 18, (39N–40E), the First Thought Mine, 28 claims—lode gold, silver; (b) to SE, in sec. 18, the First Thought Extension Mine, a principal producer—lode gold; (c) on N slope of mt. 6 mi. NE of town, in sec. 19, (39N–37E), the Gem Mine, 4 claims—lode gold; (2) E 3 mi., in SE¼ sec. 18, (39N–37E), the North Star Mine, several claims—lode gold; (3) many other area lode-gold mines, e.g., the Alice C (¼ mi. E of the First Thought), American Boy, Bald Eagle, and Blue Grass (near top of Toulou Mt. in E½ sec. 24, [39N–36E]), and dozens more.

ROCKCUT: (1) E 1 mi., on flat bench in sec. 1, (39N–36E), the Abe Lincoln Mine—lode gold, copper; (2) out 4 mi., near center sec. 19, (40N–37E), the White Elephant (Kettle R.) Mine, 6 claims—lode gold, copper, silver; (3) NE 5 mi., in secs. 19 and 30, (40N–37E), the F. H. and C. (Faith, Hope, and Charity) Mine, 13 claims—lode gold, silver.

SPRINGDALE: (1) out 2 mi., on Douglas Mt., in secs. 11 and 12, (31N–39E), the Dumbolton Mine, 10 claims—lode gold, with mercury and tin: (2) on E slope of Huckleberry Range near summit, in NW¼NE¼ sec. 36, (31N–38E), the Wells Fargo Mine, primarily antimony—lode gold, lead, silver, zinc.

VALLEY, W 9 mi. and 1 mi. N, in SE¼SE¼ sec. 9, (31N–39E), the Edna (King) Mine, 12 claims, primarily copper—lode gold, silver.

Whatcom Co.

The 63 lode-gold mines and 15 placers in Whatcom County's section of the Cascade Range produced a recorded total of 89,156 ozs. of lode gold and 2,425 ozs. of placer gold between 1903 and 1959.

AREA, extreme E part of co., in heart of the Cascade Range between lat. 48°45' and 48°50' N, long. 120°45' and 121°00' W, the Slate Cr.

district (40 area lode-gold mines, with tot. prod., 1933–52, of 29,172 gold ozs.): (1) nearly all regional crs. and streams, in bed and bench gravels and in ancient channel terraces—abundant placer gold; (2) in the Harts Pass area: (a) on Mill Cr., in secs. 30 and 31, (37N–17E), the Azurite Mine (access from the Methow Valley in Okanogan Co. and over the pass), 42 claims, silver and base metal—lode gold; (b) 5 mi. NW of the pass by rd., in sec. 33, (38N–17E), the Mammoth Mine, principal district producer, many claims—lode gold, free gold, tellurides (accompanied by lead, silver, zinc); (c) 6 mi. NW of the pass and 2 mi. N of the New Light Mill, in SW¼ sec. 26, the Chancellor (Indiana) Mine, 4 claims, lead-silver-zinc—lode gold; (d) 7 mi. NW of the pass, in center S½ sec. 27, the New Light Mine, a principal district producer, 16 claims —lode gold, silver; (e) 10 mi. NW of the pass, at head of Barron Cr., in NW¼ sec. 27, the Baltimore Mine, 22 claims—lode gold, silver, zinc; (f) on both sides of the Cascade summit about 10 mi. S of Canada, the Gold Ridge Mine, major producer—lode gold; (3) near head of Cascade Cr., in SW¼ sec. 24, (38N–16E), the Anacortes Mine, 7 claims—free gold and tellurides, silver; (4) in Allen Basin, in sec. 34, (38N–17E), the Allen Basin Mine, 14 claims—lode gold, lead, silver, zinc.

GLACIER, out 5 mi., in sec. 36, (40N–7E), the First Chance Mine— lode gold.

MT. BAKER LODGE: (1) area, the Iron Cap Mine, 7 claims—lode gold; (2) W of Wells Cr. and S of the N fork of the Nooksack R. (2 mi. by trail from the Nooksack Forest Camp), in sec. 6, (39N–8E), the Great Excelsior (Lincoln, President) Mine, 10 claims—lode gold, silver; (3) 4 mi. below lodge, near hwy., in sec. 20, (39N–9E), the Verona (Galena) Mine, 5 claims, silver and base metals—lode gold.

NOOKSACK, near E¼ cor. sec. 35, (40N–4E), the Nooksack Mine, 3 claims—lode gold.

SHUKSAN, the Mt. Baker district in N-central part of co., between lat. 48°50' and 49°00' N, long. 122°25' and 122°35' W, on N side of Mt. Baker: (1) NE, about 1 mi. SE of Twin Lakes, near head of the W fork of Silesia Cr., in SE¼ sec. 15, (40N–9E), the Lone Jack Mine, 22 claims, major producer—free gold, silver; (2) on S side of Red Mt., trail from Twin Lakes, in SW¼ sec. 4, the Gargett Mine, 7 claims, silver and base metals—lode gold; (3) N 2 mi., on Swamp Cr., in NW¼ sec. 21, the Evergreen Mine, 8 claims—lode gold (with base metals and silver); (4) 2 mi. S of Canada, in NE¼ sec. 4, the Boundary Red Mountain Mine (accessible by rd. and trail from Chilliwack, B.C.), principal producer, 11 claims—lode gold: (5) very many other lode-gold mines, including the Midas, First Chance, Henrietta, Hoosier, Gold Basin, Gold Run, Mansfield, etc.

Whitman Co.

PENAWAWA, area along the Snake R., in sec. 15, (14N–41E), the Indian Bar Placer—fine gold.

Yakima Co.

AREA, far NW cor. of co., along boundary of Pierce Co. on NE flanks of Mt. Ranier, the Summit district in T. 16 and 17 N, R. 10 and 11 E: (1) near head of Morse Cr., the Morse Cr. Placer ("good wages" made in early days)—coarse to large gold nuggets; (2) in (26N–11E): (a) in NW¼ sec. 4, the Elizabeth Gold Hill Mine—lode gold, copper, lead, silver; (b) in sec. 6, the Highland Mine—lode gold, silver; (3) in (17N–10E), in sec. 25, the Black Hawk Mine—lode gold; (4) in (17N–11E): (a) in SW¼ sec. 29, the Comstock Mine—lode gold, silver; (b) just W of the Comstock, the Crown Pt. Mine—lode gold, silver; (c) in SW¼ sec. 31, the Damfino Mine and in the S½, the Gold Hill Mine (on the Morse Cr. rd.)—lode and free gold, silver; (d) along the Cascade crest from Bear Gap to Crown Pt., in sec. 31, the Fife Mine (access by rd. up Morse Cr. from Chinook Pass hwy. and ½ mi. by trail), 14 claims —lode gold, silver; (e) in secs. 4 and 5, the Elizabeth Placer and the Gold Links Placer (altogether, 3 placer operations and 20 adjoining lode-gold claims): (f) numerous other base-metal and silver-lode mines in area, also producing lode gold.

WHITE SWAN, SW, across the Yakima Indian Reservation (get permission to enter at the agency) to the Klikitat R., in sec. 14, (8N–12E), the Surveyors Cr. Placer, productive.

WEST VIRGINIA

Unlike the other eastern Atlantic states in the general Appalachian belt, West Virginia lacks metallic minerals in commercially minable quantities. Nevertheless, the presence of small amounts of gold in Tucker County on several farms led to an abortive gold rush in 1927.

Tucker Co.

PARSONS, in Sissaboo Hollow, in outcrops of quartz over a considerable area (a few prospects in the 1920s but no mines)—assays reveal a small amount of gold in the outcrops.

WISCONSIN

Wisconsin's gold production, such as it is, comes almost entirely as a modest by-product of base-metal ore refining. Actually, more real diamonds have been found in the Pleistocene glacial drift gravels than prospector's gold, which has been so little reported as to be esteemed almost nonexistent.

Pierce Co.

ROCK ELM (Twp.), along Plum Cr., in gravel bars—placer gold, with an occasional diamond.

WYOMING

The first discovery made in Wyoming was immediately east of the Great South Pass crossing of the Continental Divide on the Old Oregon Trail in 1842, when placer gold was found in the headwaters of the Sweetwater River near the wagon train campground at the ford. However, because of the danger from Indian attacks, mining did not begin in the area until the famed Carissa Lode on the mountain above South Pass City was discovered in 1867, leading to Wyoming's first gold rush into the South Pass-Atlantic City district.

Although small amounts of gold have been mined almost every year to the present in Wyoming, the state's total production of gold through 1959 is only a very modest 82,000 troy ozs. Many of the state's stream gravels carry low-grade placer gold and, in several areas, tertiary gravel deposits show similar gold concentrations. Most of the placer gold is so finely divided that recovering it in profitable amounts is difficult to impossible.

For more information, write: State Geologist, The Geological Survey, University of Wyoming, Laramie, Wyoming 82070.

Albany Co.

AREA: (1) in the Lake Cr. district, the Albatross Claim—lode gold, with copper; (2) in SW part of co., on Rock Cr., (19N–78W), the Copper King Mine—lode gold, copper, silver; (3) extreme NE cor. of co. (just S of ESTERBROOK in Converse Co.), in NE¼ sec. 16, (28N–71W), the Three Cripples Mine, primarily iron—lode gold, some copper and silver.

ALBANY, extreme SW cor. of co., in Moore's Gulch (including Douglas Cr., the Douglas Cr. district, tot. prod., 1868–1959, between 10,000 and 11,000 gold ozs.): (1) Moore's Gulch: (a) area placer workings (first gold found in district); (b) area quartz vein mines (source of the placer gold)—lode gold, silver; (2) along Douglas Cr. and its tributaries (Lake, Muddy, Spring, Keystone, Beaver, Horse, Joe's, Dave's, Ruth's, Bear, Gold Run, and Willow crs., and in Lone Gulch): (a) in watercourse gravels, benches and terrace deposits (embracing an area 15 mi. long by 10 mi. wide), productive placers—coarse gold; (b) extending from sec. 10, (13N–80W) to sec. 2, (13N–79W), a distance of 8 mi., and on Muddy Cr. for 5 mi. from Douglas Cr. to the S line of sec. 18, (14N–78W), the Douglas Consolidated placers, very productive; (3) on top of Tenderfoot Mt., in sec. 6, (13N–79W), the Fairview Claim—lode gold; (4) in sec. 16, (14N–79W), the Albany placers (extending up cr. for 5 mi.)—fine to coarse gold, nuggets; (5) area of Blue Grass Cr., sand and gravel deposits—very finely divided placer gold; (6) W, in area of the Medicine Bow Mts., in gulches and watercourses near timberline at 10,000–11,000 ft.—colors and nuggets of native gold.

CENTENNIAL: (1) in sec. 9, (15N–78W), the Centennial Ridge Mine —lode gold; (2) the La Plata district, many area mines: (a) in the S part of the Laramie Range, the Sherman Group, Strong Mine, and the adjoining Ulcahomo Mine—lode gold; (b) in sec. 9, (15N–78W), the Utopia Tunnels—lode gold.

JELM (BOSWELL): (1) the old Cummins City area in (13N–77W): (a) the Annie Mine, primarily copper—some lode gold; (b) ½ mi. SW of the townsite, the Copper Queen Mine—minor by-product gold; (2) area of the Jelm Mts., in surface sand and gravel deposits, exceptionally rich —free-milling gold.

KEYSTONE: (1) in SW¼ sec. 22, (14N–79W), the Florence Mine— free gold, in pyrrhotite; (2) 1 mi. above town, in NW¼ sec. 22, the Keystone Mine (major producer, 40-stamp mill)—lode gold in quartz; (3) in the Holmes area: (a) the Maudem Group of mines, and (b) the Medicine Bow Mines Co., all good producers—lode gold.

LARAMIE, SE 12 mi., in the Laramie ("Black Hills") Range, the Antlers Mine, in quartz and feldspar—lode gold.

Carbon Co.

AREA, in far SE part of co., in (18N–78W), on Cooper Hill, the Cooper Hill Mine: (a) on dump—native gold, with copper and silver; (b) area float, very rich—lode gold in mineral specimens.

BAGGS, ENE, to head of Spring Cr. in (14N–86W), the Spring Cr. Mine, on dumps—free gold.

ENCAMPMENT: (1) several area lode mines, e.g., the Meta Mine— minor gold values; (2) W 4½ mi., in sec. 21, (14N–84W), the Tennant Cr. Mine—lode gold, lead, silver.

RAWLINS: (1) NE 32 mi.: (a) the Seminoe Mts. district, area mines, such as the Deserted Treasure, King, Star and Hope Gold—lode gold; (b) in SE¼ sec. 29, (26N–85W), the Sunday Morning Mine, on dumps —free gold, chalcopyrite; (2) SSE 40–50 mi., in the Sierra Madre Mts., numerous lode-gold and base-metal mines (with minor by-product gold), e.g., the Copper Gem Prospect, Dreamland King Group, Gold Coin Prospect, Independence Group, Itmay Mine, King of the Camp Prospect, North Fork Group, etc.

Carbon-Sweetwater Cos.

BAGGS (Carbon Co.), W 20 mi., along co. line (Little Snake R., Fourmile Gulch, Dry Gulch), area mesa tops in the Wasatch gravels worked by dry-wash methods—placer gold (no gold found in watercourse gravels).

Converse Co.

ESTERBROOK, on N slope of the Elkhorn Mts., E of La Bonte Cr., in NW¼NE¼ sec. 21, (29N–71W), the Snowbird Group of mines (primarily copper)—minor lode gold.

Crook Co.

SUNDANCE: (1) NW 8 mi., in the Bear Lodge Mts., the Bear Lodge district: (a) the Copper Prince Mine, on dumps—native gold (scattered small particles), with malachite and chrysocolla; (b) other area mines, on dumps—gold showings; (c) local exposures of pegmatites—free gold; (d) in area fluorite veins and outcrops—free gold; (2) in sec. 17–20, (52N–63W), the Hutchins Consolidated Gold Mining Co. mines, on dumps —native gold, tellurides; (3) the Nigger Hill placers (small but numerous workings)—abundant nuggets and grains of gold, often attached to fragments of glassy quartz (with great quantities of tourmaline, cassiterite, and small red garnets); (4) area of Warren Peaks, ½ mi. NE of the central peaks, a gold-bearing ledge extends for several hundred yds. in a NW–SE direction—gold traces (probably from iron and manganese oxides).

Fremont Co.

Fremont County has produced more gold than any other county in Wyoming, most of it coming from the ATLANTIC CITY-SOUTH PASS CITY district, in which placer gold was first noted in 1842, with lode mines opened in 1868. The total gold production through 1959 was recorded as approximately 70,000 ozs.

AREA, gravel bars, benches, and terraces along the Wind R. and the Popo Agie R.—placer gold.

ATLANTIC CITY (24 mi. S of LANDER and 1 mi. S of Rte. 28): (1) S ½ mi., the Ground Hog Mine, small quartz veins—lode gold; (2) S 1 mi., the Duncan Mine—lode gold; (3) SW 1½ mi., the Diana Mine —free gold, with arsenopyrite; (4) N 1 mi., the Midas Mine—lode gold, with arsenopyrite; (5) in secs. 31–33, (29N–98W), along Strawberry Cr. and Big Nugget Ditch, the Big Nugget Claim—placer gold (to large-size nuggets); (6) near Rongis, N of the Sweetwater R., in the Black Rock-Long Cr. area, in sec. 36, (31N–97W), the King Solomon Mine lode gold; (7) 10 mi. N of Rongis and 1 mi. NE of Tincup Springs, in SE¼SE¼ sec. 13, (31N–93W), the Anderson Mine—free gold.

DUBOIS (Wind River Indian Reservation area): (1) at Union Pass, in (41N–108W), along Warm Springs Cr., between 2 rd. crossings, in gravel bars—placer gold (very fine); (2) WNW 10 mi.: (a) along Warm Springs Cr., at Clarke's Camp in sec. 33, (42N–108W), in cr. gravel bars —placer gold (very fine); (b) S, in the Wind R. Range, in gravel deposits of Dinwoody, Dry, and Meadow crs.—placer gold.

LANDER, S 30 mi., in sec. 32, (30N–99W), the Miner's Delight Group of mines—native gold.

PAVILLION, N, in the Owl Creek Mts., in (6N–3E), along Willow Cr., area exposures of quartz veins associated with diorite dikes in pre-Cambrian granite—gold showings (obtainable by pulverizing and panning).

RIVERTON, N, in T. 1 N, R. 3 and 4 E, exposures of irregular quartz lenses and stringers in pre-Cambrian granite and schists—free-milling gold (obtainable by pulverizing and panning).

RIVERTON-KINNEAR (district, in T. 1 and 2 N, R. 2 E): (1) along the Wind R., gravel bars, especially near Neble—placer gold; (2) NE 28 mi., along the Wind R., gravel deposits—placer gold.

SHOSHONE, N, near entrance to the Wind R. Canyon, in T. 5 and 6 N, R. 6 E, area exposures of quartz veins in pre-Cambrian crystalline rocks—gold traces (obtainable by pulverizing and panning).

SOUTH PASS CITY: (1) E ¾ mi., the Pioneer Carissa Gold Mines, Inc. Mine (high on mt. overlooking the log-cabin ghost town), major producer, 1868 to present—lode gold; (2) in (29N–100W): (a) in sec. 22, 1½ mi. E of town, the B. and H. Mine—native gold; (b) in center sec. 22, the Empire State Mine—lode gold, with tourmaline and arsenopyrite; (c) in SE¼ sec. 21, the Carrie Shields Mine—lode gold; (d) in NW¼ sec. 14, the Mary Ellen Mine, quite productive—lode gold.

Fremont-Hot Springs Cos.

AREA, on headwaters of Owl Cr., in the Owl Cr. district, regional watercourse, bench, and terrace gravel deposits—placer gold.

Goshen Co.

CHEYENNE, NE 65 mi., and 6 mi. W of the Nebraska line in (19N–61W), the Goshen Hole placers, embracing about 10,000 acres of watercourse, bench, and terrace gravels—placer gold (extremely fine).

JAY ELM, W, and ½ mi. W of the highest peak in the Rawhide Buttes, at cor. of sec. 2, 3, 10, and 11, (30N–64W), the Copper Belt Mines Co. (primarily copper)—minor lode gold: (1) the Omaha Mine and (2) the Gold Hill Mine—lode gold; (3) the Emma claims—gold showings.

Hot Springs Co.

EMBAR, W, in headwaters of Owl Cr., SW of the Washakie Needles (12,495 ft. elev.), the Owl Cr. district: (1) area cr. gravels, labeled "rich placers, but lack water"; (2) area quartz vein outcrops—free gold. Discovered in the 1860s; prospectors were driven out by Indians in 1867, 1877, 1884, and 1885; little prospected since then.

Johnson Co.

BUFFALO: (1) SW, at head of Kelly Cr., the Kelly Cr. Mine, millsite (basal sandstone of the Deadwood Formation)—placer gold; (2) W 17 mi., in sec. 15, (49N–83W), the Roe Brothers Group (primarily copper mines)—assays show small amounts of lode gold.

Laramie Co.

GRANITE CANYON (in the Silver Crown district): (1) NW of the Simmon's ranch, on Middle Cow Cr., the Lenox Mine—minor lode gold in galena: (a) ¼ mi. SW of the Lenox, close to the bank of Middle Cow Cr., the Julia Lode—free-milling gold; (b) ¼ mi. NW of the Julia, the Agata Prospect, source of spectacular specimens—native gold; (2)

N 2 mi., the Great Standard Group (basically copper)—minor lode gold; (3) out 4 mi., in secs. 25 and 26, (15N–70W), the Arizona Mine (primarily copper)—free-milling gold; (4) other area mines, e.g., the King David, Colorado (Metcalf) Lode, Eureka Lode, and Copper King (Adams)—minor amounts of gold in copper minerals.

HECLA (7 mi. from GRANITE CANYON): (1) the Yellow Bird Mine, on dumps—gold traces, with pyrite; (2) in sec. 24, (14N–70W), the Hecla Mine—lode gold, with silver and copper; (3) in sec. 22, such mines as the Rambler, Coming Day, Big Elephant, and Monte Cristo (all primarily zinc)—fair values in lode gold.

LARAMIE, NE, to area 8 mi. W of the Platte-Laramie Co line (45 mi. N of the Colorado boundary), the Horse Cr. placers, in Tertiary stream gravels—small gold showings.

Lincoln Co.

AREA, extreme N part of co.: (1) in T. 34 N, R. 114 and 115 W, the Horse Cr. prospects—lode gold; (2) in NW cor. (37N–116W): (a) at mouth of Pine Cr., on S side of the Snake R., the Pine Bar diggings (in top of gravel deposit 8 ft. below barren sediments)—fine flake gold; (b) a terrace at mouth of cr., 40–50 ft. above water level and 400 to 660 ft. wide, productive placer workings; (3) E side of the Snake R., from mouth of Baily Cr. ½ mi. N, in SW¼ (38N–116W), the Davis Claim: (a) gravel bars of r. and its tributaries—very fine flour gold; (b) area terrace gravels—fine placer gold.

Natrona Co.

CASPER, S 11 mi., in (32N–79W), the Koch Deposit (close to the Koch manganese and feldspar claims), a seam of black material cuts granite, and within it occurs gold and silver.

Park Co.

AREA, along the Shoshone R. below mouth of Alkali Cr., in gravel and sand deposits—placer gold.

CLARK, vicinity, along the Clark's Fork R., in gravel deposits—placer gold.

CODY, NW 50 mi. (35 mi. SE of COOKE CITY, Montana), in the

Absoraka Mts., the Sunlight Basin district, many mines and prospects: (1) in quartz-pyrite veins (series of vugs and crustifications in fracture zones)—free gold; (2) near Lower Copper L., in complex silver-tungsten deposits—minor lode gold; (3) in argentiferous galena veins, especially on the Hoodoo Claim—minor associated gold; (4) in chalcopyrite veins —associated gold (small amounts); (5) in pyrite-chalcopyrite-gold veins, especially the Big Goose Mine—native gold, as grains; (6) in complex silver veins, especially the Tip Top Claim—minor lode gold; (7) in barren carbonate veins, e.g., the Morning Star Claim—minor lode gold; (8) many other area mines and claims—lode gold.

MEETEETSE: (1) along the Wood R., in black-sand concentrations— placer gold; (2) SW 38 mi., in (45N–104W), the Kirwin Mine—native gold.

SYLVAN PASS: (1) S, in SW¼ sec. 2, (51N–109W), the Crouch Gold Prospect (operated in the 1930s)—lode gold; (2) 2 mi. N of the Crouch Prospect, area watercourse gravels—placer gold.

VALLEY (well SW of CODY), near confluence of Needle Cr. with the S fork of the Shoshone R., the Stinking Water Region, area mines— minor lode gold.

Platte Co.

AREA, in sec. 22, (25N–66W), in pre-Cambrian rocks, a mine—lode gold, silver.

UVA, area exposures of the Arikaree Formation (its sands carry finely disseminated gold).

WHEATLAND: (1) SW 10 air mi., in the Cooney Hills, in sec. 20, (23N–69W), the Cooney Hill Prospect, primarily low-grade copper— gold traces: (2) SW 15 mi., on S side of the Cooney Hills, the Whippoor-will Claim—lode gold, with copper and silver; (3) W 22 mi., on Slate Cr., the Independence Group of mines—minor associated gold.

Sheridan Co.

SHERIDAN: (1) in (54N–87W), the Mosaic Claim—lode gold; (2) W about 60 mi. on U.S. 14 and 14 Alt., on top of the Big Horn Mts., take N-trending dirt rd. to the Bald Mt. area in (56N–91W), old site of Bald

Mt. ("Baldy") City, elev., 9,000 ft. on headwater springs of the Little Big Horn R.: (a) area old mines and prospects, N toward the Montana line, in disintegrated granite—fine-grained free gold; (b) 1 mi. W, the tumbledown log Fortunatus mill, area surrounding it contains intrusive dikes or chimneys—minor gold showings.

Teton Co.

AREA, gravel bars of the Hogback R. near its jct. with the Snake R., in small amounts—placer gold.

PART

Where to Find Gold in Canada

INTRODUCTION

The earliest recorded notice of gold existing in Canada was of placer findings in the Chaudière River Valley of Quebec, made in 1823–24. No effort to capitalize on this discovery was made for nearly twenty-five years. Indeed, only after the world excitement caused by the famous gold rush to California in 1849 and to Australia in 1851 did the gold prospecting fever strike Canadian citizens, even though productive placer mining had already begun in the Chaudière basin in 1847. Eleven years later the first placer gold was discovered in northern British Columbia, and in 1862 the first lode deposits were opened up in Nova Scotia.

In 1896 the discovery of rich placer gravels along the Klondike River in Yukon Territory stimulated the great gold rush of 1898 that was also to open up Alaska to intensive prospecting. In the seven years following the stampede over White Pass from Dyea and Skagway, Alaska, into the headwaters of the Yukon River at Whitehorse, an estimated 4,838,000 fine ozs. of gold were extracted from the gravels of Bonanza, Eldorado, Hunker, Dominion, and Sulphur creeks. Subsequently, important gold discoveries were made in other parts of Canada, particularly in the Ontario districts of Porcupine in 1909, at Kirkland Lake in 1912, and in the Lake of the Woods, with other comparatively nearby finds being made in southern Manitoba, these latter deposits all being lode-mine operations.

Placer mining in Canada reached its peak around the turn of the century and, since 1900, placer gold production has steadily declined, although it is still an important contributor to the Canadian overall gold production. At the present time the Canadian Shield, which covers the northern and eastern half of Canada with Precambrian crystalline rocks, accounts for nine tenths of the total Canadian gold production—all of it from lode mines—and places Canada second in rank among the free world's gold-producing nations, first place going to South Africa.

Between 1858 and 1972 Canada produced 195,760,091 troy ozs. of gold, reaching peak production in 1941. Ontario has been the leading producer, followed by Quebec, with most of the gold coming from quartz lode mines or as a by-product of base-metal refining. Placer gold today comes largely from the Cordilleran (Rocky Mountains) provinces, and it is in these regions where the casual gold hunter is most likely to be successful in finding raw gold.

A casual survey of the auriferous localities described in the following pages will reveal that, by and large, most of them are in remote country or far enough from urban centers to require camping, backpacking, canoeing, or air transport. Therefore, proper equipment, provisioning, knowledge of camp- and woodcraft, and means of transportation other than by automobile are vital considerations for gold hunters seriously intent on prospecting in Canada.

It is not essential to have any special clothing or general equipment. Almost any outdoorsman, worker or sportsman, already has nearly everything required for wilderness survival. The importance of proper footwear cannot be overemphasized, however, since most gold prospecting necessarily has to be done on foot. One absolute essential is some sort of water-repellant and wind-resistant coat that should not be made of rubber, plastic, or oilskin because these materials prevent the escape of perspiration when walking or working. In warm weather a light windbreaker or parka that can easily be carried when not needed is good; during cool or wet weather, a heavier, canvas hunting jacket is better.

One bit of advice pertains especially to gold hunters who are not familiar with the Canadian exigencies of climate and environments. Before planning a trip, it might be advisable to write to a merchant at the proposed takeoff point (the request may be enclosed in a letter to the local postmaster) for information on items recommendable and available, and their prices. Although most prospectors cannot afford to engage a local outfitter or guide, whenever this is feasible this procedure is a good way for a beginner, or anyone unfamiliar with the particular area or mode of transport to be used, to acquire preliminary experience that can be a life saver on subsequent prospecting trips.

Summer in most prospecting areas of Canada is similar to that in Alaska. Unlike outdoor living in the United States, the northern summers feature an unbelievable nuisance of biting insects: mosquitoes, gnats, nosee-ums, deer flies, horseflies, hornets, yellowjackets, wasps, and flies. Life is prolific in the northland in summer, even in the so-called Barren Lands, and biting insects are an annoying part of it, although none are poisonous except to the allergic. In very many areas these noxious insects are so numerous that even the most experienced woodsman has difficulty in protecting himself. In the past, wilderness travelers used helmets, headnets, gloves, and bednets at night. Today there are easily carried insect repellants available at every supply post, so that no one need hesitate about prospecting because of the seasonal pests. Therefore, on any but the lightest trip, one should carry fly spray and sprayer, or a can containing an insecticide under pressure for use in a tent.

All the standard items of equipment familiar to Alaskans can be considered necessary in most parts of Canada, including suitable hats, sweaters, and gloves; a sewing kit; a sleeping bag (preferable to blankets), a tent or small canvas tarpaulin for shelter, camp stove for cooking inside the tent during rainy weather; a saw for cutting firewood; a dunnage bag, packsack, or packboard; canvas provision bags; illumination and matches; and a first-aid kit. Fishing tackle in most regions will add nicely to the food supply and, in some areas, a firearm may be needed for protection against wild animals. A good knife is a necessity and usually an ax, along with a small whetstone. To these should be added a 50-foot length of light rope, a coil of rabbit-snare wire, a roll of friction tape, a few nails, and sunglasses for use in high country or in traversing snowfields. Provisioning, necessary on any long prospecting trip, should be adequate yet as lightweight as possible, including standard camp utensils.

For all but casual prospecting, the methods of traveling and transporting equipment and supplies are of vital concern. These vary greatly for different parts of the Dominion. While southern Canada is fairly well served by roads and railways, from which prospecting can be done by daily trips on foot or by short backpack trips, in many parts of Canada the canoe is the basic means of summer travel, or snowmobile or dog team in winter. This is particularly true in the vast Canadian Shield, which is laced with an interlocking network of lakes and streams that permits access to almost every point, at least within a few miles. The most practicable and easily repaired canoe is the canvas-covered type that, in the last half century, has replaced the older Indian birchbark crafts. Made of a framework of pliable birch wood and covered with 4-ounce canvas duck painted with a waterproof paint, these canoes are very lightweight, yet durable; they will carry surprisingly heavy loads. Repairing a hole punched in by a rock or snag is done by plastering waterproof adhesive tape over it; nothing could be simpler.

In prospecting Canada's more remote regions, especially in the Canadian Shield, most prospectors arrange to be flown in by floatplane in order to avoid a long and arduous canoe trip, then arrange to be picked up at a later specified date for their return. Charter-plane services operate from many towns or trading posts. In any case, only the most experienced men should try to travel alone in remote areas. A prospecting party, therefore, usually consists of two or more men and their equipment and supplies. All the admonitions described in books on wilderness survival, especially if one becomes lost or injured, are vitally important to know before anyone attempts to hunt for gold in the vast hinterlands of Canada.

For more information, brochures, and prospecting suggestions, write: Information Officer, Public Relations and Information Service, Department of Energy, Mines, and Resources, 588 Booth St. Ottawa, Ontario K1A OE4, Canada.

ALBERTA

Placer gold was discovered in Alberta in 1859–60 in low-water gravel bars of the North Saskatchewan River. Since then, placer mining has been carried on intermittently, chiefly by hand methods but partly by dredges. While the main production comes from the discovery river bars, placer gold is known to occur in other regional streams. All the gold is fine and occurs in association with coarse gravels. No lode mines have been operated in this province.

For more information, write: Information Officer, Department of Mines and Minerals, Petroleum Plaza, 9945 108th Street, Edmonton, Alberta, Canada.

AREA: (1) entire length of the Saskatchewan R., in low-water sand- and gravel bars (annually renewed and productive)—placer gold; (2) NE cor. of province, some 5,000 sq. mi. underlain by Precambrian rocks that elsewhere throughout the Canadian Shield region have proved to carry gold, good potential prospecting region but difficult to reach.

BATTLEFORD to EDMONTON, for 60 mi. along the North Saskatchewan R. (including all tributaries), on low-water bars—fine placer gold (no concentrations on bedrock).

LETHBRIDGE (Topo. 82 SE Cranbrook-Lethbridge), along Oldman R. (precise localities not mapped), in gravel bars—placer gold (fine).

BRITISH COLUMBIA

The first gold recognized as such in British Columbia were small visible grains in quartz outcrops on Moresby Island in the Queen Charlotte group. But not until mainland placer discoveries were made three years later did gold mining really begin. Until 1893 all gold recorded came from placers, although mention is made of the erection of arrastras and small stamp mills to work lode-gold deposits in the 1870s.

Large-scale placer mining began in 1857 in south-central British Columbia, when gold was found in abundance at the mouth of the Nicoamen River above LYTTON. A year later the rich bars at YALE were in production and, in 1860, important placers were found in the Cariboo district of central British Columbia at Quesnel forks and along Keithley and Antler creeks. The following year the celebrated Williams, Grouse, Lowhee, and Lightning creeks placers attracted hordes of eager miners. Although placer mining continues on a small scale to today, its heyday for the individual miner ended about 1885, except in the ATLIN district of far northwestern British Columbia, where rich gold deposits were not discovered until 1898.

Between 1858 and 1961 a total of 5,220,779 ozs. of placer gold were recorded, along with another 11,879,957 ozs. of lode gold. The last important gold mine operating in the province, the Bralorne-Pioneer, closed finally in 1971 to end commercial gold mining which, with the greatly increased world price of gold, can be expected to be renewed in 1975. For the modern gold hunter there still remain almost countless auriferous streams, bench and terrace gravel deposits, and unprospected and unworked areas where a man or woman can pan out colors, flakes, grains, and sizable nuggets of raw gold. Also, there are literally hundreds of old lode mine dumps carrying residual minerals that may assay out good values of gold not visible to the eye.

For information, brochures, and maps, write: Librarian, Department of Energy, Mines, and Resources, Geological Survey of Canada, Library, Sixth Floor, 100 West Pender St., Vancouver, B.C. V6B 1R8, Canada.

Mainland British Columbia

AREA: (1) along Lorne Cr., discovered in 1884, productive placers (part of old channel still unworked)—coarse gold, sizable nuggets; (2) Squaw Cr. (discovered by an Indian in 1927), gravels notable for large boulders; tot. prod., about 5,000 ozs.—placer gold (very coarse, nuggets to 4 ozs., with largest found weighing 46 ozs.).

ALEXANDRIA: (1) between town and FOUNTAIN along river, at Haskell, Big, and Island bars, productive placers; (2) between town and QUESNEL, at British (Cornish) and Ferguson (Rich) bars, productive placers.

ATLIN (district, in NW cor. of B.C., on E side of the Coast Range, a major gold-producing center discovered in 1898): (1) the Engineer Mine, principal producer—lode gold; (2) along Spruce and Pine crs. (first discoveries and most productive of region)—placer gold; (3) second most productive crs. in order of decreasing importance: Boulder, Ruby, McKee, Otter, Wright, and Birch—abundant placer gold (fine to coarse); (4) Atlin Lake, on E side: (a) in vicinity of Surprise L., in crs. named above; (b) along the O'Donnell R. and its tributaries, productive placers.

BARKERVILLE (in central B.C., the Cariboo district, one of oldest placer centers in province and scene of earliest attempts to mine lode gold): (1) the Cariboo Gold Quartz Mine and (2) opposite, the Island Mt. Mine, 1870s–1900 (reopened in 1930s), both principal producers —lode gold; (3) the whole plateau region (deeply dissected and glaciated, elev., 3,500–6,500 ft.): (a) all regional ancient stream gravels on bedrock in valley bottoms; (b) all regional rock benches or old channels (usually buried under glacial drift); (c) all interglacial stream gravels (also usually under drift); (d) all glacial outwash gravels filling stream valleys; and (e) all postglacial stream gravels (as secondary erosional deposits)— abundant placer gold.

BIG BEND AREA: (1) along French Cr. (discovered in 1865 to initiate a "rush"), rich placers; (2) along McCulloch and Camp crs., the lower part of Goldstream R., and Carnes Cr.—abundant placer gold (coarse, rough, large nuggets).

BOUNDARY (district), area mines very productive, operated to 1919— lode gold.

BRIDGE RIVER STA. (40 mi. NW of LILLOOET, q.v.): (1) W of Cadwallader Cr., about 5 mi. S of its jct. with Bridge R., the adjoining Bralorne and Pioneer mines (largest lode-gold producers in B.C., tot. prod., 1914–44, of over 1,500,000 gold ozs.), 58 claims, extensively developed—lode gold; (2) on hwy. toward the Pioneer Mine, 1 mi. below jct. of Gun Cr. with Bridge R., the Minto Mine, productive—lode gold; (3) 6½ mi. N of the Bralorne, the Wayside Mine, rich—lode gold.

CASSIAR (district, in N part of province in the Dease L. area, just N of the Arctic divide and draining N via the Dease R. into the Liard; reached from Milepost 648.9 on the Alcan Hwy. and 143 mi. on the Cassiar Rd. Tot. prod. of placer and lode gold, mostly during the 1870s, over $5,000,000): (1) along Dease, Thibert, and McDame crs. (primary placer operations): (a) bed gravels, very productive; (b) adjoining benchs and old-channel gravels (probable source of gold in existing stream gravels), very rich—placer gold; (2) E of Dease L., and along the Stikine R. above Telegraph Cr., all gravels of bed and benches—placer gold (not in commercial quantities).

EAST KOOTENAY: (1) along Wild Horse R., most productive of regional streams (tot. prod., above $1,000,000), placers—coarse gold; (2) along the Bull and Moyie rivers, bars and bench gravels—placer gold; (3) on Perry, Weaver, Palmer Bar, Findlay, Boulder, and Quartz crs., productive placers (hydraulic and drift-mining operations).

FORT HALKET (abandoned fort on the Liard R.): (1) at McCullough's Bar, rich placers, discovered in 1872; (2) at mouth of Rabbit and Fort Nelson rivers and on bars along the upper Liard R., productive placers (no gold found below Devil's Portage).

GRAND FORKS (in the Similkameen district immediately N of the U.S. boundary, the Greenwood-Grand Forks district): (1) numerous area mines, productive—lode gold; (2) the Brooklyn Claim (at PHOENIX, reworked 1937–40, tot. prod. of 6,336 gold ozs.)—lode and by-product gold (from major copper-silver ores); (3) the Dentonia Claim (near Jewel L., tot. prod. of 4,178 ozs.)—lode and by-product gold; (4) W, at ROCK CR.: (a) along Boundary Cr., and (b) in Rock Cr. tributary of the Kettle R.—placer gold, platinum (minute grains).

GREENWOOD: (1) at Camp McKinney, one of earliest gold camps in B.C., the Cariboo-Amelia Mine (tot. prod., 1894–1903, of more than $1,000,000)—lode gold; (2) out 8 mi., at Jewel L. (7 mi. from the Eholt RR sta.), the Dentonia (Jewel-Denoro) Mine (cf. also under GRAND FORKS), 24 claims, discovered in 1896 and extensively developed—lode gold.

HAZELTON (district), mines on Rocher Déboulé Mt. (primarily silver-copper)—by-product gold.

HEDLEY (district in S part of B.C. in the Osoyoos mining division), on Nickel Plate Mt., the Nickel Plate Mine, largest area producer, tot. prod., 1903–30, of about 532,000 ozs.—lode gold.

HOPE: (1) area: (a) along Siwash Cr., well-worked placers; (b) along the Coquihalla R., extensive placers; (2) 25 mi. below, on the Frazer R., at Maria Bar (nearly opposite mouth of the Chilliwack R.)—flour gold; (3) between town and PRINCETON: (a) along the Similkameen R., in bed and bench gravels—placer gold; (b) along Granite Cr., in gravels—placer gold, platinum (nuggets to ½ oz.); (c) gravels along Lockie (Boulder) Cr.—placer gold (to large nuggets); (d) along the Tulameen R., in bed and bench gravels—placer gold, platinum nuggets; (e) in gravels along Cedar, Slate, and Lawless (Bear) crs.—placer gold, platinum; (4) between town and Spences Bridge, along the Frazer R. from Boston Bar to Sisco Flat, rich placers—"heavy" gold; (5) between town and YALE, q.v., at Cameron, Emery, and Texas bars, rich placers.

HOWE SOUND (on W-central B.C. coast), the Britannia Mine, discovered in 1898 and one of largest gold mines in B.C. before 1942—lode gold.

KAMLOOPS: (1) N about 50 mi., the N Thompson R. district: (a) 1½ mi. E of Dunn L. (8 mi. out of Chua Chua sta. on RR), the Windpass Mine, 36 claims, productive—free and lode gold, tellurides, pyrite, native bismuth grains; (b) just S of the Windpass, the Sweet Home Claim (tot. prod., 1934–40, of 33,324 ozs.)—lode gold; (2) in the Kamloops L. area, along Criss Cr., near its mouth on the Deadman R., in bed and bench gravels—placer gold, platinum.

KASLO, NW of the N end of Kootenay L.: (1) along the Lardeau R., in bed and bench gravels, and (2) along Five-Mile and Canyon crs. placer gold.

KITSUMGALLUM (district), area quartz vein exposures—free gold, with sulfides.

KOOTENAY L., out 17 mi. and just E of the YMIR, q.v., district, the Bayonne Property (tot. prod., 1935–42, of 40,643 ozs.)—lode gold.

LILLOOET (district, with Bridge R.): (1) along the Frazer R., from FOSTER'S BAR to well above FOUNTAIN: (a) river and bench gravel deposits, very productive placers (discovered in late 1800s)—coarse gold; (b) near town, in river bars—gold nuggets (to 6 ozs.) found in years past; (2) gravel deposits of the lower Frazer R.—flour gold; (3) along Bridge R., in gravels—abundant placer gold; (4) along Cayoosh Cr.: (a) at mouth of cr., rich placers; (b) the Bonanza (Golden Cache) Mine, principal producer—lode gold; (c) on the Frazer R., the Grange Mine —minor lode-gold showings; (5) on Cadwallader Cr., the Coronation (Bras d'Or) Mine and other area mines—lode gold.

LYTTON: (1) near jct. of the Thompson R. with the Frazer R., in bar and bench gravels, rich placers; (2) halfway to FOSTER'S BAR, extending to well above FOUNTAIN (cf. also under LILLOOET), rich placers—gold nuggets to considerable size; (3) between town and FOUNTAIN, at Mormon, Foster, Great Falls, Lillooet, and Upper Mormon bars, rich placers; (4) between town and YALE, q.v., at Sailor, Nicaragua, and Boston bars, rich placers.

McLeod R. Area, along the Little McLeod R., in shallow gravels on low-lying benches (discovered in 1932)—placer gold (medium nuggets), platinum.

Midway, W 32 mi. (and 8 mi. N of the U.S. boundary), at Camp McKinney, the Cariboo-Amelia Mine (cf. under Greenwood).

Nelson: (1) area mines, e.g., the Fern Gold, Queen, Nugget, Mother Lode, Perrier, Second Relief, Kootenay Belle, Ymir-Wilcox, and Yankee Girl—lode gold (tot. prod. of 20,114 ozs., free-milling); (2) S 6 mi., on Toad Mt., the Athabasca Mine, productive—lode gold; (3) W 6 mi., on Eagle Cr., the Granite-Poorman (Poorman) Mine, extensive quartz workings, tot. prod., 1890–1919, of over $1,000,000—lode gold; (4) S 27 mi., on Wild Horse Cr., the Ymir Mine, largest gold mine in Canada, 1900–6—lode gold; (5) between town and Slocan City, q.v., many auriferous prospects along veins and quartz outcrops in the Nelson granite—free and lode gold.

Nicoamen, along the Thompson R., gravel bars and benches—placer gold.

North Bend, area site of the old Beatty dredge, in bouldery ground below water level—coarse gold.

Okanagan Area (in S-central B.C.): (1) area streams: (a) entering the North Thompson R. and (b) entering Okanagan L.—minor placer gold; (2) along such streams as: (a) Scotch Cr., for 8 mi. above its mouth; (b) Tranquille Cr., worked for many years by Chinese; (c) Mission Cr., richest part 7 mi. above its mouth; (d) Cherry Cr., producer of coarse gold and nuggets; and (e) Harris Cr.—abundant placer gold (coarse to nuggets), with considerable platinum.

Oliver (in S part of the Okanagan Valley), just NW, the Fairview Camp mines, especially the Stemwinder Claim, Morning Star, and Fairview mines (tot. prod., around 11,000 ozs.)—lode gold.

Omineca Area: (1) all regional streams, especially: (a) along Vital Cr., discovered in 1869, rich placers; (b) along Germansen Cr., rich placers, discovered in 1870; (c) along the Manson R., productive placers, discovered in 1871 (tot. prod. of all areas above, $1,500,000)—coarse gold, nuggets to 24 ozs.; (2) along Slate and Tom crs., dragline shovel operations—coarse placer gold; (3) along Harrison, Silver, Rainbow, and Philip crs., rich placers—gold, platinum; (4) N part of district: (a) in crs. N of the Omineca R., e.g., along Jim May, McClair (McLaren), and McDonnell, rich placers—nuggets over 1 oz.; (b) along the Ingenika R., at jct. of McConnell Cr., placers (so rich as to have caused small "rushes" in 1932; tot. prod., about 1,100 ozs.)—placer gold, platinum.

OSOYOOS (camp and mine, operated 1935-40 with tot. prod. of 14,638 ozs.)—lode gold.

PEACE R. AREA: (1) along the Parsnip R., about 20 mi. above its mouth, rich placers discovered in 1861; (2) on the Finlay R., 3 mi. above its jct. with the Parsnip R., at FINLAY FORKS, on Pete Toy's Bar, "rich diggings"—placer gold; (3) 26 mi. above the Peace R. Canyon, at Branham Flat, productive placers (worked by dragline in early 1920s).

PREMIER (N of Stewart, which is at the head of Portland Canal on the NW coast), the Salmon R. district: (1) the Silbac-Premier (Premier) Mines, an early "fabulous bonanza," now exhausted, tot. prod., 1918–44, of 1,728,545 ozs.—lode gold, with 40,927,968 ozs. of silver; (2) nearby, the Big Missouri Mine (tot. prod., 1938–42, of 56,946 ozs.)—lode gold.

PRINCETON (cf. under LILLOOET, tot. prod. of creeks named, 1885–1961, between 10,000 and 20,000 ozs.)—placer gold, with abundant platinum (often more abundant than gold, nuggets to ½ oz.).

QUESNEL: (1) along Scottie Cr., in bed and bench gravels—placer gold, platinum; (2) 7 mi. above, at Long Bar, rich placers; (3) 13 mi. above, at Spanish Bar—abundant placer gold; (4) from 20 mi. above town to below HOPE, q.v., along the Frazer R. at intervals, extensive placer workings, annually renewed by regional erosion, rich; (5) between town and LILLOOET, q.v., along the Frazer R., in low-water gravel bars and adjoining bench deposits—placer gold; (6) cf. the associated Cariboo district at BARKERVILLE and (a) gravel bars and benches along the Cottonwood, Quesnel, and Horsefly rivers, rich placers; (b) Hixon, Government crs.—abundant placer gold.

REVELSTOKE (cf. BIG BEND AREA).

ROSSLAND (camp in the Trail Cr. division of the West Kootenay district, 6 mi. W of the Columbia R. and 5 mi. N of the U.S. boundary; tot. prod., 1894–1930, of 2,868,227 ozs. of lode and by-product gold (from silver-copper ores): (1) the Rossland Mine, principal producer, 1891–1928 (40 per cent of provincial gold production); (2) out 2½ mi., the O.K. and I.X.L. mines, very productive—lode gold.

SALMO: (1) on Sheep Cr. (NELSON area), the Yellowstone Mine, productive—lode gold; (2) W 3 mi., on Erie Cr. (13 rd. mi. from RR at ERIE), the Second Relief Mine, intermittently productive, 1900–40 —lode gold; (3) S 7 mi., on Sheep Cr., the Kootenay Belle Mine, 200 acres, quite productive, 1900–40—lode gold; (4) out 10 mi., in valley of Waldie (Wolfe) Cr., just above its jct. with Sheep Cr., the Queen Mine, very productive, 1902–32—lode gold; (5) out 15 mi., to head of Fawn Cr. (30 mi. S of NELSON): (a) the Reno Mine, 16 claims, staked 1912–13 (mill burned, 1932)—lode gold; (b) adjoining mines of Motherlode and Nugget, productive—lode gold.

SAVONA, N 43 mi., on shore of Vidette L. near headwaters of Deadman Cr., the Vidette Mine, 22 claims, 1,300 acres (tot. prod., 1934–39, of 28,256 ozs.)—lode gold.

SHUSWAP LAKES AREA, along Scotch Cr. (flows into Shuswap L. from the N), in bed and bench gravels—placer gold.

SKEENA (mining division), along Douglas Cr., discovered in 1886 and revived during 1930s, productive gravels—placer gold (nuggets to 6 ozs.).

SLOCAN CITY (primarily a lead-silver-zinc camp), area mines—by-product gold.

SMITHERS (mainly a silver district), on Hudson Bay Mt., area mines —lode and by-product gold, with arsenic and bismuth.

SQUAMISH (district, at head of Howe Sound on the N-central coast): (1) N, the Squamish R. district, the Ashloo Gold Mines, tot. prod., 1936–39, of 6,323 ozs.—lode gold; (2) S, on Howe Sound, the Britannia Mine (cf. under HOWE SOUND).

TATLA L., on Blackburn Mt. and Perkins Peak, near E contact of Coast Range intrusives, area mines and prospects—lode gold.

TRANQUILLE, N, along the Tranquille R., in bars and bench deposits, productive—placer gold, platinum.

TULSEQUAH (Taku R.) district, in far NW cor. of province, NW of the Taku R. near the Alaskan boundary, the Polaris-Taku Mine (tot. prod., 1937–42, of 89,311 ozs.)—lode gold.

USK: (1) the Columario Consolidated Mines and (2) the Zymoetz claims, in quartz veins—free and lode gold.

VANCOUVER, N 120 mi., on W coast on Phillips Arm, the Dorothy Morton Mine (first cyanide plant constructed in B.C., 1898), rich but small pockets—lode gold.

VERNON: (1) in gravel beds and benches of Whiteman, Bouleau, Siwash, Equesis, Moffat, Newport, Winfield, Harris, and Cherry crs.—placer gold; (2) all other regional watercourses, bench and terrace gravels —placer gold.

VERONA (NE of VANCOUVER on the Kettle Valley RR), N 1 mi., in the HOPE, q.v., district, the Aurum Property, pockets and seams in serpentine 10 mi. long by ½–1 mi. wide—free gold (films, grains, plates, thin wedges, wires, leaves), with pyrite, arsenopyrite.

WELLS (55 mi. E of QUESNEL, q.v., and 5 mi. W of BARKERVILLE, q.v., on nearby Cow Mt. on E side of Jack of Clubs L., the Caribou Gold Quartz Mine (cf. also under BARKERVILLE), 74 claims, 2,800 acres along hwy. between towns, extensively developed—lode gold.

WEST KOOTENAY: (1) area mines, very productive before 1900—lode gold; (2) the Poorman, O.K., and Fern mines, principal producers—lode gold; (3) along Boundary, May, and other crs. tributary to the Kettle R., productive placers; (4) along Forty-nine Cr. and (5) along the Salmo, Pend d'Oreille, Lardeau, and Duncan rivers, many productive placers.

YALE: (1) area: (a) the Strathyre Mine, major producer before 1900 —lode gold; (b) various other area mines, productive before 1900—lode gold; (2) nearby, on Hills Bar at foot of Frazer Canyon, rich placers (producer of more gold than any other Frazer R. locality); (3) below, along the Frazer R., many extensively worked placer deposits, with greater average annual gold production than elsewhere along r.—coarse gold, nuggets; (4) 16 mi. above, on Frazer R., from a few mi. below Boston Bar to Sisco Flat (just below LYTTON, q.v.), a distance of 25 mi., rich placers, extensively worked.

YMIR (Sheep Cr.) district: (1) 10 highly productive area mines, including (a) in the YMIR area, the Ymir Consolidated (major producer),

Wesco, Second Relief, etc.; and (b) in the Sheep Cr. area, the Kootenay Belle Gold Mine, Gold Belt Mine, Reno Gold Mine, Sheep Cr. Gold Mines (operating also the Queen Mine), Arlington, etc., all very productive—lode gold; (2) E 1½ mi., the Ymir Yankee Girl Mine, discovered in 1896 (17 mi. S of NELSON, q.v.), on N slope of Bear Cr. Valley, tot. prod. to 1918 of $470,000—lode gold, silver.

Pacific Coast Islands

PORCHER Is. (SW of PRINCE RUPERT), on NW coast near Surf Pt., the Porcher Is. Mines, tot. prod., 1933–38, of 17,244 ozs.—lode gold.

PRINCESS ROYAL Is., 1 mi. E of Bear L. (7 mi. above head of Surf Inlet), the Belmont-Surf Inlet Mine, large-scale 1917–26, tot. prod. to 1942 of 386,599 ozs.—lode and by-product gold, with large quantities of silver and copper.

GRAHAM Is. (in the Queen Charlotte group), along the NE coast, in black-sand beach deposits—placer gold, platinum.

TEXADA Is. (off S coast of B.C.), the Marble Bay, Copper Queen, and Cornell mines, tot. prod., 1897–1919, of 13,371 ozs.—lode gold.

VANCOUVER I.: (1) on E side of is. in the Clayoquot area, along the Oyster R., in sand- and gravel bars—placer gold; (2) S part of is.: (a) all area streams once worked—placer gold; (b) in gravels of China and Loss crs.—placer gold; (c) in bed and bench gravels along Leech, Gordon, Jordan, Sooke, Sombrio, and San Juan rivers, once-productive placers; (2) 70–80 mi. NW of VICTORIA, along the Bedwell R., productive placers; (3) at Wreck Bay, in black-sand deposits along area beaches—placer gold, platinum; (4) at ZEBALLOS: (a) along the Zeballos R. and its tributaries, bed and bench gravel deposits—placer gold; (b) follow trail along r., then along the Nomash R., and at an elev. of 410 ft. a pack-horse trail branches up Curly Cr. to the Golden Horn Group of claims (elev., 1,460 ft.) on S side of Nomash Valley—flake gold (excellent specimens).

MANITOBA

The history of gold production in Manitoba is short but rather spectacular, for during the 1930s the province ranked fourth among Canada's gold-producing provinces. However, most of Manitoba's gold has come from refining the extraordinarily rich copper-zinc-gold ores of the great Flinflon and Mandy mines, discovered in 1915. No placer gold has been found in commercial quantities in this Canadian Shield province.

BERESFORD L. (village): (1) the Diana and Beresford L. mines (tot. prod., 1934–40, of 11,137 ozs.)—lode gold; (2) W ½ mi., the Gunnar Gold Mines, tot. prod., 1936–42, of 101,463 ozs.—lode gold.

FLIN FLON: (1) on shore of Flinflon L., the Flinflon Mines (copper-zinc-gold), discovered in 1915, tot. prod., 1931–44 (after copper refining began), of 1,910,812 ozs.—lode and by-product gold; (2) SE 3½ mi., the Mandy Mines (primarily copper), tot. prod., 1915–19, of 6,000 ozs. —by-product gold.

HERB L. (settlement), just N: (1) the Laguna Gold Mines, tot. prod., 1936–39, of 52,463 ozs.—lode gold; (2) the nearby Gurney Mine (tot. prod., 1937–39, of 25,164 ozs.)—lode gold.

NORWAY HOUSE (N end of L. Winnipeg), N 174 mi. by rd. or air to Island L., in far NW part of province, the Island L. Mines, discovered in 1931, 150 claims on Gold I., extensively developed (tot. prod., 1934–35, of 6,390 ozs.)—lode gold.

SHERRIDON (on Kississing L., 97 mi. by RR from THE PAS, q.v., and 45 mi. N of the Flinflon Mines), the Sherritt-Gordon Mine (copper-zinc)—minor by-product gold.

THE PAS: (1) access via RR 91 mi. to FLIN FLON, q.v.; (2) out 287 mi. on the Hudson Bay RR to ILFORD, then 130 mi. to God's L. (usually served by airplane from WINNIPEG, NORWAY HOUSE, and ILFORD), on N shore of Elk I., the God's L. Mines, discovered in 1932, 149 claims, 6,400 acres, tot. prod. to 1942 of 160,531 ozs.—lode gold.

WINNIPEG, NE 125 mi.: (1) on N shore of Rice L., the San Antonio and Forty-four mines, major producers—lode gold; (2) in T. 22, R. 16 E (within 10 mi. of the Ontario line): (a) the Central Manitoba Mine, 1,500 acres, productive 1924–34—lode gold; (b) 16 mi. SE, the Diana (Gem Lakes Mines, Ltd.) Mine, 52 claims, 2,080 acres astride the interprovince boundary—lode gold; (c) 17 mi. NW, the Gabrielle Mine (adjoining the San Antonio, q.v.), first claim staked in the Rice L. area, 1911—lode gold.

NORTHWEST TERRITORIES

An auriferous region extends from the north shore of Lake Athabasca to a line about halfway between Great Slave Lake and Great Bear Lake, with a gap of 400 miles separating Lake Athabasca from the next gold-bearing region to the southeast. These remote territories are served almost exclusively by airplanes. Prospecting requires careful attention to wilderness survival techniques.

For information, write: Resident Geologist, Department of Indian Affairs and Northern Development, Yellowknife, N.W.T., Canada.

GOLDFIELDS (on N shore of Athabasca L.): (1) just W, the Box Property (with 1,000-ton mill), tot. prod., 1939–42, of 64,066 ozs.—lode gold; (2) S 1 mi., the Athona Property, minor producer from low-grade ores—lode gold.

GREAT SLAVE L. AREA, along the South Nahanni R. and its Flat R. tributary, in gravel bars and bench deposits—placer gold.

THOMPSON L. (post on an is. in Thompson L.), near S shore of E arm of lake, the Thompson-Lundmark Gold Mine (presently inactive)—lode gold.

YELLOWKNIFE (NE of the N arm of Great Slave L., in basin of the Yellowknife R.): (1) area, such mines as (a) the Con-Rycon and Negus mines, major producers to 1942; (b) the Giant Mine, all in quartz lenses in chlorite schist zones—aurostibnite and free gold (nuggets, plates, blebs, films, leaves, spongiform masses, distorted octahedral crystals); (c) other properties of the Akaitcho Yellowknife Gold Mines Ltd. (currently under development)—lode gold; (d) the Ptarmigan Mine, closed in 1942—lode gold; (2) SE 55 mi., on the Outpost islands (most westerly is. in the E arm of Great Slave L.), the Slave L. (Philmore) Mine, worked in 1950s, a good producer—lode gold, with copper and tungsten.

NOVA SCOTIA

The province of Nova Scotia has the longest continuous record of lode-gold production of any province in Canada, gold having been mined from quartz veins every year since 1860, even though the total output through 1934 of 930,539 troy ozs. was less than that produced in other provinces.

The earliest gold was panned from beach sands at several places as early as 1849, and black-sand concentrations still yield gold colors. Actual mining, however, did not begin until 1860, when lode veins were found in the Mooseland district. By World War II more than 100 localities had been developed into gold producers to greater or lesser degrees. These localities varied in size from less than 1 square mile to 3 square miles in extent along a stretch of land 10 to 75 miles wide and 275 miles long.

The gold-bearing series of Nova Scotia is a monotonous succession of quartzites and slates more than 30,000 feet thick. These series occupy that half of the province lying along the Atlantic Coast and extending the full length of the Nova Scotia Peninsula, characterized by anticlines more or less regularly spaced about 3 miles apart. Some anticlines have been traced for 100 miles. Gold-quartz veins occur near the crests of plunging anticlines, often passing completely across a crest from one limb to the other.

Nova Scotia gold occurs commonly in the native form, and much of it is very coarse. The auriferous veins of the province are far from being exhausted, because in the early days the only claims granted were extremely small, from 20 to rarely more than 150 feet along a particular vein. Therefore, large numbers of short shafts and stopes dot the mining areas. In fact, only the more readily mined surface and subsurface gold ores were removed.

For information, write: Information Officer, Department of Mines, Provincial Building, 1649 Hollis St., Box 1087, Halifax, Nova Scotia, Canada.

Guysborough Co.

AREA, the Richardson Mine, tot. prod., 1893–1910, of 53,835 ozs.—lode and native gold.

Halifax Co.

CARIBOU, the Lake Mine (fissure type), rich and very productive—lode gold.

OVENS (W of Halifax, on the S coast), area black sands along beaches, very productive in early days—placer gold.

Queens Co.

BROOKFIELD, the Libbey Mine (deepest mine in province at 1,997 ft.), very productive—lode and native gold.

ONTARIO

One day in the summer of 1866 Marcus Powell was digging along what he thought was a seam of copper at what came to be known as the Richardson location in Hastings County, when he broke through into a solution cavity 12 feet long by 6 feet wide and 6 feet high. The flickering light of his miner's lamp glittered back at him from walls coated with leaves, nuggets, and ragged crystals of raw, native gold, the first gold found anywhere in Ontario. The gold fever that struck all who soon heard the story opened up a broad region to excited prospecting, especially for a distance of 70 miles through Hastings, Peterborough, Addington, and Frontenac counties, where gold-bearing quartz veins in abundance crop out along contacts of diorite or granite.

Gold production really began in Ontario in 1866, and it began to boom in the early 1900s, when the rich lode deposits of the Lake of the Woods district reached full production. In 1909 came the astonishing discovery of the Porcupine gold field and, from 1911–12, the impressive Kirkland Lake fields, followed by important gold finds in the Rainy River area of the northwestern part of the province. All of Ontario's total production of 2,155,518 troy ozs. through 1933 came only from lode mines. Although many streams in northern Ontario have gravel deposits that pan out placer gold, no commercial placer operations have ever been developed.

For information, write: Information Officer, Ministry of Natural Resources, Whitney Block, 99 Wellesley St. West, Toronto, Ontario, Canada.

ATIKOKTAN, area, the Elizabeth, Hammond Reef, Sunbeam, and other mines, good producers—free gold, with pyrite, lead, chalcopyrite.

BELMONT TWP. (Hastings Co.), in lot. 20, con. 1, the Cordova Mine, tot. prod., 1898–1940, of about 20,000 ozs.—lode gold, pyrite.

COBALT, WNW about 65 mi. (about 60 mi. due S of the Porcupine district), in the West Shiningtree district, the Ronda Gold Mines, tot. prod., 1929 only, of 2,727 ozs.—lode gold.

DRYDEN: (1) vicinity, numerous mines; (2) around Wabigoon L., several productive mines—lode gold.

EAGLE L., the Grace, Golden Eagle, and other area mines, in quartz —free and lode gold.

EMPIRE (BEARDMORE) STA. (130 mi. E of FORT WILLIAM on the Longlac-Fort Williams branch of the Canadian Nat'l. RR), the Northern Empire Mine, discovered in 1925, many claims near sta., sporadic development but productive—lode gold.

FAVEL STA. (on the Canadian Nat'l. RR), N 75 air mi.: (1) at SE angle of Red L. in the Patricia portion of the KENORA, q.v., district, the Howey Mine, discovered in 1925, one of most famous mines in Ontario, very productive—lode gold; (2) 4 mi. N of the Howie, the McKenzie-Red L. Mine, 400 acres on N part of McKenzie I., opened in 1928, productive —lode gold.

GALBRAITH TWP., in lot. 12, con. III, the Havilah Mine, discovered in 1899, intermittently worked—lode gold.

GERALDTON STA. (on the Canadian Nat'l. RR), S 2 mi., on Little Long L. in the THUNDER BAY, q.v., district, the Little Long Lac Mine, 35 claims, discovered prior to 1916, in quartz—free gold.

GOUDREAU STA. (on the Algoma Central RR, 25 mi. NE of MICHIPICOTEN, q.v., and 170 mi. N of SAULT STE. MARIE), E 5 mi., in T. 49, R. 27 E: (1) the Algoma Summit (McCarthy-Webb) Mine, 7 claims, 322 acres, quite productive—lode gold; (2) the Algold and Cline L. mines—lode gold.

HARKER, HOLLOWAY TWPS., S of Abitibi L., the Lightning R. claims —minor free gold.

HISLOP TWP., in N½ lot. 1, con. II, the Ross Mine, several levels in depth, good producer—lode gold.

KENORA: (1) out 7 mi. to L. of the Woods, on Sultana I., the Sultana Mine, very productive, 1891–1906—lode gold; (2) on E shore of l., near

N end (about 10 mi. S of the Sultana), the Wendigo Gold Mines, tot. prod., 1936–43, of 67,414 ozs.—lode gold; (3) SW 25 mi., on shore of Shoal L.: (a) the Mikado Mine, major producer—lode and free gold (highly spectacular specimens); (b) on Cameron I. in l., the Duport Mine, tot. prod., 1935–36, of 4,572 ozs.—lode gold.

KINGHORN STA. (on the Canadian Nat'l. RR), NW 12 mi., at Atigogama L., the Dikdik Mine, discovered in 1933, good producer—lode gold.

KIRKLAND L. (315 air mi. N of TORONTO), the Kirkland Gold Belt, including the outlying areas of BOSTON CR., LARDER L., SWASTIKA, and MATACHEWAN, with all chief producing mines in an area 4 mi. long by ¾ mi. wide along a single main "break" of ore zone, highly productive 1906 to present: (1) in town: (a) the Kirkland L. Gold Mine, 1 mi. deep, tot. prod., 1918–44, of 602,409 ozs.—lode gold; (b) immediately W, the Macassa Mine, 242 acres, tot. prod., 1933–44, of 527,304 ozs.—lode gold; (2) E 5–7 mi., near center of Lebel Twp.: (a) the Bidgood Mine, 753½ acres, opened in 1920, productive—lode gold; (b) adjoining, and close to the Nipissing Central RR, the Moffatt-Hall Mine, 15 claims (tot. prod., 1920–34, of 2,525 ozs.)—lode gold; (3) on S shore of Kirkland L., the Teck-Hughes, Wright-Hargreaves, Sylvanite, Toburn (Tough-Oakes), and L. Shore mines, tot. prod. all these mines, 1913–44, of 14,111,856 ozs.—lode gold; (4) SSE 12 mi., at BOSTON CR., the Barry-Hollinger (Patricia) Mine, 360 acres, tot. prod., 1916–36, of 68,-883 ozs.—lode gold, some tellurides.

LARDER L. (Kirkland L. district): (1) the Chesterville, Kerr Addison, Laguerre, and Yama mines (combined tot. prod., 1937–44, of 758,069 ozs.)—lode gold; (2) NW 6 mi., the Argonaut Mine, tot. prod., 1919–35, of 37,932 ozs.—lode gold.

MANITOU L. (1) area lode mines, and (2) vicinity of MINE CENTER, many gold-quartz deposits and such mines as the Foley, Golden Star, and Laurentine, etc., all quite productive—lode gold.

MARMORA TWP. (Hastings Co.), in lot. 9, con. VIII, the Deloro Mine, tot. prod., 1899–1903, about 90,000 ozs.—lode gold (with small grains of free gold).

MATACHEWAN (in the Kirkland L. district, in Bannockburn and Argyle twps.): (1) the Matachewan Consolidated Mines, 22 claims, 880 acres, discovered in 1917, minor producer—lode gold; (2) from ELK L.: (a) by rd., the Ashley Mine, productive—lode gold; (b) NE 26 mi., on the Montreal R., the Young-Davidson Mine, 5 claims, 163 acres, open-pit operations—lode gold.

MATHESON, E 10 mi.: (1) on S shore of Painkiller L., the Amalgamated Goldfields (Blue Quartz) Mine, 600 acres, intermittently productive, 1907–28—lode gold; (2) in Munro Twp., the Croesus Mine (a small deposit of probably the richest gold ore in Ontario), tot. prod., 1915–19, of 12,616 ozs.—lode and free gold (2,274 troy ozs. from original 765 lbs. of ore).

MICHIPICOTEN (district, discovered in 1897 with sporadic operations ever since), area mines—lode gold.

MOSS TWP. (WNW of FORT WILLIAMS and E of the L. of the Woods and Patricia groups of mines), the Moss Mine, tot. prod., 1883–1941, of 54,282 ozs.—lode gold (first gold discovery site in western Ontario).

OSWAY, HUFFMAN TWPS., the Jerome Gold Mine (source of spectacular native-gold specimens), tot. prod., 1941–43, of 56,879 ozs.—native gold.

PORT ARTHUR, W 90 mi., in the Thunder Bay district of NW Ontario and 18½ rd. mi. from the Tip Top siding on the Ft. Frances branch of the Canadian Nat'l. RR, the Ardeen (Moss) Mine, 900 acres, productive 1871–1930s—lode gold.

RATHBUN TWP., on E side of Wanapitei L., the Crystal Mine, discovered in 1898, productive—lode gold.

RED L. (1) the Howie Mine, a famed producer, 1930–41, of 421,322 ozs.—lode gold; (2) the Gold Eagle, Hasaga, Madsen Red L., and Red L. Gold Shore mines, all very productive from large quantities of generally low-grade ores—lode gold.

SAULT STE. MARIE, N about 100 mi., on Wawa L. in the Michipicoten district (reached via HAWK JCT. and the Algoma Central RR), the Minto Mines, several properties worked as early as 1899 and intermittently productive—lode gold.

SAVANT L. STA. (on the Canadian Nat'l. RR), S 12 mi., in the Thunder Bay district, on W shore of Couture L. near the North Bay of Sturgeon L., the St. Anthony Mine, discovered in 1900 and productive through the 1930s—lode gold.

Sioux Lookout (on the Canadian Nat'l. RR): (1) due N 100 air mi., at Summit L. in the Patricia portion of the Kenora district, q.v., the Casey Summit Mine, discovered in 1931—lode gold; (2) N 125 mi., in the Pickle-Crow area a few mi. N of the E end of L. St. Joseph (in the Patricia part of the Kenora district, q.v.), the Central Patricia Mine, 90 claims, 3,837 acres, staked in 1928—lode gold; (3) out by airplane or boat (access also from Hudson) to area of Woman and Confederation lakes, the Jackson-Manion Mine (40 mi. from the Howie Mine), 56 claims, 2,600 acres, productive—lode gold.

Sudbury: (1) on House L. in Mongowin Twp., about 1½ mi. W of West R. Sta. on the Algoma Eastern RR, the McMillan Mine, discovered in 1926, 34 claims, minor producer—lode gold; (2) SW 8–14 mi., near SW end of Long L., in timber berth 69, the Lebel Oro (Long L.) Mine, 9 claims, tot. prod., 1909–16 and 1937–39, of 54,708 ozs.—lode gold; (3) SW about 35 mi., near Howie Cr., an area about 6 mi. long by 1 mi. wide, numerous gold-bearing quartz exposures in the Cobalt Series (much metamorphosed) although little mined—free and lode gold (low-grade generally).

Swayze (80 mi. SW of Porcupine, cf. under Timmins): (1) in NE cor. Swayze Twp., the Kenty Mine, productive—native gold, with tourmaline, calcite, specularite, galena, chalcopyrite; (2) numerous other area mines—native and lode gold.

Thunder Bay (district, E of L. Nipigon, in the 55-mi. stretch between it and Little Long Lac): (1) 11 productive regional mines—lode gold; (2) near E end of district, in Errington and Ashmore twps.: (a) the Little Long Lac Mine, oldest and largest area mine, productive—lode gold; (b) 2 mi. SE, the Hard Rock and McLeod-Cockshutt mines (adjoining properties)—free gold, pyrite, abundant arsenopyrite; (6) 6 mi. W of the Hard Rock Mine and SW of Magnet L., the Bankfield and adjoining Tombill properties, productive—lode gold.

Timmins (about 450 mi. due N of Toronto), the Porcupine Gold Belt, an area E of town 5 mi. long by 3 mi. wide comprising most of

Tisdale Twp. and N part of Deloro Twp., discovered in 1909 and one of most productive gold-mining regions of Canada: (1) (a) the Hollinger Mine, tot. prod., 1910–44, of 13,416,667 ozs.—lode gold; (b) adjoining, many claims covering 560 acres, extensively worked to considerable depths (tot. prod., 1911–34, over $15,000,000)—lode gold; (2) the McIntyre Mine, tot. prod., 1912–44, of 5,403,342 ozs.—lode gold; (b) NE of the McIntyre (as an extension of the same vein), the Coniaurum Mine, 626 acres, discovered in 1909, tot. prod., 1912–34, of $7,600,000 —lode gold; (3) E 4 mi.: (a) the Dome Mine, 11 claims, 438 acres, tot. prod., 1909–44, of 4,972,837 ozs.—lode gold; (b) 2 mi. SW of the Dome, the Buffalo-Ankerite and Marbuan (March Gold) mines, very productive—lode gold; (4) many other area mines, especially the Paymaster Consolidated, West Dome, Porcupine Paymaster, and Vipond (Anglo-Huronian), all productive—lode gold; (5) SE cor. of Tisdale Twp., the Preston East Dome Mine (tot. prod., 1938–44, of 372,971 ozs.)—lode gold.

TYRRELL, KNIGHT TWPS., straddling the intertwp. boundary, the Tyranite Mines (tot. prod., 1939–42, of 30,352 ozs.)—lode gold.

WOMAN LAKE: (1) the Uchi Gold Mines, producer of 114,467 ozs., 1939–43—lode gold; (2) the Hudson Patricia and J. M. Consolidated mines, productive—lode gold.

QUEBEC

The province of Quebec has a long history of gold production, since placer gold was first observed in the gravels of the Chaudière River Valley in 1824. Until 1900 all gold production came from placers, but from 1901 to 1914 the principal producers were lode-gold mines. Since then, practically all gold has been derived from hard-rock mining. From 1877 to 1896, a total of 1,808,711 ozs. of placer gold were recovered, and there are very many areas where the modern gold hunter may still find placer gold.

Precambrian rocks underlie all of Quebec north of the St. Lawrence and Ottawa rivers, except in a narrow strip between MONTREAL and QUEBEC CITY. Although gold prospects appear throughout the Precambrian region, the area most favorable for gold hunting occurs directly east of the Porcupine and Kirkland Lake fields of Ontario, for about 100 miles east of the interprovincial boundary.

For more information, write: Information Officer, Mineral Deposits Service, Department of Natural Resources, 1620 boul. de l'Entente, Quebec City, P.Q., Canada.

AMOS: (1) in NW¼ of Bourlamaque Twp., the Lamaque Mine, 44 claims, 2,452 acres (reached by water via the Harricanaw R.), productive —lode gold; (2) SW 30 mi. by rd. or the Harricanaw R., the O'Brien Mine, 10 claims, 455 acres, staked in 1944, quite productive—lode gold; (3) out 42 mi. by the Harricanaw R., on is. in DeMontigny L. (an expansion of the r.): (a) the Siscoe Mine, discovered in 1912, 1,174 acres, productive—lode gold; (b) 1½ mi. SE, on SW shore of l., the Sullivan Consolidated Mine, 890 acres, staked in 1911, minor producer—lode gold.

BEAUCHASTEL TWP., in SW part just NE of L. Renaud, the Arntfield Gold Mines and the adjoining Francoeur Gold Mines (both low-grade ores)—lode gold.

CADILLAC TWP. A chain of important gold deposits known as the Cadillac "break" runs westerly through Cadillac and the middle of Bosquet Twp., turning SE into Malartic Twp. and possibly extending W into Joannès Twp.; along the "break," the Thompson Cadillac, O'Brien (spectacular shoots of native gold), Central Cadillac, Amm, Lapa-Cadillac, and Pan-Canadian mines—free, native, lode gold.

LA SARRE, due S 20 mi., in Duparquet Twp., near N end of L. Duparquet, the Beattie Mine (22 mi. NW of NORANDA), developed mainly in late 1920s, quite productive—lode gold.

MALARTIC, adjoining on immediate E: (1) the Canadian Malartic, Sladen Malartic, and East Malartic mines; and (2) 3 mi. farther E, the Malartic Goldfields Mine—native gold (usually too disseminated to be visible).

NORANDA (in NW Quebec, in Rouyn Twp.), area, the Noranda (Horne) Mine, 1,509 acres, staked in 1920 near shore of Osisko L., very productive—lode gold (tot. prod., 1927–39, of 2,817,951 ozs.).

QUEBEC CITY, SE 40–50 mi., in the Chaudière R. Basin, all area streams, richly productive placers, chiefly: (1) along Meule Cr. tributary of Mill R. (flowing into the Chaudière R. at BEAUCEVILLE, extensively hydraulicked from 1911–12)—placer gold; (2) regional buried channel placers (overlain by barren glacial drift, difficult to mine, probably productive)—placer gold.

ROUYN (TWP.): (1) some 34 area mines that are minor producers lode gold; (2) ESE 4½ mi., the McWatters Gold Mines (spectacular specimens erratically distributed)—native gold; (3) S 4½ mi., in SW¼ of twp., the Granada Gold Mines, 6,000 acres, tot. prod., 1929–35, of 51,286 ozs.—native gold, with tourmaline and hornblende in quartz; (4) E 25 mi., and about 45 mi. from AMOS, q.v., in N part of Fournière Twp., the Canadian Malartic Mine (cf. also under MALARTIC), 443 acres —lode gold; (5) E 30 mi., in Dubuisson Twp., the Green-Stabell Mine, 327 acres, staked in 1914, minor producer—lode gold.

VAL D'OR, center of an area containing numerous productive lode-gold mines: (1) the Siscoe (cf. under AMOS), Sullivan Consolidated, Green-Stabell (cf. under ROUYN), Cournor (Bussières), Perron, and Payore Holdings—native gold (occasionally in pockets to 1,000 ozs.); (2) SE: (a) NW cor. of Bourlamaque Twp., 1 mi. from W boundary and 3½ mi. from N boundary, the Lamaque Gold Mines (cf. also under AMOS), second-largest gold mine in Quebec, tot. prod., 1935–44, of 944,686 ozs.—lode gold; (b) adjoining on the N, the Sigma Mines (tot. prod., 1937–44, of 472,008 ozs.)—native gold.

VALLEY JCT. to BEAUCEVILLE, along the Chaudière R. and its tributaries, rich placer deposits (extensively worked, 1875–85)—placer gold (coarse, angular nuggets, many attached to bits of quartz).

SASKATCHEWAN

The first gold found in Saskatchewan was in 1932. However, the total commercial production of the province has been as a by-product of copper refining from the ores of the Flin Flon mines in Manitoba, with overlap into Saskatchewan.

For more information, write: Information Officer, Department of Mineral Resources, Mineral Records Branch, Government Administration Bldg., Albert Street, Regina, Saskatchewan S4S OB1, Canada.

REGION, about 251,700 sq. mi. (one third of entire province) is underlain by Precambrian crystalline rocks that have proved auriferous elsewhere in the Canadian Shield; hence modern gold hunters might expect favorable prospecting throughout this immense region.

YUKON TERRITORY

Canada's production of gold is best exemplified by the great gold discoveries in Yukon Territory that led to the last of the great gold rushes in the world, in 1898 to the Klondike. From all over the world, prospectors and miners swarmed in from San Francisco and Seattle to Skagway, Alaska, then via the Dyea trail over White Pass to Whitehorse and by raft, rough-hewn boat, and canoe down the turbulent Yukon River to Dawson. Actually, prospecting for placer gold had been carried on for at least 15 years prior to the discovery of the Klondike in 1896, but from this one field alone the main production of placer gold in Canada has come.

With tens of thousands of prospectors and miners working hundreds of creeks, streams, and river valleys, placer gold production reached a maximum in 1900, when it exceeded $20,000,000. The richest of the creeks was Bonanza Creek, a comparatively small stream near its mouth, where it measures only 15 feet in width and 3 to 4 inches in depth. Bonanza Creek flows through a valley flat 300 to 600 feet wide. The creek gravels were 4 to 8 feet thick and extended across the valley, overlain by a few feet of frozen muck. For 13 miles along its length the valley proved to be the most highly productive placer area in the Klondike, through its middle section producing more than 50 ozs. ($1,000) a running foot. The fact that the creeks, rather than the river valleys, were so auriferous suggests that any search for new fields should be along the regional creeks draining into the main rivers. Probably no such rich creeks as the Bonanza remain to be discovered.

There are three main types of auriferous deposits that characterize all the gold-bearing stream valleys in the Klondike: low-level gravels (gulches, creeks, rivers); terrace gravels at intermediate levels; and high-level gravels, especially the White Channel deposits. Of all these, the creek gravels have proved most auriferous, flooring the bottoms of their valleys to depths of 4 to 10 feet. They rest on bedrock, and the gold in many places extends down into cracks and joints for 2 to 3 additional feet; hence, bedrock must also be excavated to such depths in order to recover all the available gold.

In almost all areas the auriferous ground must be thawed before it can be profitably mined. In 1919 it was discovered that thawing frozen ground with cold water was half as costly as thawing by steam, with a great saving in the cost of dredging. In areas where surficial muck is not interbedded with sand and gravel, it can be excavated without being thawed, using a pick, and then panned, sluiced, or run through a rocker. Otherwise, thawing is necessary, with heated rocks covered with moss or a blanket to retain the heat, by means of wood fires, or with hot or cold water using "points" or pipes driven to bedrock, with water or steam forced into the ground. Other special techniques for getting to placer gold can be learned from resident old-timers in the various areas.

For information, write: Resident Geologist, Department of Indian Affairs and Northern Development, Whitehorse, Yukon, Canada.

Dawson-Mayo District

DAWSON: (1) a few mi. S, on Bonanza Cr. (where the great gold rush of 1898 began), in a triangle about 4 mi. wide enclosed among Eldorado Cr., Upper Bonanza Cr., and Victoria Gulch, richest placers of Yukon Territory, discovered in 1896; (2) all area crs., extensively placered, e.g.: Eldorado, Bear, Hunker, Quartz, Dominion, Gold-Run, Sulphur, All Gold, Eureka, Caribou, Granville, Adams, Bedrock, Gold Bottom, Miller, Kirkman, Henderson, Pan, Hidden, Cash (Gold), Thistle, and Indian—abundant placer gold; (3) along the Klondike and Sixtymile rivers, in gravel bars, benches, and terrace deposits, very productive—placer gold (coarse, nuggets); (4) practically all other regional crs., benches, and terraces of the Yukon R. drainage clear back to glaciated crests of the Cordillera—placer gold (colors, grains, coarse, ragged, nuggets, etc.).

MAYO: (1) along Dublin Gulch, rich placers—gold, with scheelite, cassiterite, garnets; (2) along Haggart, Highet, Johnson, Duncan, Steep, Minto, Ledge, Scroggie, Barker, Congdon, Canadian, and Rude crs.—abundant placer gold (often very coarse), in some crs. with cassiterite and scheelite; (3) along the Stewart R.: (a) above mouth of the Mayo R. and (b) along all tributaries leading to Hess and Lansing rivers—abundant placer gold.

Kluane-Whitehorse District

AREA, almost all regional stream and river gravels, worked at one time or another—placer gold.

CARMACKS: (1) along Nansen, Victoria, Discovery, Webber, Back, Seymour, and Stoddart crs., rich—placer gold; (2) E fork of Nansen Cr., rich placers; (3) NW 28 mi., on Freegold Mt.: (a) area mines (two types of gold deposition)—free and lode gold; (b) the Laforma Group of claims, opened in 1939—minor lode gold.

KLUANE, DEZADEASH LAKES AREA: (1) in gravels along the Tatshenshini, Bates, Alsek, and Primrose rivers, extensive placers; (2) along Shorty, Beloud, Silver, and Squaw (Dollis) crs.—placer gold; (3) in gravel deposits along Iron, Sugden, Kimberley, Victoria, Goat, Mush, Marshall, Granite, Sandpiper, Wolverine, Bullion, Sheep, Fourth of July, Ruby, Gladstone, Cultus, Printers (New Zealand), McKinley, and Dixie crs.—abundant placer gold, often with native copper nuggets (to 28 lbs.); (4) in Burwash and Arch crs.—coarse placer gold (with nuggets of native silver and platinum (in Burwash Cr.); (5) along Kletson Cr. (near international boundary)—placer gold, with native copper.

LABERGE: (1) along the Lewis and Teslin rivers, in low-water bars and bench gravel deposits, extensive placers; (2) along Livingstone, Summit L., Cottoneva, Little Violet, Mendocina, and D'Abbadie crs.—placer gold.

Peel River Area

AREA: (1) along the entire length of the Peel R., in bed, bench, and terrace gravels—placer gold; (2) along the Bonnet Plume R., placers; (3) along the Wind R., especially at mouth of the Little Wind R., productive placers; (4) along Hungry Cr.—placer gold; (5) gravel deposits along a stream nearly opposite the mouth of the Rackla R.—rather abundant placer gold.

Pelly River-Watson Lake Area

AREA: (1) along the Pelly R., from its mouth to Campbell Cr., in sand- and gravel bars—placer gold; (2) between Hoole Canyon and the Hoole R., in sand- and gravel bars (most productive of region)—placer gold; (3) regional tributaries of the Pelly R. joining it from the S—abundant placer gold.

Index